Corrosion Preventive Materials and Corrosion Testing

S.K. Dhawan
CSIR-National Physical Laboratory
New Delhi, India

Hema Bhandari
Chemistry Department
Maitreyi College, University of Delhi
New Delhi, India

Gazala Ruhi
Chemistry Department
Maitreyi College, University of Delhi
New Delhi, India

Brij Mohan Singh Bisht
National Test House, Ministry of Consumer Affairs
Ghaziabad, Uttar Pradesh, India

Pradeep Sambyal
CSIR - National Physical Laboratory
New Delhi, India

CRC Press
Taylor & Francis Group
Boca Raton London New York

CRC Press is an imprint of the
Taylor & Francis Group, an **informa** business

A SCIENCE PUBLISHERS BOOK

Cover credit: Cover illustrations reproduced by kind courtesy of the authors.

CRC Press
Taylor & Francis Group
6000 Broken Sound Parkway NW, Suite 300
Boca Raton, FL 33487-2742

First issued in paperback 2021

© 2020 by Taylor & Francis Group, LLC
CRC Press is an imprint of Taylor & Francis Group, an Informa business

No claim to original U.S. Government works

Version Date: 20200106

ISBN 13: 978-0-367-50456-4 (pbk)
ISBN 13: 978-1-138-11875-1 (hbk)

Publisher's Note
The publisher has gone to great lengths to ensure the quality of this reprint but points out that some imperfections in the original copies may be apparent.

Visit the Taylor & Francis Web site at
http://www.taylorandfrancis.com

and the CRC Press Web site at
http://www.crcpress.com

Preface

The design and production of *"Corrosion Preventive Materials and Corrosion Testing"* is entirely different from other handbooks dealing with the same subject. While other corrosion handbooks have been generally the state-of-the-art information on corrosion engineering by a principal author, this book is a result of collective efforts of many authors who have a clear understanding on corrosion and its implementation in different areas. Although four author's names appear on its cover, this handbook is indeed the result of the cumulative efforts of many generations of scientists and engineers and their understanding and preventing the effects of corrosion, one of the most constant foes of human endeavors.

The design and construction of this handbook are made for the new millennium with latest information processing techniques presently available. This book deals with prominent topics, for example, how conducting polymer-based composite coatings have addressed and tackled corrosion problems. The conducting polymer systems, selected as supportive examples, have been chosen for the performance of coatings in the marine environment. The conducting polymer-based paints/coatings have found an immense application in its fight against corrosion. These coatings overcome the drawbacks of conventional anti-corrosive coatings, and offer following advantages like its self-healing ability (it helps in passivating the regions of scratch, pores, etc.), are environmental friendly and based on green technology (free from heavy metal ions and hazardous chromates), have a long service life, are economically feasible and also have additional anti-static properties.

This book is aimed at practicing engineers, providing a comprehensive guide and a reference source for solving corrosion problems by new generation coatings based on conducting polymers. During the past decades, progress in the development of materials capable of resisting corrosion has been significant. There have been substantial developments in the area of protective coatings. This book should prove to be a key information source concerning numerous facets of corrosion damage, from detection and monitoring to its prevention and control.

The book is divided into six main chapters and is followed by a short chapter on the future directions of this area accompanied by supporting materials. Each chapter is relatively independent and can be consulted without having to go through the previous chapters. The first chapter contains fundamental principles governing aqueous corrosion and covers the main environments causing corrosion such as atmospheric, natural waters, seawater environments, inspection, monitoring and testing. This section also provides elements for understanding protective coatings, corrosion inhibitors, cathodic protection and anodic protection.

The second chapter gives an overview of conducting polymers that are considered as futuristic electronic materials with promising properties. It has attracted the attention of many material scientists because of its boundless applications in corrosion protection and other areas like energy storage, EMI shielding, sensors, batteries, anti-static, supercapacitors, organic solar cells, light emitting diodes, field emission transitions, electrochromic devices, etc., that can be attributed to its tailorable electrical conductivity, ease of synthesis, low processing cost, corrosion resistant, lightweight and good environmental stability.

The third chapter covers the development of conducting copolymer composites, to be used as the corrosion protection with hydrophobic properties, greater durability and high wear resistance that is expected to encourage a major revolution in the world of corrosion. It describes the development of highly resilient, inexpensive and environment friendly hydrophobic epoxy coating based on conducting copolymer composite. SiO_2 particles were incorporated in conducting copolymer based on aniline and pentafluoro aniline. The synthesized poly(aniline-co-pentafluoro aniline)/SiO_2 composites were formulated with epoxy resin. Different physico-mechanical properties as well as anti-corrosion performance of epoxy formulated copolymer-silica composites coatings is demonstrated by using mild steel substrate in 3.5 wt.% NaCl medium.

The fourth chapter is devoted to epoxy coating modified with conductive copolymer nanocomposites based on poly(AN-co-PFA)-zirconia, which is an excellent corrosion resistant material under the harsh marine environment. This may also disclose a new opportunity in various technological applications for marine engineering materials, which requires very high salt resistance ability.

The fifth chapter discusses the chemistry of polypyrrole and mentions different synthesis routes of the polypyrrole composites. Various polypyrrole based composites, containing fillers like SiO_2, TiO_2, clay, ZnO, Al_2O_3, MWCNTs, grapheme, etc., and their designing as corrosion resistant coatings over ferrous and non-ferrous substrates are presented in this chapter. The corrosion characterization of the polypyrrole-based composite coatings is the core content of the chapter. An interesting section of this chapter is on the self-healing property shown by the polypyrrole/SiO_2 composite coatings and presents the FESEM images of the artificial defect on the coated surfaces and their self-healing tendency with the passage of time. The present chapter concludes that the polypyrrole-based composite coatings offer advanced corrosion protection to the metal surfaces.

Next, how biopolymer conducting polymer hybrid coatings can effectively control corrosion forms the basis of the sixth chapter. Elaborate discussions on two different composite coating systems, namely: chitosan/polypyrrole/SiO_2 and Polypyrrole/gum acacia are presented in this chapter. In order to develop highly durable and efficient conducting copolymer based anti-corrosion material, a significant effort has been presented in this book. Finally, what is the future scope and how self-healing coatings can be designed using conducting polymer coatings form the basis of the seventh chapter. An appendix appears at the end of the book which explains all the terminologies used in the chapters.

In this book, an attempt is made to present data on highly durable, self-healing and hydrophobic conducting copolymers based nanocomposites and evaluation of such copolymers nanocomposites for the use of corrosion protection of iron in a corrosive environment.

S.K. Dhawan
Hema Bhandari
Gazala Ruhi
Brij Mohan Singh Bisht
Pradeep Sambyal

Acknowledgments

The book *Corrosion Preventive Materials and Corrosion Testing* is an outcome of the extensive research carried out by my students Dr. Hema Bhandari, Dr. Gazala Ruhi, Dr. Brij Mohan Singh Bisht and Dr. Pradeep Sambyal along with my other students whose contribution can never be forgotten. In fact, Ms. Ridham Dhawan is the researcher, whose contribution in designing some of the chapters have been very fruitful and her input has been a part and parcel of many sections of this book.

Our acknowledgments also go to the staff working in the conducting polymer group, who are the key figures in designing the pilot plant, powder coating units and preparation of samples. As mentioned in the Preface, this book attempts to summarize the present state of our knowledge of the corrosion phenomena and their impact on our societies. Many of the opinions expressed in the book have come either from our work with corrosion team at NPL or more often from our study of the work of other corrosion engineers and scientists. Of the first kind, we are particularly indebted to Mr. Brijesh Sharma, who is a pioneer in setting up lab reactors and other experiments that were carried out by us in the laboratory. We are also thankful to various science and engineering pillars responsible for the present state of our knowledge in the field of corrosion. Our utmost gratitude goes to Dr. D.K. Aswal, Director of National Physical Laboratory, for his pragmatic vision of the quantification of corrosion damage. Heartfelt thanks to National Test House team as well, where our samples were tested in the Salt Spray Chamber under corrosive conditions.

Contents

Introduction to Corrosion

"Keep up your bright swords, for the dew will rust them."

A sarcastic quote by Othello in William Shakespeare's famous play *Othello* emphasized the menace of corrosion.

1.1 What is corrosion?

A broad definition of corrosion is "the deterioration of the useful properties of materials when they interact with their environment". The substances that are used as construction materials, process equipment including the products manufactured from them, are termed as materials. These materials can be metals and their alloys, composites, polymers, ceramics, etc. The term "deterioration" specifically refers to the undesirable chemical reaction between these materials and their environment containing liquids, gases, salts, etc. Corrosion is commonly known as "rust". However, the word "rusting" is more appropriately reserved for corrosion of iron. So, it can be stated that the terms "corrosion" and "rusting" are synonyms to each other.

Iron and its alloys are the most commonly used materials opted to design various structural/functional components, and thus are largely affected by corrosion. Corrosion occurs and progresses because of the spontaneity of its process. Nature allows materials to stay in their lowest possible energy state (or most stable state), and thus most of the metals/alloys have a tendency to corrode (combine with water/oxygen present in the environment) so that they can achieve this state. In one concluding statement, we can say, almost every metal has a tendency to convert into its stable oxide state. Corrosion is a phenomenon of global interest as losses due to corrosion are huge, and it adversely affects the economy of organizations in particular and the countries in general. Just like any other natural calamity, corrosion can also cause catastrophic damage to bridges, buildings, gas/petroleum pipelines, structural components of industries, automobiles, home appliances, drinking water systems, etc. (Yamashita and Uchida 2002; Mohebbi and Li 2011; Walker 1976).

The interaction of metals with their environment has vital importance in the construction of materials of superior performance. For example, iron/steel when exposed to an industrial atmosphere forms a layer of iron oxide commonly known as "rust". These rust layers are generally heterogeneous, porous and flaky in nature, and has almost no mechanical integrity to protect the steel from corrosive environments. In addition to this, it is also devoid of self-healing properties like passive oxide films.

The corroding steel surface evidences formation of iron oxides/hydroxides, basically consisting of α-, β-, γ-FeOOH, γ-Fe$_2$O$_3$, Fe$_3$O$_4$ and amorphous oxy-hydroxides in the rust layers (Evans 1976; Yamashita et al. 1994; Leigraf and Graedel 2000). These rust layers behave as a good protective barrier, providing corrosion protection to the steel surface during their early stages of exposure to the environment. However, the rust components have a tendency to show some level of affinity towards water/electrolyte. Therefore, with the lapse of time, the protective film deteriorates due to the formation of defects/pores. The corrosion process commences from the defect sites and propagates further. As a concluding remark, rust layer can be mechanically dense and tightly adherent to the metal surface initially but is not sufficient to form a compact/robust film to protect the metal surface from water and oxygen. On the other hand, copper forms a well adherent and compact layer of CuSO$_4$.3Cu(OH)$_2$, which is highly protective in nature and isolates metal from the atmosphere. This is the reason why copper sheets on roofs tend to last for well over 100 years.

1.2 Problems due to corrosion

Corrosion is detrimental to structural materials and sometimes even causes catastrophic failures resulting in a huge monetary loss as well as loss of lives. Researchers throughout the world have done extensive research to understand the mechanism of corrosion phenomena and the corrosion resistant properties of various metals and alloys in order to use them for different applications. They have made simultaneous efforts to control corrosion by developing superior alloys, changing the microstructure of the alloys, applying corrosion resistant coatings, altering the environments and so forth. Their ultimate aim was to improve the service life of materials to the maximum extent. All these efforts are aimed to mitigate the effect of corrosion. Some of the major harmful effects of corrosion are as follow:

- Reduction of metal thickness resulting in loss of mechanical integrity and structural failure or breakdown of metallic components.
- Severe casualties or loss of human lives due to structural failure or breakdown (e.g., aircraft, cars, bridges, etc.).
- Monetary losses due to the shutdown of industrial plants.
- Economic losses due to fluid contamination because of vessel corrosion.
- Leakage of pipelines causing the release of hazardous chemicals in the environment.
- Mechanical damage to valves, pumps, etc., or blockage of pipes by solid corrosion products.
- Additional expenditure on maintenance of corroded components and redesigning equipment that can withstand corrosion.
- Loss of technically important surface properties (electrical conductivity, surface reflectivity, ease of fluid flow on the surface, etc.) of metals.

The global cost of corrosion, estimated by NACE's International Measures of Prevention, Application and Economics of Corrosion Technology (IMPACT) study, has risen to \$2.5 trillion, which is equivalent to 3.4% of the world's GDP. The data provided by NACE of the global cost of corrosion is mentioned in Table 1.1.

Table 1.1: Global cost of corrosion (Billion US $) (copyright NACE international)

Economic Regions	Agriculture CoC US $ Billions	Industry CoC US $ Billions	Services CoC US $ Billions	Total CoC US $ Billions	Total GDP US $ Billions	CoC % GDP
United States	2.0	303.2	146.0	451.3	16,720	2.7%
India	17.7	20.3	32.3	70.3	1,670	4.2%
European Region	3.5	401	297	701.5	18,331	3.8%
Arab countries	13.3	34.2	92.6	140.1	2,789	5.0%
China	56.2	192.5	146.2	394.9	9,330	4.2%
Russia	5.4	37.2	41.9	84.5	2,113	4.0%
Japan	0.6	45.9	5.1	51.6	5,002	1.0%
Four Asian Tigers + Macau	1.5	29.9	27.3	58.6	2,302	2.5%
Rest of the World	52.4	382.5	117.6	552.5	16,057	3.4%
Global	152.7	1,446.7	906.0	2,505.4	74,314	3.4%

From the above data, it can be noted that the economic losses due to corrosion are huge in developed countries like the United States, European region, China, etc. Massive industrial growth is the main reason behind it. The major cost could be because of the industrial shutdown, damage of structural components of industries, frequent replacement of damaged metallic components and maintenance of refinery plants and so on. Apart from this, loss of efficiency, pollution, expulsion or undesirable failure of machinery, loss of production during shutdown of plants, etc., are some of the additional, indirect economic losses due to corrosion. Furthermore, loss of human lives during accidents caused by corrosion like damage of pipelines, the bursting of boilers, sudden leakage of gas pipelines are always beyond interpretation in terms of currency.

India with a GDP of approximately $2 trillion loses as much as $100 billion every year on account of corrosion. Some of the recent accidents around the world due to corrosion are mentioned here. On November 22, 2013, the Donghuang II oil pipeline suddenly exploded in Qingdao in eastern China. The blast killed 62 people and left 136 people injured. In 2009, a 50-foot-tall, high-pressure crystal production vessel, at NDK Crystal manufacturing facility in Belvidere, Illinois, exploded, injuring by standers and killing a trucker at a nearby gas station. On May 20, 2000, as hundreds of NASCAR fans left a stock car racing and were crossing a pedestrian bridge an 80-foot section of the concrete and steel walkway snapped in half. Pedestrians fell 17 feet below to the highway beneath. The bridge failure injured 107 people. On August 19, 2000, a 30-inch natural gas pipeline owned by El Paso Natural Gas exploded, leaving an 86-foot-long, 46-foot-wide and 20-foot-deep crater. Twelve people died in that 1,200-degree fireball. The cause of all of these accidents was either pipe or structural failure due to corrosion. These accidents give a clear indication that several times corrosion is proved catastrophic, taking a toll on human lives. Typical examples of corrosion are shown in Figure 1.1.

Figure 1.1: Typical examples of corrosion

1.3 Classification of corrosion

Corrosion can be classified in a number of ways like low-temperature corrosion and high-temperature corrosion. It can also be categorized as wet corrosion and dry corrosion. The preferred classification here is uniform corrosion and localized corrosion. Uniform corrosion is a common type of corrosion that occurs evenly on the surface of the metal substrate. The failure of a protective coating or barrier film on the metal structure results in uniform corrosion. It causes thinning of the metal as a result of chemical or electrochemical reactions that occurs on the metal surface. Initially, it damages the surface and leads it to a point of failure. It is a type of corrosion which is predictable, controllable and preventable, so it is often considered to be a safe form of corrosion. The most common examples of corrosion are oxidation and tarnishing, atmospheric and immersed corrosion of metals. It can be protected by various protection methods like cathodic protection, anodic protection and applying of paints, etc. Localized corrosion occurs at selective sites on the metal substrate. It causes severe degradation of metal in comparison to uniform corrosion. Localized corrosion is a dangerous form of corrosion because it is very difficult to detect and occurs usually without any warning. Corrosion action at localized sites depends upon various factors like the exposure time, defects in barrier coatings and variation in electrolytes, etc. Localized corrosion has been classified into following forms.

1.3.1 Pitting corrosion

Pitting is the most destructive form of localized corrosion. It results in the formation of small pits or cavities in the substrates as shown in Figure 1.2(a and b). Pitting occurs due to the collapse of

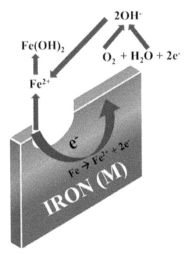

Figure 1.2(a): Schematic presentation of pitting corrosion

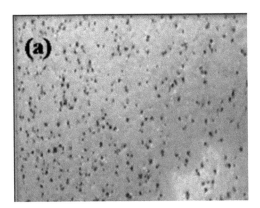

Figure 1.2(b): Pitting corrosion on exposure to atmospheric conditions

the barrier coating in the presence of aggressive anions. It occurs in a very small area of the metal surface while the remaining surface remains unaffected. Pit region becomes anodic and the rest of the portion becomes cathodic.

It initiates the galvanic reaction, which results in an increase in the pH inside the pit. This acidified electrolyte inside the pits hinders repassivation of the metal and promotes propagation of pits.

It is very difficult to predict pitting due to their small size and sometimes pits may be covered with corrosion product. Pitting results in minimal weight loss in the metal substrate, which can lead to a complete failure of the structure. Pitting corrosion is believed to be the cause of the 1967 collapse of the U.S. Highway 35 bridge between Point Pleasant, WV and Kanauga, OH, when that structure suddenly fell into the Ohio River. According to Jane Alexander (2017), investigators determined the cause of this disaster had begun decades earlier with a small crack that formed during the casting of the bridge's I-beams. The I-bar subsequently broke under the compounding stress from the corrosive environment and the newer, heavier vehicles crossing the bridge.

1.3.2 Crevice corrosion

It is a special type of pitting with the geometry of crevice, and generally occurs adjacent to the gap or crevice between the two joining metal surfaces. It usually occurs in engineering structures like

Figure 1.3: Crevice Corrosion of steel die mold

between the bolted joints, below flanges or between the flanges, at nuts and rivets' heads, etc. (see Figure 1.3). The size of crevice is quite narrow to maintain the stagnant zone and wide enough to permit access of liquid. Its initiation depends upon various factors like variation in the concentration of oxygen, pH and concentration of constituents. The concentration of oxygen and pH is quite high in bulk solution in comparison to inside of the crevice, this step up the electrochemical cell. Inside, the crevice oxidation of iron takes place, Fe converts into Fe^{2+}. At cathode, reduction of oxygen takes places that results in the formation of passive layer $Fe(OH)_2$ at the mouth of the crevice. After initiation, propagation mechanism is quite similar to the pitting corrosion mechanism.

1.3.3 Galvanic corrosion

Galvanic corrosion results when two different metal couples in the presence of the electrolyte or corrosive conditions. It is also known as bimetallic corrosion. The essential conditions for galvanic corrosion to occur are two electrochemically different metals should be present, there must be electrical contact between them and both metals should be exposed to the electrolyte. The driving factor for this type of corrosion is the potential developed between the dissimilar metals (Figure 1.4). The metal, which is more active or less noble, acts as an anode and tend to corrode faster. However, a substrate, which is less active or noble, acts as a cathode and corrodes at a slower rate. The electrolyte provides a mean of transfer of an ion from an anode to a cathode. Most of the bimetallic corrosion occurs in the sea environment due to salt water effectiveness as an electrolyte. The electrolyte acts as a conduit for ion migration, moving metal ions from the anode to the cathode. The anode metal, as a result, corrodes more quickly than it would otherwise, while the cathode metal corrodes more slowly and in some cases the rate of corrosion is very small. The position of metals in the galvanic series is responsible for galvanic corrosion. Hence, the effect of galvanic corrosion can be minimized by selecting metals close to each other in the galvanic series. Typical examples of galvanic corrosion are galvanic corrosion of the body of the ship in contact with bronze or brass propellers; in the heat exchangers between the tubes and tube sheet; defects in the copper coating on the surface of steel coated with copper; steel pipe fitted with brass fittings, etc.

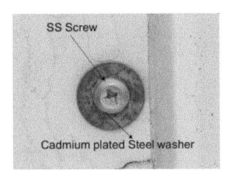

Figure 1.4: Galvanic corrosion in water pipelines and SS screw fixed on steel washer

1.3.4 Erosion corrosion

Erosion corrosion is a combined effect of corrosion or erosion which occurred due to the relative movement between a fluid and metal substrate surface. This type of corrosion mainly occurs in the pipeline, the main reason for deterioration is turbulence of the fluid. The rate of erosion corrosion depends upon velocity and the physical condition of the fluid. The combined effect of corrosion and erosion leads to aggressive pitting in the substrate. High shear stress is commonly observed in this type of corrosion. The presence of abrasive particles in the fluid causes the depletion of the outer layer because of the relative movement of the solid with respect to the surface. Cavitation is a special case of erosion corrosion that is caused by the collapse of vapor bubbles in liquid contacting a metal surface.

1.3.5 Intergranular corrosion

Intergranular corrosion is a special form of corrosion which generally occurs at grain boundaries or region next to its boundaries. It is also known as the intergranular attack or interdendritic corrosion. The main reasons for intergranular corrosion are the formation of precipitates and segregate in the specific region of grain boundaries. The presence of precipitates and segregates make the grain boundaries physically and chemically different from the grains, causing selective dissolution of grain boundaries or the region close to the grain boundaries. Intergranular corrosion is generally confined in a very small area, but in some cases, the complete grain gets dislodged due to the total destruction of the boundaries. It severely affects the mechanical properties of the metal substrates. The well-known example of intergranular corrosion is the sensitization of stainless steels or weld decay. In this case, chromium gets precipitates at grain boundaries that lead to a depletion of Cr concentrations in the region next to these precipitates, making these areas susceptible to corrosive attack. Identification of this corrosion is usually done under microscopic examination, but in some cases, it is even visible to the naked eyes.

Intermetallic compounds such as Mg5Al8, formation at the grain boundaries form a galvanic cell with an alloy matrix in the corrosive marine environment and severe intergranular corrosion occur. It

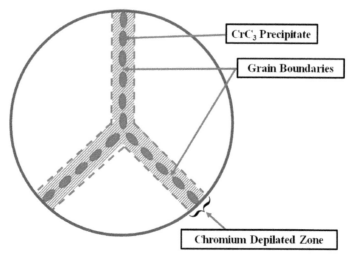

Figure 1.5: Intergranular corrosion in SS

can also occur in many alloy systems such as stainless steel, Ni and Al alloy. Small amounts of iron in aluminum have been shown to segregate the grain boundaries and cause intergranular corrosion.

Even depletion of chromium in SS results in IC. In steel, depletion of chromium occurs. If chromium in steel is less than 10%, then relatively low corrosion resistance is approached. Usually, SS 304 contains 0.06 to 0.08% of carbon and so carbon reacts with chromium leading to the formation of chromium carbide as shown in Figure 1.5. This phenomenon is observed when steel is heated in the sensitization temperature range (950–1450 °F). If metal is cross-sectionally cut and examined with a scanning electron microscope, the corroded area of CrC_3 will be observed as a deep narrow trench.

1.3.6 Cracking corrosion

Cracking corrosion is another class of corrosion that leads to sudden failure of the structure. Cracking corrosion generally occurs when ductile metal substrates is subjected to stress at elevated temperature. It includes corrosion fatigue, stress corrosion cracking and hydrogen-damage types of corrosion.

1.3.7 Stress corrosion cracking

Stress corrosion cracking arises due to the combined effect of tensile stress and a corrosive environment. Either external stress or residual stress inside the material can also cause stress corrosion cracking as shown in Figure 1.6. It causes unexpected failure of the metal structure. It commonly occurs in areas of high stress, pressure vessels, pipelines and reactors buried under the earth.

Apart from metal, materials like ceramics, pure materials and polymeric materials are susceptible to stress cracking corrosion in certain corrosive environment. A corrosive environment that initiates cracking mainly constitutes aggressive corrosive ions like Cl^-, NH_3 in the medium. The mechanism of stress corrosion cracking as shown in Figure 1.7 is still not well understood, but it is thought to be caused by stress, aggressive corrosive environments and susceptible microstructures and can be correlated as shown.

It is possible that in some cases beginning of crack may have been associated with the formation of a brittle film at the surface of the material, which could lower ductility due to a different metal composition than the bulk material. The most obvious identifying characteristic of stress corrosion

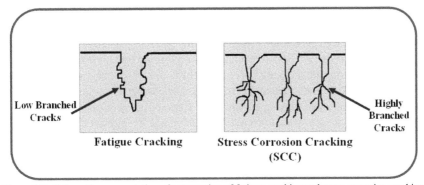

Figure 1.6: Schematic representation of propagation of fatigue cracking and stress corrosion cracking

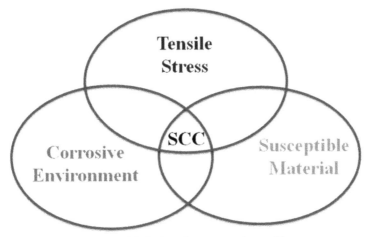

Figure 1.7: Mechanism of stress corrosion cracking

cracking in pipelines, regardless of pH, is the appearance of patches or colonies of parallel cracks on the external surface of the pipe. Pitting is commonly associated with stress cracking corrosion phenomenon. Aluminum and steel are two metals which are more susceptible to SCC. Stress corrosion cracking in pipelines begins when small cracks develop on the external surface of buried pipelines. In severe cases, it can end in a structural failure.

1.3.8 Fatigue corrosion

Fatigue corrosion arises due to the simultaneous effect of cyclic stress and corrosive environments. The collective effect of these two processes is much more dangerous than alone. It generally occurs at pits, surface defects or irregularities. Fatigue corrosion is similar to stress corrosion cracking in many ways, except it can take place in any environment. In fatigue corrosion, trans-granular propagation is generally observed and does not exhibit branched propagation as observed in the stress corrosion cracking (Figure 1.2). The extent of fatigue corrosion depends upon factors like environment, loading and metallurgical. Morphology and size of cracks depend on the frequency of stress. High frequency stress results in fine cracks and low frequency stress develops broad cracks. In order to understand the complex corrosion processes, its electrochemistry needs to be discussed first. The following section mentions an elaborate discussion on the electrochemistry of corrosion.

1.4 Electrochemistry of corrosion

The mechanism of corrosion can be explained chemically/electrochemically by which the deterioration of metal surfaces take place. The basic mechanism is the ionic transport processes that take place at the metal/electrolyte interface during corrosion. These processes are multi-step and the slowest step governs the rate of the overall corrosion reaction. The corrosion rates can be evaluated in terms of weight loss/gain of the metal during corrosion process, build up of the corrosion product layer, changes in the mechanical or physical properties of the metal, changes in the surface appearance of the metal, etc. The mechanism of corrosion is explained electrochemically by measurement of the corrosion potential and corrosion current. The electrochemical corrosion involves the expulsion of ions from the metal surface to the environment and simultaneous movement of electrons within the material. The mechanism is valid if the environment possesses ions and the material is conducting in nature (metals). The corrosion of metals in an aqueous environment is the best example of an electrochemical mechanism in which metal atoms go into the solution in form of metal ions and electrons migrate to the corroding site where they are consumed by the species in contact with the metal. The most acceptable electrochemical theory of corrosion is given by Whiteny (Whiteny 1903). In addition to this, Evans, Bockris, Frunkin, Parsons and Vetter have made significant contributions in the elucidation of the mechanism of electrochemical reactions (Bokris and Reddy 1970; Bokris and Yang 1991). During the aqueous corrosion process, the surface of the metal is distributed into several anodic and cathodic regions. If we take the example of iron, from the anodic regions, the iron (Fe) goes into solution as ferrous (Fe^{2+}) ions, constituting the anodic reaction. Under this condition, the oxidation of Fe atoms to Fe^{2+} ions takes place with the simultaneous release of electrons, whose negative charge would quickly build up in the metal and prevent further anodic reaction or corrosion. The schematic representation of corrosion is shown in Scheme 1.

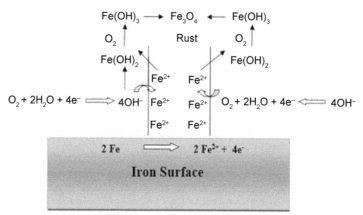

Scheme 1: Mechanism of corrosion on the iron surface

The anodic reaction will continue if the electrons released during the anodic reaction can be moved to a site on the metal surface where a cathodic reaction is possible. At a cathodic site, the electrons react with some reducible components of the electrolyte and are themselves removed from the metal as shown in the following equations:

$$Fe \rightleftharpoons Fe^{2+} + 2e^- \tag{1}$$

Hydrogen evolution

$$2H^+ + 2e^- \longrightarrow H_2 \text{ (in acidic solution)} \tag{2}$$

Oxygen reduction

$$O_2 + 4H^+ + 4e^- \longrightarrow 2H_2O \text{ (in acidic solution)} \tag{3}$$

$$O_2 + 2H_2O + 4e^- \longrightarrow 4OH^- \text{ (in neutral and alkaline solution)} \tag{4}$$

Metal deposition

$$M^{n+} + ne^- \longrightarrow M \tag{5}$$

Metal reduction

$$M^{n+} + e^- \longrightarrow M^{+(n-1)} \tag{6}$$

According to Faraday's Law, the rates of the anodic and cathodic reactions must be equivalent, which is determined by the total flow of electrons from anodic regions to cathodic regions and known as, corrosion current (i_{corr}). Since the corrosion current must also flow through the electrolyte by ionic conduction, the conductivity of the electrolyte will influence the way in which corrosion cells operate. The corroding piece of metal is described as a "mixed electrode" since simultaneous anodic and cathodic reactions are proceeding on its surface. The mixed electrode is a complete electrochemical cell present on the metal surface. Metals have their own tendency to corrode depending on the free energy difference (ΔG) between the metal and its corrosion products. Hence, ΔG gives the tendency of a metal to corrode. The change in free energy is related to the electrode potential by the following relation

$$\Delta G = -nFE \tag{7}$$

Here, n = number of electrons involved in the reaction.
F = Faraday's constant
E = Electrode potential

The magnitude of the change in free energy (ΔG) of a particular corrosion reaction is the measure of the spontaneity of the reaction. It gives an idea about the extent to which a reaction will proceed before the attainment of equilibrium. The negative value of ΔG ($\Delta G \ll 0$) shows the tendency of a metal to react readily with the species in the solution, but whether the reaction proceeds to any extent will depend on the kinetic factors. On the other hand, if the value of ΔG is positive ($\Delta G > 0$) then the metal shows its stability in the environment. If the value of $\Delta G = 0$, then the system will not proceed in any direction (equilibrium state). The value of free energy change (ΔG) and equilibrium constant (K) of corrosion reaction are calculated by using standard electrode potential (E^0) or the standard chemical potentials. The value of electrode potential is calculated from the Nernst equation.

$$E = E° - RT/nF \ln [\text{reduced}]/[\text{oxidized}] \tag{8}$$

Here, E°   = Standard electrode potential
R   = Gas constant (8.314 JK^{-1}mol^{-1})
T   = Absolute temperature
F   = Faraday's constant (96,500 Cmole^{-1})
n   = number of electrons transferred in the reaction
[oxidized] = concentration of oxidised species (mole/l)
[reduced]   = concentration of the reduced species (mole/l)

When a metal or alloy is immersed in a corrosive electrolyte, the metallic surface gets divided into several anodic and cathodic sites which create potential differences across the metal surface. This ultimately results in the differences in the free energy in these regions. The commencement of oxidation and reduction reactions in these regions causes the metal surface to act as localized galvanic cells and

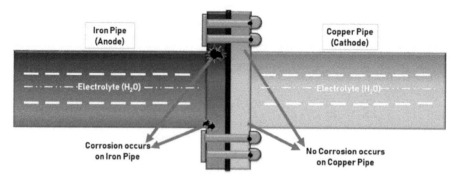

Figure 1.8: Mechanism of corrosion when two metals with different electrochemical potentials are joined

the corrosion reaction proceeds further. Therefore, measurement of oxidation and reduction potential has a great significance in the prediction of corrosion of the metal in various corrosive systems. These potentials are also called as half cell potentials or redox potentials. Standard reduction potentials are used to evaluate the electrochemical stability of various metals. Hydrogen is chosen as a standard for reference and its electrode potential is given as zero volt. The electrode potentials of other elements are compared with that of hydrogen and are called a standard electrode potential for that particular element. So, the metal having potential more negative than the standard hydrogen electrode (SHE) tends to corrode in acidic solutions. Corrosion occurs when two metals with different electrochemical potentials are in contact with a corrosive electrolyte or water. Like copper pipe in contact with an iron pipe will lead to formation of the galvanic cell leading to the formation of corroded surface at the iron side of the joint because copper has positive electrode potential compared to iron; thereby iron pipe is working as the anode and copper pipe as the cathode which can be depicted in Figure 1.8. Positive ions from Fe anode move towards Cu cathode through the aqueous medium. H^+ ions are released at Cu cathode. Fe^{2+} move towards the cathode and OH^- ions towards anode leading to the formation of ferrous hydroxide and towards the formation of rust at the joints towards the iron pipe.

These include metals like tin, nickel, iron, chromium, aluminum, etc., together with other metals with negative potential. On the other hand, metals like copper, mercury, silver, palladium, etc., and other metals with potential more positive than the SHE and will remain passive in acidic solutions. The above description shows the significance of oxidation–reduction potential in predicting the occurrence of corrosion. The standard oxidation-reduction (redox) potentials at 25°C (EMF Series) is given in Table 1.2. The EMF series as mentioned in Table 1.2 is widely used to make an initial selection of possible corrosion resistant alloys for different corrosive media. However, electrochemical (EMF) series has its own limitations. The electrode potentials listed in the EMF series is measured under standard conditions and is independent of other factors of the environment. Further, it does not take any consideration about the alloys/composites which are frequently used materials nowadays.

In order to overcome these limitations of EMF series, a new series known as galvanic series (Table 1.3) has been suggested. In galvanic series metal and alloys are arranged in accordance with their measured potentials in a given environment. The potentials listed in galvanic series are the free corrosion potentials, and the wider the separation between two metals in the series the more severe is the corrosion of more active metal among them. The order of metals listed in galvanic series alters by the change in the environment. For example, aluminum is anodic to zinc according to EMF series, whereas, zinc is shown to be anodic to aluminum in galvanic series. It has been actually observed in the marine environment that zinc is anodic to aluminum.

Table 1.2: Standard oxidation–reduction (Redox) potentials at 25°C (EMF/V)

Au	$\rightarrow$	$Au^{3+} + 3e^-$	-1.42	(+) Noble
Pt	$\rightarrow$	$Pt^{2+} + 2e^-$	-1.2	
$O_2 + 4H^+ + 4e^-$	$\rightarrow$	$2H_2O$	-1.23	
Pd	$\rightarrow$	$Pd^{2+} + 2e^-$	-0.83	
Ag	$\rightarrow$	$Ag^+ + e^-$	-0.799	
2Hg	$\rightarrow$	$Hg_2^{2+} + 2e^-$	-0.798	
$Fe^{3+} + e$	$\rightarrow$	Fe^{2+}	-0.771	
$O_2 + 2H_2O + 4e^-$	$\rightarrow$	$4\ OH^-$	-0.401	
Cu	$\rightarrow$	$Cu^{2+} + 2e^-$	-0.34	
$Sn^{+4} + 2e^-$	$\rightarrow$	Sn^{2+}	-0.154	
$2H^+ + 2e^-$	$\rightarrow$	H_2	-0.000	
Pb	$\rightarrow$	$Pb^{2+} + 2e^-$	0.126	
Sn	$\rightarrow$	$Sn^{2+} + 2e^-$	0.140	
Ni	$\rightarrow$	$Ni^{2+} + 2e^-$	0.23	
Co	$\rightarrow$	$Co^{2+} + 2e^-$	0.27	
Cd	$\rightarrow$	$Cd^{2+} + 2e^-$	0.402	
Fe	$\rightarrow$	$Fe^{2+} + 2e^-$	0.44	
Cr	$\rightarrow$	$Cr^{3+} + 3e^-$	0.71	
Zn	$\rightarrow$	$Zn^{2+} + 2e^-$	0.763	
Al	$\rightarrow$	$Al^{3+} + 3e^-$	1.66	
Mg	$\rightarrow$	$Mg^{2+} + 2e^-$	2.38	
Na	$\rightarrow$	$Na^+ + e^-$	2.71	
K	$\rightarrow$	$K^+ + e^-$	2.92	(–) Active

Table 1.3: Galvanic series in seawater (3% NaCl) of some commercial metals and alloys

Cathodic (noble)

Platinum
Gold
Graphite
Silver
18-8-3 Stainless steel, Type 316 (passive)
18-8 Stainless steel, Type 304 (passive)
Titanium
13% Chromium stainless steel, Type 410 (active)
67 Ni-33Cu alloy
76 Ni-16Cr-7Fe alloy (passive)
Nickel (passive)
Silver solder
Bronze
70–30 Cupro-Nickel
Silicon Bronze
Copper
Red Brass
Aluminum Brass
Tin
Lead
Cast iron
Wrought iron
Mild steel
Aluminum
Galvanized steel
Zinc
Magnesium alloys
Magnesium

Anodic (active)

1.4.1 Thermodynamics aspects of corrosion process (Pourbaix diagram)

Pourbaix diagram (Potential–pH diagram) explains the thermodynamic basis of the corrosion reaction. Pourbaix correlated the dependence of pH and the potential of the electrode upon the condition of the electrode (Pourbaix 1966). In the Pourbaix diagram, there is a diagrammatic presentation of metal; corroding, not corroding and passivated in aqueous solution. The diagram has been divided into domains of immunity, corrosion and passivation. Each domain indicates a region in which one species is most thermodynamically stable. When a metal is in the most stable species, then it is considered as immune to corrosion but if a soluble ion is the most stable entity then the metal will corrode. On the other hand, if in a region the insoluble product is the most stable species, then it is considered to be passive. Figure 1.3 illustrates the potential—pH diagram for iron exposed to water. When iron corrodes in water following possible reactions take place:

$$Fe \rightleftharpoons Fe^{2+} + 2\,e^- \qquad\qquad (9)\ \text{Corrosion reaction}$$

$$Fe^{2+} \rightleftharpoons Fe^{3+} + e^- \qquad\qquad (10)\ \text{Oxidation reaction}$$

$$2Fe^{2+} + 3H_2O \rightleftharpoons Fe_2O_3 + 6H^+ + 6e^- \qquad\qquad (11)\ \text{Precipitation reaction}$$

$$Fe^{3+} + H_2O \rightleftharpoons Fe(OH)_3 + H^+ \qquad\qquad (12)\ \text{Hydrolysis reaction}$$

$$2Fe + 3H_2O \rightleftharpoons Fe_2O_3 + 6H^+ + 6e \qquad\qquad (13)\ \text{Corrosion reaction}$$

$$Fe + 2H_2O \rightleftharpoons HFeO_2^- + 3H^+ + 2e^- \qquad\qquad (14)\ \text{Corrosion reaction}$$
(Hypoferrite)

$$HFeO_2^- + H_2O \rightleftharpoons Fe(OH)_3 + e^- \qquad\qquad (15)\ \text{Precipitation reaction}$$

$$Fe^{2+} + 2OH^- \rightleftharpoons Fe(OH)_2 \qquad\qquad (16)\ \text{Precipitation reaction}$$

Here, reaction (9), (10) and (15) are independent of pH and is represented by straight horizontal lines, while reactions (11), (13) and (14) are dependent on both pH and potential and is represented on the potential–pH plot by sloped lines. Reactions (12) and (16), which depend on pH, are represented by vertical lines. Oxygen is evolved above but not below the line "cd", in accordance with the reactions.

$$H_2O \longrightarrow \tfrac{1}{2}\,O_2 + 2H^+ + 2e^- \qquad\qquad (17)$$

Hydrogen is evolved below but not above the line "ab", in accordance with the reaction.

$$H^+ \longrightarrow \tfrac{1}{2}\,H_2 + e^- \qquad\qquad (18)$$

It can be noticed in Figure 1.9 that the redox potential of the hydrogen electrode (line ab) lies above the immunity region along all the pH scale. This shows that it may be dissolved with the evolution of hydrogen in aqueous solutions of all pH values. In the pH range of 9.4 to 12.5. However, a passivating layer of $Fe(OH)_2$ is formed (reaction (16)). At low pH, iron is corroded with the formation of Fe^{2+} ions (reaction (9)). At higher pH, soluble hypoferrite ions can form within a restricted active potential range. At higher redox potential in corroding medium, the passivating layer consists of $Fe(OH)_3$ or $Fe_2O.nH_2O$ or Fe_2O_3 or Fe_3O_4. Soluble ferrate (FeO_4^{2-}) ions can form in alkaline solution at a very noble potential. Though the potential–pH diagram is quite useful in showing the conditions of potential and pH under which the metal will corrode, there are several limitations regarding the use of this diagram in practical corrosion problems. Since the data in the potential–pH diagram are thermodynamic, they do not convey any information regarding the rate of the reactions.

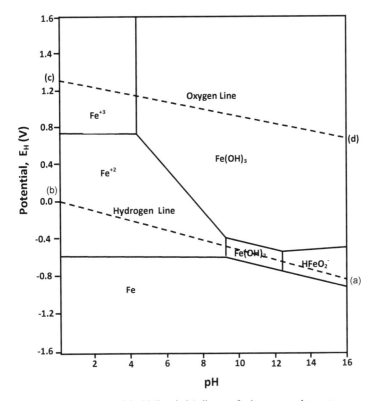

Figure 1.9: Potential–pH (Pourbaix) diagram for iron exposed to water

1.4.2 Kinetic aspects of the corrosion process

1.4.2.1 Polarization methods of determination of corrosion rate

One of the widely accepted laws of electrochemistry is given by Tafel, Bockris and Reddy (Bokris and Reddy 1977). This law is most frequently used to study the kinetic aspects of the corrosion process. There are basically two methods that are based on polarization measurement to measure the corrosion kinetics (corrosion rate). The first one is Tafel slope extrapolation and the second one is the linear polarization resistance method (Stern Geary method). Tafel explained that for an electrochemical reaction taking place on a corroding metal surface, the logarithm of the current density varies linearly with the potential of the electrode. Here, the potential of the electrode is also polarized from the open circuit potential (OCP) values. The Tafel polarization curve is composed of anodic polarization curve where metal oxidation takes place, and cathodic polarization curve where reduction of the oxidant from the solution takes place. Under a condition when the metal surface is largely polarized and either the anodic or cathodic corrosion reaction predominates, the potential (E) versus logarithm of current density (log i) curve gives a straight line (Tafel lines). The potential versus log i curve is shown in Figure 1.10. The linear regions of the Tafel plot are extrapolated to intersect at the point coordinate, which gives the value of corrosion potential (E_{corr}) and corrosion current density (i_{corr}).

The slope of the linear anodic and cathodic regions gives the values of anodic (β_a) and cathodic (β_c) Tafel constants respectively. At potential away from the corrosion potential (E_{corr}), say in the anodic region, the current density exhibits the kinetics of the anodic corrosion reactions.

When, potential (E) >> corrosion potential (E_{corr}) and anodic current density (i_a) >> cathodic current density (i_c), the Tafel slope of anodic reaction is obtained, i.e., the value of βa is obtained and is expressed by the relation:

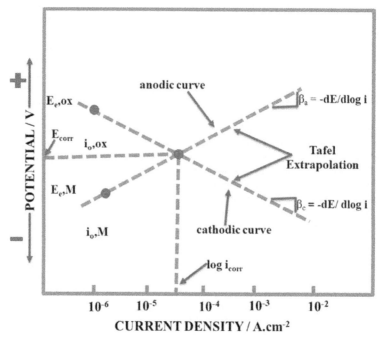

Figure 1.10: Tafel polarization curve for the corroding system

$$\beta a = [dE/dlogi]_{E \gg Ecorr} = 2.3RT/\alpha_a zF \tag{19}$$

Similarly, for the βc, the relation is expressed as

$$\beta c = -[dE/dlogi]_{E \ll Ecorr} = -2.3RT/\alpha_c zF \tag{20}$$

Here, α_a and α_c are the charge transfer coefficients of anodic and cathodic reactions, respectively. Here, z is the valency of the metal and F is Faraday's constant. The Tafel extrapolation method is a widely used method of corrosion rate determination. However, one of the limitations of this method is the various values of βa and βc can be obtained by poor selection of slopes which ultimately results into the erroneous values of corrosion potential (E_{corr}) and corrosion current density (i_{corr}) (Rochini 1993). In order to minimize the possibilities of errors, two important points should be considered: firstly, the anodic/cathodic extrapolation should be done 50–100 mV away from the E_{corr} and secondly, either the anodic or the cathodic polarization curve show linear, semi-logarithmic behavior (Tafel behavior) over a range of one decade of current density. Along with this linear region, an accurate extrapolation can be done. If these points are not considered, the Tafel extrapolation can cause the measurement of several orders of magnitude high/low corrosion rate as compared to the actual corrosion rate. The linear polarization resistance (LPR) or the Stern and Geary equation is a widely used method of corrosion measurement. The variation of current density is found to be almost linear with respect to the applied potential in a region within a few mili-Volts from the E_{corr}. The expression for the current density can be written as:

$$i = i_{corr}[\{1 + \alpha_a (zF/RT) (E-E_{corr})\} - \{1 - \alpha_c (zF/RT) (E-E_{corr})\}] \tag{21}$$

$$i = i_{corr} [(\alpha_a zF/RT + \alpha_c zF/RT) (E-E_{corr})] \tag{22}$$

Replacing $\alpha_a zF/RT$ with $2.3/\beta a$ and $\alpha_c zF/RT$ with $2.3/\beta c$ and $E-E_{corr}$ with ΔE, equation (22) can be written as:

$$i_{corr} = 1/2.3 \ \beta a.\beta c/\beta a + \beta c \ (\Delta i/\Delta E)_{Ecorr} = B/Rp \tag{23}$$

{here, B = 1/2.3 βa.βc/βa+βc} and Rp = (ΔE/Δi) (E–Ecorr)$^{-\infty}$.

Here, R_p is the polarization resistance, which is given by the linear portion of the Tafel line (dE/di) at t = ∞ and ΔE = 0. Equation (xxiii) is the Stern Geary equation or linear polarization resistance method that correlates the polarization resistance with the corrosion current density of the corroding metal surface. The abovementioned methods enable us to measure the instantaneous rates of corrosion. However, these methods are susceptible to production of erroneous results.

1.4.2.2 Electrochemical Impedance Spectroscopy (EIS)

Electrical methods are used to perform *in situ* studies of electrochemical systems. The reason behind is that the kinetics of the surface reactions occurring in electrochemical systems can be evaluated using electrical response. Electrochemical Impedance Spectroscopy (EIS) is a non-destructive technique which provides accurate kinetic and mechanistic information of a corroding system (Walter 1991; Singh 2003; Moutarlier et al. 2005; Souto et al. 2003; Wang et al. 1996). It is one of the powerful tools to study the corrosion rate (Pebere et al. 1989; Cáceres et al. 2007; Aoki et al. 2001). The impedance technique has been widely used to evaluate the interaction of the metal-coating system with the corrosive environments (Mansfeld et al. 1986), degradation of highly resistive organic coatings (Tsai and Mansfeld 1993; Grundmeier et al. 2000; Kendig and Scully 1990), corrosion rates at the defects and porosity in the coatings (Zayed and Sagues 1990; Deflorian et al. 1994; Bonora et al. 1996).

Impedance is measured by applying a sinusoidal potential to an electrochemical cell and then the sinusoidal current is measured. The study is carried out by sweeping the frequency between a particular ranges and measuring impedance at each point. Impedance measurement is carried out by applying small amplitude perturbation signals because the polarization of the electrochemical system can demonstrate non-linear behavior. This approach facilitates the confinement to the almost linear part of the polarization curve. Hence, for this linear/pseudo-linear system, for the sinusoidal potential, the current will oscillate at the same frequency but shift in phases (Figure 1.11). In order to determine the cell impedance, the input and output signals are analyzed. The input potential signal is expressed in terms of polar and Cartesian variables as mentioned below:

$$E_t = E_o \sin(\omega t) \tag{24}$$

Here, E_t is the oscillating potential at time t, E_o is the signal amplitude and ω is angular frequency. The angular frequency (ω expressed in radians/second) and the frequency (f expressed in Hertz) are related to each other by the following expression:

$$\omega = 2\pi f \tag{25}$$

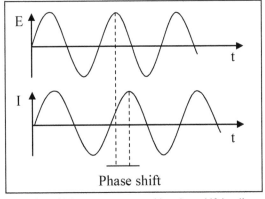

Figure 1.11: Sinusoidal current response with a phase shift in a linear system

In a linear system, the responding current signal I_t has the different amplitude (I_o) and is phase shift (ϕ), the expression relating I_t and I_o is written as follows:

$$I_t = I_o \sin(\omega t + \phi) \tag{26}$$

For the calculation of impedance, an expression analogous to Ohm's law is applied which is:

$$Z = E_t/I_t = E_o \sin(\omega t)/I_o \sin(\omega t + \phi)$$

$$Z = Z_o \sin(\omega t)/\sin(\omega t + \phi) \tag{27}$$

Impedance can be expressed as a complex function. Expressing sinusoidal potential E(t)

$$Et = E_o \exp(j\omega t) \tag{28}$$

$$I_t = I_o \exp(j\omega t - \phi) \tag{29}$$

Here, j is an imaginary number $= \sqrt{-1}$. The impedance (Z), when presented as a complex number, follows Ohm's law and related to potential and current signals as mentioned below:

$$Z(\omega) = E/I = |Z| \exp(j\phi) = |Z| (\cos\phi + j\sin\phi) = Z_r + jZ_j \tag{30}$$

Here, Z_r and Z_j are the real and imaginary parts of the complex impedance, respectively. Impedance data is presented in the form of Nyquist plot and Bode plots. The Nyquist plot is obtained by plotting real part of the impedance along the horizontal axis and the negative of the imaginary part of the impedance along the vertical axis. In the Nyquist plot, each point is the impedance at a particular frequency. Interestingly, the low frequency region of the Nyquist plot lies on the right side and the high frequency region on the left side.

In the Nyquist plot, the impedance is presented as an arrow of length |Z|. The angle between this arrow and X-axis is known as a phase angle and is represented by ϕ = arg Z. However, one limitation of the Nyquist plot is that at any point of the curve, the value of frequency is not known. The impedance data is interpreted by applying different equivalent circuit models using resistors, capacitors, diffusion element, inductors, etc. For the Nyquist plot shown in Figure 1.12 (which has one semicircle showing one time constant), the equivalent circuit known as Randle's circuit is used. The Randle's circuit consists of a resistor connected to a parallelly connected resistor and capacitor (Figure 1.13). This circuit is a simple model consisting of solution resistance, charge transfer resistance and double layer capacitance.

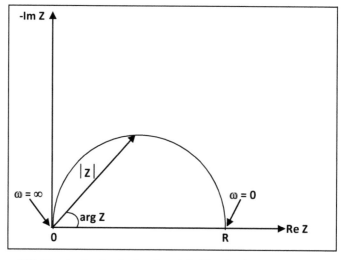

Figure 1.12: Nyquist plot showing imaginary (−Im Z) and real (Re Z) part of impedance

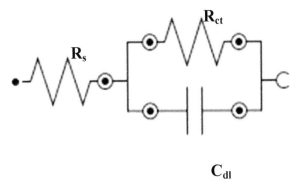

Figure 1.13: Simple Randle's circuit with one time constant having circuit elements like electrolyte resistance (R_s), charge transfer resistance (R_{ct}) and double layer capacitance (C_{dl})

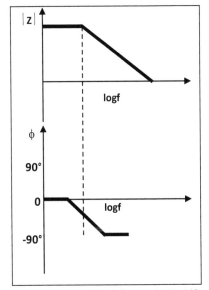

Figure 1.14: Bode plots; impedance (|Z|) versus log f and phase angle (ϕ) shift versus log f for one time constant

The value of impedance from the Randle's cell is calculated as follows:

$$Z = R_s + R_{ct}/1 + j\tau\omega \tag{31}$$

Here, τ is the time constant associated with the RC circuit. Another EIS plot is presented by the Bode plot. It consists of two plots; modulus of impedance, i.e., |Z| versus log of frequency plot and phase angle shift versus log of frequency plot. As compared to the Nyquist plot, Bode plots are more informative since it gives the values of frequency during the whole scan. The Bode plot corresponding to the Nyquist plot (Figure 1.12) is shown in Figure 1.14.

Some common equivalent circuit models on which the Nyquist plots are fitted to extract the EIS data are mentioned below:

1. **Purely inductive coating:** This circuit model is applied for undamaged coating with high impedance. The circuit contains a resistor (due to the electrolyte) and capacitor (for coating capacitance) in series (Figure 1.15a).

2. **Simple Randle's Cell:** It is the most common cell model and the basis of the more complex circuit systems. It contains solution resistance, double layer capacitor and polarization resistor

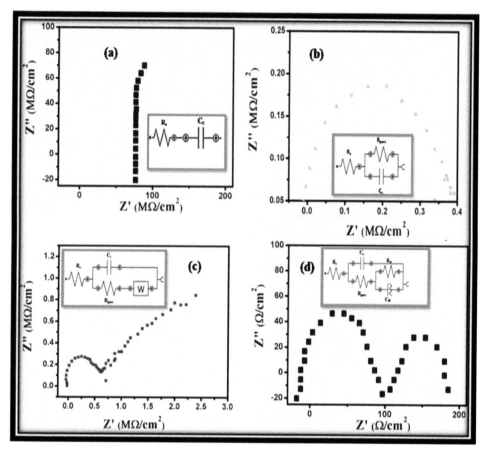

Figure 1.15: Electrical equivalent circuits of (a) Purely inductive coating, (b) simple Randle's cell circuit, (c) coating showing diffusion of electrolyte and (d) coatings in contact with the electrolyte (degradation in the process)

(Figure 1.15b). A semicircle is obtained for the simplified Randle's cell. The intercept at x-axis in high frequency yields solution resistance value. While intercept in low frequency region gives the sum of polarization resistance and solution resistance. The diameter of the semi-circle is equal to the polarization resistance.

3. **Coating showing diffusion of electrolyte (mixed control circuit):** This equivalent circuit is generally applied to the coatings showing diffusion of the electrolyte in the intact coatings. A diffused semicircle is generally obtained for this type of coating in Nyquist plot. In this circuit new impedance is measured that arises due to the diffusion known as Warburg Impedance (Figure 1.15c).

4. **Equivalent Circuit for Failed Coating:** When the coating is exposed to the electrolyte after certain time, electrolyte diffuses into the coating and causes undercoating corrosion process to commence. The process leads to the failure of the protective property of the coating. For this type of coating, we get a two-time constant plot in the Nyquist plot (Figure 1.15d).

1.5 Metal corrosion and passive film

Passivity is an inherent property of many structural metals, alloys, etc., which are active in the EMF series but corrodes regardless at a very low rate under specific environmental conditions. In the beginning of eighteenth century, H.H. Uhlig observed that iron corrodes rapidly in dilute HNO_3 but

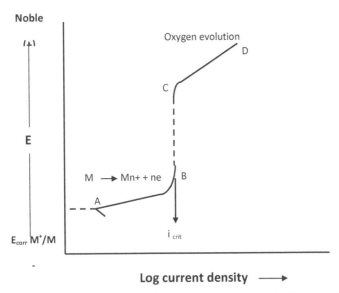

Figure 1.16: Galvanostatic anodic polarisation curve for metals exhibiting passivity

is almost unattacked in conc. HNO_3 (Uhlig 1978). Later on, Wagner defined the corrosion resistant state of iron as its "passive" state and proposed an extended definition of passivity (Wagner 1965). According to him, a metal is passive if on increasing the electrode potential towards more noble value, the rate of anodic dissolution in a given environment under steady-state conditions becomes less than the rate at some less noble potential. Metals such as chromium, iron, nickel, titanium and alloys containing these elements show essential inertness and behave like noble metals (Pt, Au); though their oxidation and reduction potential indicates that they should possess corrosion tendency in acidic solutions.

When electrode potential of metal is polarized towards more anodic potential, its dissolution rate increases. However, the polarization of metals showing the active/passive transition to higher anodic potentials causes reduced corrosion rates. This is due to the formation of a passive film on the surface of the metal. The electrochemical description of passivity is explained through galvanostatic polarization (at constant current) and potentiodynamic polarization (at varying potential) method.

The measurement of polarization of the electrode at constant current (galvanostatic polarization) does not give enough information regarding the mechanism of passivity. As shown in Figure 1.16, after increasing the applied current density beyond I_{crit}, there is a sharp increase in the potential followed by oxygen evolution (the region between points B and C). The region between points B and C is called a passivation range of the metal. One of the major drawbacks of this technique is that it gives no information about the behavior of metal between points B and C. On the other hand, potentiodynamic polarization technique is used to explain the mechanism of anodic passivation of metals and alloys. The anodic polarization curve for an active-passive electrode is given in Figure 1.17. Point A gives the value of the equilibrium potential of a metal at particular environmental conditions. The region between point A and B show the anodic polarization of the corroding metal. At Point A, the rate of oxidation is equal to that of reduction and the potential corresponding to this is called corrosion potential (E_{corr}). The polarization of the electrode potential of a metal in a positive direction causes an increase in the dissolution of rate. The curve AB corresponds to the Tafel dependence of the electrode potential with respect to current density during the dissolution of the metal.

$$M \longrightarrow M^{2+} + 2e^- \tag{32}$$

Point X, known as Flade-potential, is an equilibrium potential at which the formation of the protective oxide film is thermodynamically possible.

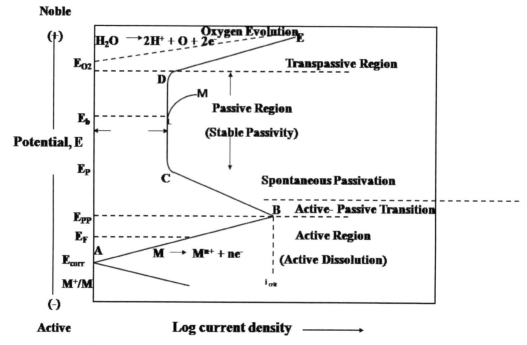

Figure 1.17: Potentiostatic anodic polarization curve for metals exhibiting passivity

$$M + H_2O \longrightarrow MO + 2H^+ + 2e^- \tag{33}$$

However, the rate of the formation of a protective film at this particular potential is very small due to the dissolution of the film. The rate of dissolution continues with the increase of electrode potential in a positive direction. At Point B, the accelerated rate of metal dissolution is equal to the retardation of this process. It is due to the formation of a protective film on the surface of the metal. Thus, a limiting passive current known as critical current density (I_{crit}) is achieved. The potential at this point is called a primary passive potential. On further polarization, the current density decreases sharply till a complete passive potential is reached (Point C). The reduced current density established over a range of potential is known as passive current i_p. It corresponds to the existence of a very thin film with a high dielectric property. The region between point C and D is called a passive region. Here, the rate of anodic reactions become independent of the electrode potential and is determined by the rate of chemical dissolution of the protective film. In this region thickening of the oxide layer with the increase of potential occurs. It involves the movement of M^{n+} outwards and further combination of these cations with O^{2-} and OH^- at the film solution interface. The end of the passive range (Point D) corresponds to the oxygen evolution reaction as follows:

$$4OH^- \longrightarrow O_2 + 2H_2O + 4e^- \tag{34} \text{ in neutral or basic solution}$$

$$2\,H_2O \longrightarrow O_2 + 4H^+ + 4e^- \tag{35} \text{ in acidic solution}$$

The curve DE shows the transpassive region, causing metal dissolution and oxygen evolution.

1.6 Tests for corrosion protection

1.6.1 Surface studies

Surface studies include the analysis of the surface of the metal before and after exposure to the aggressive corrosive environment in order to estimate the rate and the mechanism of corrosion.

Techniques like X-ray diffraction, scanning electron microscope and Transmission Electron Microscope are used to study the structure and chemical composition of the corrosion product formed on the surface of the metal. Field emission scanning electron microscope also has been employed to get surface morphology and chemical composition of the powder coated mild steel specimens.

1.6.2 Salt spray test

Salt spray testing is an accelerated corrosion test that produces a corrosive attack to coated samples in order to evaluate the coating for use as a protective finish. The appearance of corrosion products such as rust is evaluated after a pre-determined period of time. Test duration depends on the corrosion resistance of the coating; generally, the more corrosion-resistant the coating is, the longer the period of testing before the appearance of corrosion/rust. The salt spray test is one of the most widespread and long-established corrosion tests. ASTM B117 was the first internationally recognized salt spray standard, originally published in 1939. The apparatus required for salt spray (fog) exposure consists of a fog chamber, a salt solution reservoir, a supply of suitably conditioned compressed air, one or more atomizing nozzles, specimen supports, provision for heating the chamber and other necessary means of control. Salt spray chamber is made of a material resistant to corrosion by the use of sprayed solution. The photograph of the salt spray chamber is shown in Figure 1.18.

The most common test for steel based materials is the neutral salt spray test, which reflects the fact that this type of test solution is prepared to a neutral pH of 6.5 to 7.2. Each test surface shall be placed in the cabinet, facing upwards at an angle of 15 degrees to the vertical. The salt solution was prepared by dissolving 561 parts by mass of sodium chloride in 95 parts of water conforming to Type IV water in Specification D1193 (except that for this practice, limits for chlorides and sodium may be ignored). The exposure zone of the salt spray chamber shall be maintained at $35 \pm 2°C$ and at 65% relative humidity. Specimens for the evaluation of paints and other organic coatings should be suitably cleaned. To determine the development of corrosion from an abraded area in the paint or organic coating, a scratch or scribed line was made through the coating with a sharp instrument so as to expose the underlying metal before testing. Coated mild steel panels were prepared by coating on fully finished mild steel specimen dimension 15.0 cm × 10.0 cm × 0.12 cm. The coated panels were exposed to the salt spray of 5.0% NaCl solution for 60 to 120 days as per ASTM B117 method. All the coated mild steel panels were provided with a scribe mark across the panel and placed in the salt spray chamber. Figure 1.19 shows the photographs of epoxy coated (EC) and epoxy with copolymer

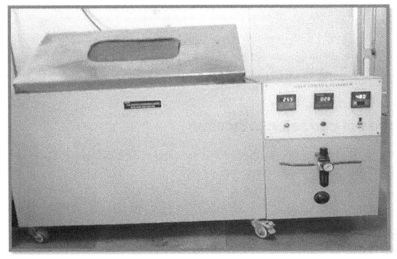

Figure 1.18: Image of a salt spray chamber

Figure 1.19: Images of test panels kept in salt spray chamber

composite coated mild steel panels which are kept in the salt spray chamber for 60 to 120 days. Anti-corrosion performance of coating is evaluated by visual observations, by means of the presence of spread of corrosion along the scribe mark after exposure to salt spray chamber.

1.6.3 Weight loss method

The weight loss method can be applied for corrosion study, if the specimens are of the same size and have been tested for the same span of time. Weight loss is expressed as the loss in the weight per unit area or per unit area per unit time. If we introduce the density of metal, the loss in the thickness of metal per unit time can be calculated. Corrosion Rate (C.R.) in mm/year can also be calculated by weight loss method as follows:

$$C.R.(mm/year) = \frac{(w_0 - w_1) \times 87.6}{a.t.d} \tag{36}$$

where w_0 = initial weight of the sample (g); w_1 = weight of the sample after corrosion (g);

a = area (cm^2); t = time (h) and d = density of metal or polymer (g/cm^3)

1.7 Methods of corrosion protection

A number of corrosion prevention methods have been reported in literature. There are mainly four primary ways to control corrosion. These are:

- Materials selection
- Cathodic and anodic protection
- Corrosion inhibitors
- Corrosion resistance coating

1.7.1 Materials selection

The most common and important method of controlling the corrosion is the selection of the right and proper materials for the particular corrosive environment. Corrosion behavior of each metal and alloy is unique and inherent, and corrosion of metal and alloy has a strong relation to the environment to which it is exposed. The rate of corrosion directly depends upon the corrosivity of the environment and is inversely proportional to the corrosion resistance of the metal. Hence, the knowledge of the nature of the environment to which the material is exposed is very important. Moreover, the corrosion resistance of each metal can be different in different exposure conditions. Therefore, the right choice of the materials in the given environment (metal corrosive environment combination) is very essential for the service life of equipment and structures made of these materials. Selection of correct material depends on the various factors such as their cost, availability, physical, mechanical and chemical properties. In addition, there are many factors which can affect the performance of the material. Such as corrosion and wear characteristics is an important factor, hence improper selection of materials without attention of their corrosion behavior in a corrosive environment can lead to the failure of components and plant shutdown; therefore in order to avoid failures due to corrosion, the material should be compatible with environment and it must have sufficient resistance to corrosion for the designed life. In addition to that an appropriate preventive maintenance practices must also be employed. The selection of material must be based on an extensive knowledge of the service environment. The material selected has to meet the criteria for mechanical strength, corrosion and wear resistance for specific service conditions. For example, high performance corrosion resistant alloys are specified for valves and piping systems in corrosive environments encountered in refineries and chemical processing plants.

Mechanical properties of the materials are also affected by the environment. For example, the yield strength of steels may be significantly reduced by saline water over a period of time. It is possible to reduce the corrosion rate by altering the corrosive medium. The alteration of the corrosive environment can be brought about by lowering temperature, decreasing velocity, removing oxygen or oxidizers and changing the concentration.

1.7.2 Cathodic and anodic protection

There are two ways to provide cathodic protection to metals namely using sacrificial anode and impressed current method. Cathodic protection of ferrous materials is done by using sacrificial anodes. Active metals like magnesium, zinc, aluminum are used as the sacrificial anode (Figure 1.20) in environments of low conductivity (e.g., soil), where the magnitude of iR drop between the electrode is substantial. A large potential difference between two metals (to be protected and sacrificed) is an absolute necessity. On the other hand, in highly conducting environment like water, a low potential drop is more economical to use. Among other variables, the relative positions of electrodes, polarization characteristics of electrodes and applied current density play a crucial role in cathodic protection. The impressed current method for cathodic protection is cheaper as compared to the sacrificial anode method, and permits greater control over the system. The method involves the use of an external source of current, such as D.C. rectifier, the battery in combination with a counter electrode (e.g., graphite, platinum, etc.) in order to polarize the metals to be protected cathodically.

Cathodic protection of big ship hulls, storage tankers, harbor structure, jetties, off-shore platforms, etc., is done by attaching more reactive metals like Mg or Al, wherein iron becomes the cathode and Mg/Al becomes anode. The reactive metal anode dissolves first and is sacrificed to protect iron as shown in Figure 1.20a.

Another important method of cathodic protection is by impressed current cathodic protection as shown in Figure 1.20b. The schematic representation of protection of a structure using aluminum anode shows how a structure of a steel structure can be protected in marine environment conditions if the sacrificial anode is attached. Aluminum metal is used more often than any other type of galvanic

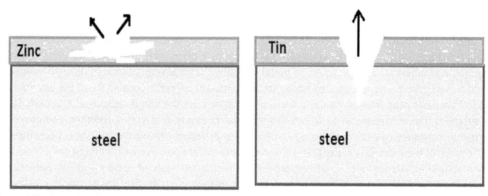

Figure 1.20: Example of cathodic protection using zinc and tin on steel, zinc makes a protecting layer that saves the steel from corrosion while tin is protected by corrosion of steel

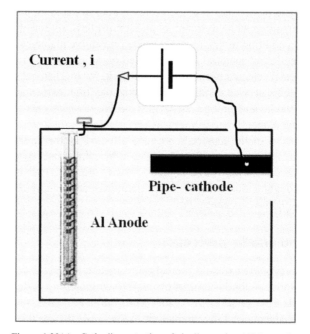

Figure 1.20(a): Cathodic protection of pipelines using Al/Mg anodes

metal in the marine environment due to its relatively low consumption rate which is approximately 7 lbs per amp year. Nowadays, the alloy composition of anodes is used because they have high open circuit potential. A typical alloy being used in industry for cathodic protection is Al 99.2%, Zn 0.3–0.6% and Hg 0.03 to 0.06%.

Anodic protection is also one of the recent developments in the field of corrosion control. The most extensive use of anodic protection was found in protecting, storage and handling of sulfuric acid. It can be used to control the corrosion of metals that exhibit the tendency to form a protective passive film. When anodically polarized from its free corrosion protection, considering the potential of the metal is maintained in the range which leads to the formation of the passive layer on its surface, then the corrosion current density becomes very low and stable. The passive layer formed on the surface of metals show electrical resistivity, and is relatively insoluble in the chemical environment. The failure of electrical supply leading to the depassivation is one of the major disadvantages of this method. Conducting polymers have also been used as anode materials for piplines with deteriorating

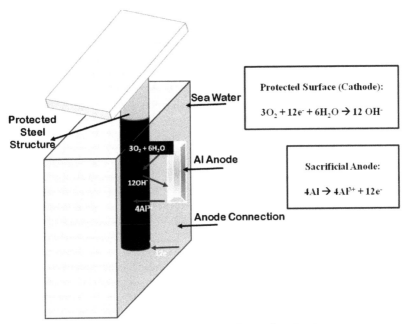

Figure 1.20(b): Schematic representation of protection of steel structure

coatings. Gibson et al. (1993) have elaborated the use of conductive polymeric cable anodes that eliminate the problems of conventional groudbeds.

1.7.3 Corrosion inhibitor

Corrosion phenomena are well-known in the petroleum industry, which leads to maximum damage to oil field equipment. Remarkable corrosion protection efforts have been made by petrochemical industries to prevent structural damage. An addition of corrosion inhibitors in small quantities can significantly retard the corrosion process. The definition of corrosion inhibitor as given by the National Association of Corrosion Engineers (NACE) is "a substance which retards corrosion when added to an environment in small concentrations". Inhibition usually involves the adsorption of organic species onto the metal surface to form a monomolecular barrier. As with barrier coatings, the adsorbed molecules can restrict the anodic and/or cathodic corrosion reactions and slow the corrosion rate. Santos et al. (1998) described the inhibitor action as involving the displacement of adsorbed water from the metal surface by soluble organic species (Org.)

$$Org_{sol} + nH_2O_{ads} \longrightarrow Org_{ads} + nH_2O_{sol} \tag{37}$$

The corrosion inhibitors for protection of metals in an aggressive environment are selected on the basis of their solubility or dispersibility in the solutions. For example, in a hydrocarbon system, a corrosion inhibitor soluble in hydrocarbon is used. Two-phase systems composed of both hydrocarbons and water, utilize oil soluble water-dispersible inhibitors. Corrosion inhibitors are used in many areas such as petroleum industry, packaging industry, potable water system, construction industry, engine coolants, chemical reactors and sour gas system, etc. Inhibitors are divided into two major classes: inorganic and organic inhibitors. The anodic type of inorganic inhibitors consists of chromates, molybdates, nitrites and phosphates, and the cathodic type inorganic inhibitors include polyphosphate and zinc. The organic inhibitors which have film-forming ability include amines, amine salts and imidazolines-sodium benzoate mercaptans, esters, amines, ammonia derivatives and so on. A number of studies have indicated that monomeric aniline and functionalized aniline derivatives are effective corrosion inhibitors for iron and steel. The most effective and efficient inhibitors are organic compounds having π-bonds in their structures. The efficiency of an organic compound as an inhibitor is largely dependent on its adsorption on the metal surface, which consists of replacement of water by an organic inhibitor at the interface. The adsorption of these materials is influenced by the presence of functional groups such as –NH, –N-N–, –CHO, R-OH, R-R, etc., which are available on the inhibitor molecules.

The inhibiting action applied by organic compounds on the dissolution of metallic species is related to interactions by adsorption between the inhibitors and the metal surface (Sathiyanarayanan et al. 2007; Yagan et al. 2006; Bhandari et al. 2008). High electron density of the heteroatom such as nitrogen and sulfur atoms present in hetero-compounds facilitates the organic molecule to get adsorbed onto the metal surface (Yagan et al. 2006). Moreover, the high-quality inhibitors should have properties like cost-effectiveness, high inhibition efficiency, low toxicity and easy production and better solubility in common organic solvents. Conducting polymers such as polyaniline, polypyrrole and their copolymers are found to be highly efficient inhibitors for protection of metal from corrosion due to the presence of π electrons, quaternary nitrogen atom and large molecular size (Kumar et al. 2013; Ruhi et al. 2014; Sambyal et al. 2015).

The corrosion inhibitor functions by one or more of these mechanisms:

1. By adsorption as a thin film on the surface of a corroding material
2. By inducing the formation of a thick corrosion product
3. By forming a passive film on the metal surface

Conducting polymers are conjugated systems, having alternate single and double bonds and therefore are rich in electrons and hence can get effectively adsorbed on metallic surfaces. In the case of polyanilines, for effective adsorption the electrons are not only available from aromatic nuclei but also from the chemically flexible –NH groups flanked either side by phenyl rings to ensure strong adsorption of polymeric cations on metallic surfaces. Studies on corrosion inhibition behavior of iron in HCl medium were studied by a new class of conducting polymer, poly phenetidine or poly(2-ethoxy aniline) by Dhawan et al. (1995). Attachment of ethoxy group at the ortho-position of aniline gives a polymer which is soluble in common organic solvents like CH_3OH, C_2H_5OH and chloroform that has been effectively used as a corrosion inhibitor for iron in acidic medium. Co-existence of delocalized π-electrons in the polymer backbone and quaternary ammonium nitrogen in the polymer facilitates its strong adsorption on the iron surface which leads to better corrosion inhibition response. Undoped and doped forms of Poly(2-ethoxy aniline) are shown in Figure 1.21.

It was observed that poly(2-ethoxy aniline) gave corrosion inhibition efficiency of > 80% even with 25 ppm of polymer, whereas monomer 2-ethoxy aniline gave a corrosion inhibition efficiency of ~ 54% at 20,000 ppm for iron in HCl medium. The inhibitive action of ortho-methoxy substituted polyaniline, a new class of conducting polymer, on the corrosion of iron in acidic chloride solution has been evaluated by Electrochemical Impedance Spectroscopy (EIS), Linear Polarization Resistance (LPR), Weight Loss (WL) and by the Logarithmic Polarization Technique (LPT) (Sathiyanarayanan et al. 1994). Inhibition efficiencies of nearly 80–88% have been observed even at 25 ppm concentration. Double-layer capacitance studies indicate a strong adsorption of polymer, following the Temkin adsorption isotherm, is largely responsible for its inhibitive action. It was also observed that undoped form of poly(2-methoxy aniline) gave better corrosion inhibition efficiency than the doped form of poly(2-methoxy aniline). The reasons for the undoped polymer to offer better efficiencies is that a neutral polymer has an imine structure and hence has better adsorption than a doped polymer, and the reason for obtaining better efficiencies for SMA doped polymer can also be attributed to the presence of the $-NH_2$ group in sulphamic acid.

Even corrosion inhibition performance of soluble self-doped copolymers of aniline and 4-amino-3-hydroxy-napthalene-1-sulfonic acid (ANSA) (Figure 1.22) were studied in HCl medium by Bhandari

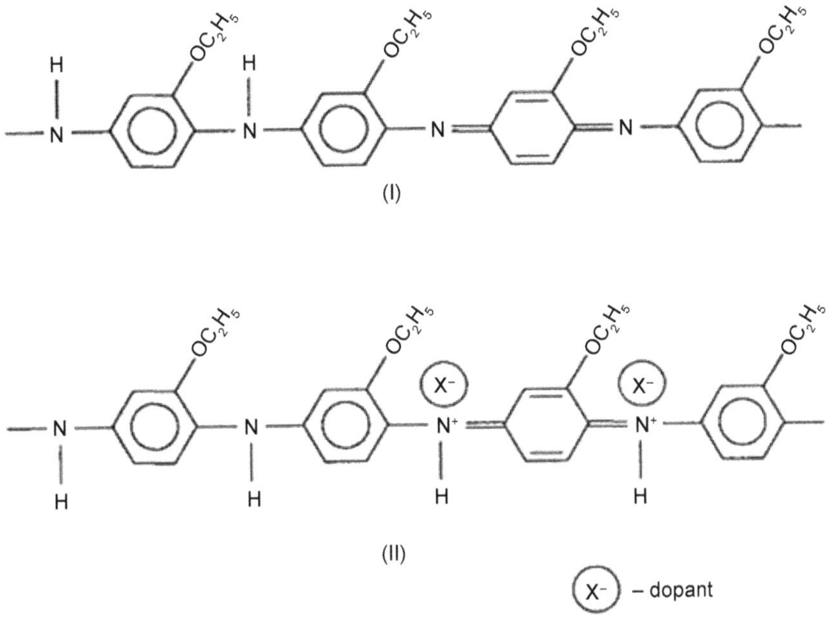

Figure 1.21: Undoped and doped forms of poly(2-ethoxy aniline)

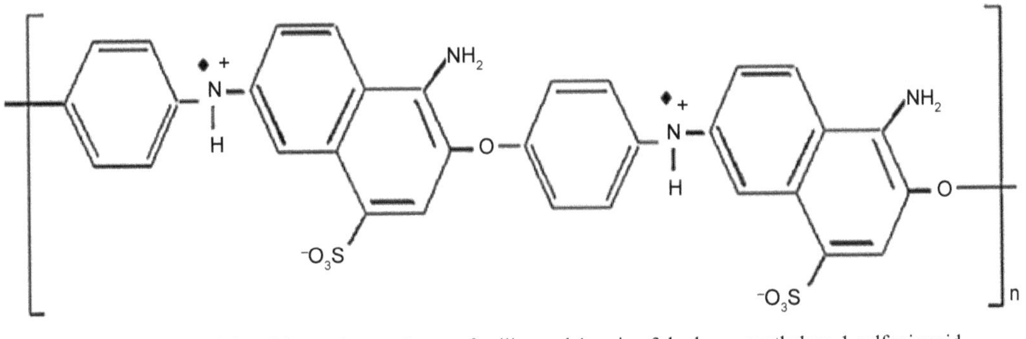

Figure 1.22: Self-doped form of a copolymer of aniline and 4-amino-3-hydroxy-napthalene-1-sulfonic acid

et al. (2011). ANSA is tri-functional monomer and have three functional groups (i.e., $-NH_2$, $-OH$ and $-SO_3H$) along with two fused benzene rings. Copolymerization of ANSA with aniline in the absence of any external dopant leads to the formation of self-doped copolymers, where the polymerization proceeds through the linkage of $-OH$ group resulting in the formation of naphthalene oxide type of structure and the color of copolymers to be brownish-red. During the polymerization of ANSA, two functional groups ($-NH_2$ and $-OH$) may take part in the polymerization depending on the nature of the reaction medium. In copolymerization of ANSA with aniline in the absence of any external dopant, polymerization reaction occurred selectively via $-OH$ group. Moreover, in such medium, the $-SO_3H$ group of ANSA unit facilitated the polymer to acquire intrinsic protonic doping ability leading to the formation of highly processable, soluble and stable, electrically conducting polymer. Corrosion inhibition response was evaluated using Tafel Extrapolation method and Electrochemical Impedance Spectroscopy. The results showed that copolymer exhibited a drastic shift in corrosion potential (E_{corr}) and corrosion inhibition efficiency were found above 90% in HCl medium. Moreover, copolymer showed a larger degree of surface coverage onto the iron surface leading to better corrosion inhibition behavior.

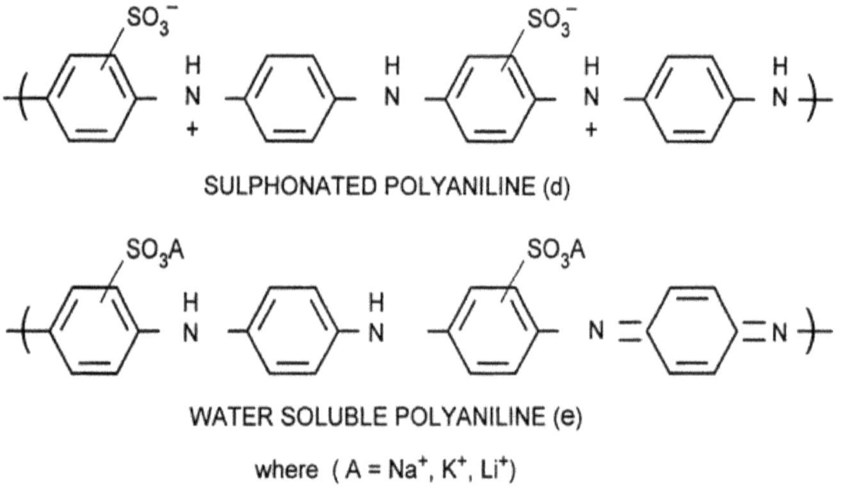

SULPHONATED POLYANILINE (d)

WATER SOLUBLE POLYANILINE (e)

where (A = Na⁺, K⁺, Li⁺)

Figure 1.23: Compensated sulphonated polyaniline

Even sulphonated polyaniline as water-soluble polymer has been synthesized by Koul et al. (2001) which is found useful for corrosion inhibition for iron in a corrosive medium. The sulphonated polyaniline has the structure as shown in Figure 1.23. The sulphonated group attached to the polymeric backbone remains facing the electrolyte to prevent the corroding ions from reaching the iron surface and to give increased corrosion inhibition efficiencies.

Even our studies have indicated that water soluble nigrosine, whose structure is shown in Figure 1.23a can be used as an inhibitor for ion in HCl medium.

100 ppm of the compound was found to give an efficiency of 95.6% in HCl medium. Even impedance analysis of the water-soluble nigrosine has given efficiency of more than 90% and thus can be effectively used as a corrosion inhibitor.

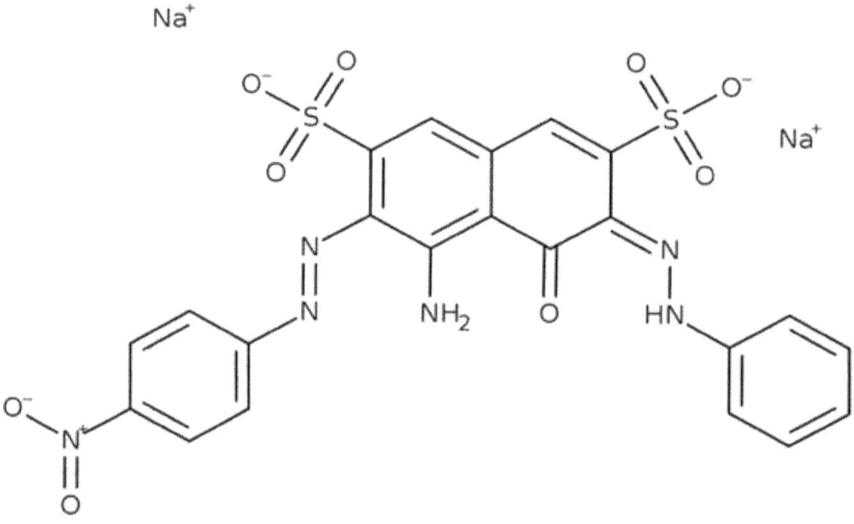

Figure 1.23(a): Water soluble nigrosine

1.7.4 *Corrosion resistant coating*

Protective coatings are the most generally used method for preventing corrosion. The function of a protective coating is to provide a satisfactory barrier between the metal and its environment. Coatings can be broadly classified into three types such as metallic coatings, inorganic coatings and organic coatings. The basic requirements for corrosion protection are good environment resistance, long term stability of the protective layer and strong adhesion of coating to the substrate. The functioning of any protective coatings is based upon three basic mechanisms: (i) Barrier Protection, (ii) Chemical inhibition and (iii) Galvanic (sacrificial) protection. Completely isolate metals and alloys from its environment to achieve barrier protection. Protection of metals through chemical inhibition is achieved by adding inhibitor molecules into the coating system. Corrosion resistant coating can be applied to the metals in the form of paints, metallic coating, inorganic coating, conversion coating, organic coating, etc.

1.7.5 *Paints*

Paint is any liquid or a dispersed solution in a binding medium that provides a thin layer of a barrier to the substrate. From the electrochemical theory of corrosion, it follows that corrosion can be stopped by suppressing either the cathodic or the anodic reaction or by inserting between the cathodic and anodic areas a large resistance, which impedes the movement of ions. Calculations indicate that paint films are so permeable to water and oxygen that they cannot suppress the cathodic reaction. Paints can inhibit corrosion by modifying the anodic reaction; for this to occur the pigment must be either metallic, basic or soluble. In general, paint films protect by virtue of their high electrolytic resistance; they readily acquire a charge, consequently, they are relatively impermeable to ions. They are normally used to protect color and offer texture to the substrate. Paints mainly constitute of pigments, binders, solvents and additives, etc., materials like metal oxide (TiO_2, Fe_2O_3, Pb_3O_4, etc.) or $ZnCrO_4$, $PbCO_3$, etc., are present in pigments depending upon the requirements of the color. Various properties like flexibility, gloss, durability and toughness depend upon binders. It includes both synthetic and natural materials such as silanes, polyester, acrylics and natural oils. In the automobile industry various type of paints are used (resins, varnishes and lacquers) for the protection of components from water.

Recent findings by a team at The University of Manchester (Nair et al. 2014) has revealed that a thin layer of graphene paint can make impermeable and chemically resistant coatings, which could be used for packaging to keep food fresh for longer duration and protect metal structures against corrosion. Graphene paint can be applied to practically any material, independently of whether it is plastic, metal or even sand.

The degree of corrosivity of metals in normal to marine environment has been classified by the International Organisation for Standardisation into five degrees of corrosivity from C1 to C5, in order of increasing corrosivity and while selecting a particular paint, this classification has to be considered:

(i) **C1 – Very low corrosion risk:** Buildings, houses with normal air, interiors only. This is suitable for houses, offices in a city environment, school and colleges, etc.

(ii) **C2 – Low corrosion risk:** Homes and buildings in dust environment—this is applicable for cold storage, sports arenas, homes exposed to high degree of PM 2.5, garage, etc.

(iii) **C3 – Moderate corrosion risk:** This is for high rise buildings in coastal area, and industries exposed to different gases and solvents.

(iv) **C4 – High corrosion risk:** Chemical plants and industrial units exposed to a corrosive environment.

(v) **C5 – Very high corrosion risk:** Offshore platforms, bridges, water lines and oil pipelines.

Five important steps which must be followed for preparing any metal surface for painting to prevent it from corrosion are:

(i) **Clean the surface:** To properly prepare new metal surfaces, use kerosene to remove grease and use a mild detergent to remove any dust on the surface. Then, apply a rust-inhibitive primer. If painting is to be done on a painted metal surface, remove dust with a clean, dry cloth and then clean the surface with a sand paper. Finally, clean the surface with spirit solution to get better adhesion of paint.

(ii) **Remove loose and peeling paint:** If the painted metal surface is in poor condition, then clean the surface by wire brushing or using sand paper. Even the power cleaning of the surface can also be done, which will help in removing the paint.

(iii) **Remove rust:** Before painting metal, it is important to check if any rust is present on the surface. If rust is present, remove it by brushing or using a sand paper, and then apply a good quality rust-inhibitive primer, which can be used to cover rusted spots to turn them into non-rusting, paintable surfaces.

(iv) **Repair small holes and dents:** If metal surface that needs to be painted has small holes or dents, then apply a proper epoxy on the holes and apply a primer on the surface. For larger holes, make a mesh of the size of the hole and paste it on the hole encapsulated with epoxy, and then apply primer on the surface. Even pin-holes should be properly covered and filled with epoxy. The detection of pin holes on the metal surface can be done by salt droplet indicator solution test. The test consists of placing a salt solution containing phenolphthalein indicator and potassium ferricyanide ($K_3Fe(CN)_6$) on the iron surface. If a blue color is formed on the center of the droplet, then pin-holes are present on the metal surface. This blue color is due to reaction of ferricyanide ions with ferrous ions at the metal surface leading to formation of ferriferrocyanide complex at the anodic area on the iron surface. Also, a pink colored ring is formed on the edge of the droplet, which is due to the formation of OH^- ions on the iron surface. This is because indicator phenolphthalein turns pink in the presence of OH^- ions. The whole reaction process of detection of pin-holes can be shown in Figure 1.24.

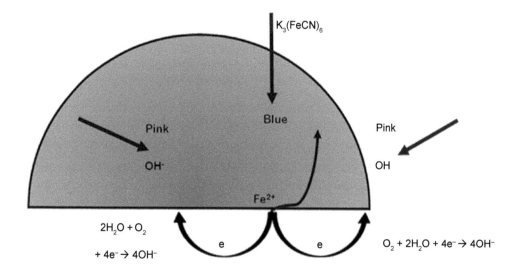

Figure 1.24: Detection of pin-holes at iron surface

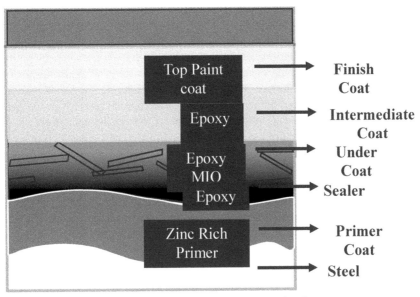

Figure 1.25: Protective paint steps on metal surface

(v) Applying the right primer on the surface: For painting on any metal surface, proper selection of primer is very important. Water-based (latex) primers should not be used on metal surfaces, as moisture can seep through and cause paint to fail within weeks or months. Zinc primers are suitable for applying on the surface which will give better adhesion to the surface. These points can be seen in Figure 1.25, where the sequence of applying paint is given.

1.7.6 Metallic coatings

The two most commonly used methods of applying metallic coatings to structural steel are hot-dip galvanizing and thermal(metal) spraying. In general, the corrosion protection afforded by metallic coatings is largely dependent upon the choice of coating metal and its thickness, and is not greatly influenced by the method of application Metals like zinc, cadmium, tin, aluminum, nickel, lead and chromium, etc., are used as a corrosion resistant coating. These coatings are applied by hot-dipping, metal spraying, electro-deposition, etc. Zinc and cadmium are anodic to steel, and provide sacrificial protection when used as a coating. Tin coating applied by hot-dipping or electro-deposition are widely used in the electrical and food industry. Aluminum coating reduces the corrosion rate of metals on which it is applied by forming a thin oxide layer. Chromium coatings are used for engineering purpose because of its hardness and excellent wear resistance. Metallic coatings like lead and cadmium cannot be used in contact with drinking water or food products because of its toxic nature. Metallic coatings provide an external thin layer of coating to the metal substrate. This coating can be applied for various reasons such as enhancement in appearance, corrosion resistance, adhesion, wear resistance, etc. Although these metals exhibit excellent performance as coating materials, they have certain shortcomings. Tin coating is generally used for ferrous and non-ferrous metal substrates. It is widely used in the food industry due to their non-toxic behavior and anti-corrosive properties. But tin coated food containers are susceptible to contamination due to the hydrogen evolution. Nickel coating is commonly used in automobile industry for a decorative and anti-corrosion purpose. The major problem associated with this coating is tarnishing and slow, superficial corrosion. Chromium coatings are usually used due to their hardness and wear resistance properties. However, application of these coating is discouraged due to their toxicity.

Metallic coatings can be applied to steel to provide a layer of protection, thereby allowing steel to be used in various corrosive environments. Metallic coatings are generally used for two types of protection: barrier protection and galvanic protection. When a metallic coating such as zinc is applied to steel, it dries and hardens it and forms an impervious barrier that prevents moisture intrusion. This removes one of the essential components needed for corrosion to occur. Without moisture or humidity, oxidation cannot occur and therefore rust cannot form. Another important aspect of barrier protection is called corrosion film protection. Some metals, such as aluminum, react with oxygen to form a protective aluminum oxide film on its surface. This oxide film is resilient and firmly adheres to the surface of the aluminum to prevent moisture intrusion and further corrosion. This makes aluminum an ideal material for sheet metal.

Zinc, most commonly used to coat structural steel, reacts with O_2 and moisture in the atmosphere to form corrosion products that creates a defensive layer and protects the underlying steel. Freshly exposed zinc reacts with oxygen to form zinc oxide, and with water to form zinc hydroxide. When zinc hydroxide reacts with carbon dioxide in the atmosphere, the resulting product is zinc carbonate. These corrosion film products are resistant to water intrusion and tightly adhere to the steel surface so that it does not easily peel off like the corrosion formed on iron. Zinc, however, is a reactive metal and will slowly corrode and erode over time. The degradation rate of zinc is still several times lesser than that of steel and will therefore significantly prolong the service life of the steel it is meant to protect. The most common method of applying a zinc coating to structural steel is by hot-dip galvanizing. The galvanizing process involves the following stages: (i) Any surface oil or grease is removed by suitable degreasing agents. (ii) The steel is then usually cleaned of all rust and scale by acid pickling. This may be preceded by a blast cleaning to remove scale and roughen the surface, but such surfaces are always subsequently pickled in inhibited hydrochloric acid. (iii) The cleaned steel is then immersed in a fluxing agent to ensure good contact between the steel and zinc during the galvanizing process. (iv) The cleaned and fluxed steel is dipped into a bath of molten zinc at a temperature of about 450°C. At this temperature, the steel reacts with the molten zinc to form a series of zinc/iron alloys integral with the steel surface. (v) As the steel work piece is removed from the bath, a layer of relatively pure zinc is deposited on top of the alloy layers. The schematic presentation of hot dip galvanizing is shown in Figure 1.26.

The second method by which metallic coatings protect steel is by providing galvanic protection of the underlying steel by allowing the metallic coating to corrode preferentially to the steel, thus acting as a sacrificial coating. Nanan (2017) has elaborated various aspects of metallic coatings which gives a better understanding of the subject.

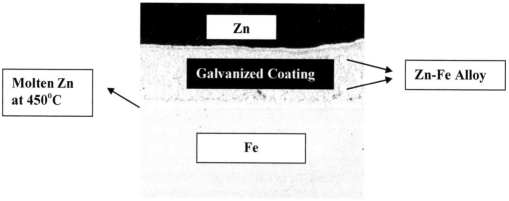

Figure 1.26: Hot-dip galvanizing process

1.7.7 Inorganic coatings

Inorganic coatings provide a barrier to metal substrate against the highly aggressive corrosive conditions. Inorganic coatings like vitreous, porcelain enamels and glass linings can be used on steel against corrosive environments. It includes a different process such as anodizing, enameling, surface conversion, phosphating, etc. These coatings when applied and heated on the metal surface gets soften and bonded with the metal. Enamel coated steels have a very long life. Vitreous enamels are used to protect against high temperature gases. However, failure occurs in these coatings by forming a network of cracks through which rust appears.

Anodizing involves the electrolytic oxidation of a surface to produce a tightly adherent oxide scale which is thicker than the naturally occurring film. Anodizing is an electrochemical process during which aluminum is the anode. The electric current passing through an electrolyte converts the metal surface to a durable Al_2O_3. The difference between plating and anodizing is that the oxide coating is integral with the metal substrate as opposed to being a metallic coating deposition. The oxidized surface is hard and abrasion resistant, and it provides some degree of corrosion resistance. However, anodizing cannot be relied upon to provide corrosion resistance to corrosion prone alloys, and further protection by painting is usually required. Fortunately, the anodic coating provides an excellent surface both for painting and adhesive bonding. Anodic coatings break down chemically in highly alkaline solutions (pH > 8.5) and highly acid solutions (pH < 4.0). They are also relatively brittle and may crack under stress, therefore supplementary protection such as painting is particularly important with stress corrosion prone alloys.

Anodic coatings can be formed in sulfuric acid, chromic acid, phosphoric acid or oxalic acid solutions. Chromic acid anodizing is widely used with 7000 series alloys to improve corrosion resistance and paint adhesion, and unsealed coatings provide a good base for structural adhesives. The thickness of aluminum oxide coating produced by anodizing varies from 2 mm to 25 mm. In industry hard coat anodizing takes place where anodization is carried out in H_2SO_4 medium because it provides durable coating and resistance to abrasion; and corrosion resistance is needed and bulk anodizing for carrying out anodization on small, irregular shaped articles, which are processed in perforated aluminum is performed. Anodizing processes are also carried out on titanium and magnesium alloys. The process produces a smooth coating with a uniform texture and appearance, and a uniform blue to violet color.

Phosphating or phosphate conversion coating

In this process, metal is immersed in an ortho-phosphoric acid bath containing other metal salts such as zinc phosphate and manganese phosphate, giving a layer of phosphate on the steel surface. During the steel forming process, phosphating is often carried out in order to improve the surface properties of the sheet. The main components of a phosphating solution are:

- Ortho-phosphoric acid (H_3PO_4);
- Bivalent metal cations: Zn^{2+}, Mn^{2+}, Fe^{2+};
- Accelerator—an oxidizing reagent (H_2O_2, KNO_3, etc.) increasing the coating process rate and reducing the grain size of the deposit.

When iron is immersed in a zinc phosphate solution, the following chemical reactions start: Iron dissolves in the phosphoric acid solution:

$$3Fe + 6H^+ + 2PO_4^{3-} \longrightarrow 3Fe^{2+} + 2PO_4^{3-} + 3H_2 \tag{38}$$

Phosphate ions combine with zinc ions leading to the formation of zinc phosphate and its deposition on the substrate surface:

$$3Zn^{2+} + 2PO_4^{3-} \longrightarrow Zn_3(PO_4)_2 \tag{39}$$

Zinc phosphate coating is applied when increased corrosion resistance is required. Zinc phosphate can withstand more than 200 hours of the neutral salt test.

Nitriding

Steels containing nitride forming elements such as chromium, molybdenum, aluminum and vanadium can be treated to produce hard surface layers providing improved wear resistance, improved fatigue and corrosion fatigue resistance. Gas nitriding can be applied to unalloyed steels and irons to produce a corrosion and wear-resistant nitride layer. In alloy steels that contain nitride-forming alloy elements (Cr, Mn, Mo, V, W, Al, Ti) a deeper diffusion layer develops.

1.7.8 Conversion coatings

Conversion coatings are the coatings that develop on the metal surface by the chemical or electrochemical action between the electrolyte solution and a metal surface. It is widely applied to metal or alloys substrates such as steel, magnesium, tin, zinc and aluminum, etc. Conversion coatings are mainly classified into three categories, i.e., oxide coatings, phosphate coatings and chromate coatings. Oxide coatings are the corrosion by-product that shows good adhesion with the substrate. Some examples of oxide coatings are formation of black oxide, anodizing, etc. Phosphate coating provides good adhesion to the surface and is widely used for cast iron and low carbon steel. Chromate coating is extensively used in a commercial scale. Chromate coating is fabricated due to the interaction of chromate salt and water. A passive layer of coherent chromium oxide film with good corrosion resistance is generated on the metal surface. However, this coating is banned by the environment safety agencies due to their toxicity and adverse effect on humans. Recently Bagal et al. (2018) developed the chemical zinc phosphate coatings on low carbon steel by using nano-TiO_2 in the standard phosphating bath. The results showed that the corrosion rate of nano-TiO_2 phosphate coated samples were four times less than the bare, uncoated low carbon steel.

1.7.9 Organic coatings

Organic coatings are one of the most extensively applied methods for protection of metals in the corrosive environment. Organic coatings comprise of three components such as the binder, pigments and additives like dryers, stabilizing agents, hardening agents and dispersion agents, etc. The binders are generally low molecular weight polymers, which control the basic physical and chemical properties of the coating. The addition of pigments can remarkably affect the properties of the coating. They provide color and also act as a barrier for corrosive species in an aggressive medium.

It could be shown that for typical organic coatings used for corrosion protection the diffusion rate of H_2O and O_2 far exceeds the diffusion limited value for oxygen reduction (Feser and Stratmann 1990; Thomas 1989). The roles of organic coatings, which provide corrosion protection, are as follows:

 (i) The barrier for ions leads to an extended diffusive double layer.
(ii) Adhesion of the coating.
(iii) Blocking of ionic paths between local anodes and cathodes along the metal/polymer interface.

1.7.10 Mechanism of protection

For the corrosion protection, three mechanisms have been proposed such as physical barrier effect, anodic protection and self-healing mechanism.

1.7.11 Barrier protection

Conducting polymer, when coated on the metal substrates, acts as a physical barrier between the metal and corrosive environment. The polymer coating works as a barrier against the penetration of oxidants and aggressive anions, and protects the metals from corrosion. The coating blocks the diffusion of the ions and corrosion products, thus preventing the formation of galvanic coupling between the local anodes and cathodes.

1.7.12 Ennobling mechanism

Wessling in 1996 proposed the corrosion protection of metals by an ennobling mechanism using conducting polymers. The conducting polymers have the ability to induce the oxidation of metal surface to form its metal oxide and maintain the metal in its passive or "noble" form. In order for this to happen, the oxidation potential value of the polymer should be lower than that of metal to be protected. The oxide layer is also called as a passive layer which inhibits further corrosion in two ways; first, it acts as another layer of barrier against diffusion of corrosive species and secondly, it also acts as an insulating layer which inhibits the flow of electrons. Conducting polymer also increases the passivity of the oxide layer and constrains the cathodic and anodic delamination of the coating.

1.7.13 Self-healing mechanism

The self-healing coatings automatically repair and prevent the extension of corrosion. The self-healing mechanism is based on the assumption that the conducting polymer releases the doping anions present in the conducting polymer matrix when defect on the coating appears. The doping anions diffuse into the defect site and reduce the rate of corrosion. The passive oxide layer then re-forms between the metal substrates and polymer coating, brought about by the oxidative proficiency of the conducting polymers (Kamaraj et al. 2015).

1.8 Testing methods of coatings

During the life cycle of a coating product, testing, evaluation and analysis are required to ensure product quality and regulatory compliance requirements are met. There is an urgent need to develop alternative smart coatings for an existing application due to considerations such as green compliance goals or need to improve performance and prolong their durability. The chemical composition and physical characteristics of coatings have a direct impact on their performance. Accordingly, it is essential to conduct comprehensive coating testing and analysis programs. These often require state-of-art analytical equipment and techniques, and should be conducted by analysts who have experience in industry coating application and can give special insight on this area. During product development, there should be application studies such as raw material substitution, substrate surface interactions, detailed analysis of formulations, processing and then end-use performance through exact procedures such as confirmation of molecular structure, conductivity, thermal stability, anti-corrosive evaluation, physico-mechanical property like adhesion, flexibility of coating, scratch resistance, abrasion-free and accelerated weathering product, failure analysis test, etc.

International regulatory bodies established guidelines for quality assurance, measurements and quality check through a certain set of standard operating modules, e.g., American Society for Testing Materials (ASTM), British Standards Institution (BS), DIN, SI, GB, IS, ISO, IEC 17025 and the Commission of European Communities, etc. These standards and systematic operating procedures are used for quality check of raw and finished products and for their end-use applications, evaluation, performance and significance, etc.

In the present time, it may be necessary to seek more up-to-date information than that available in literature which have been consulted, and this will necessitate making use of review publications, abstracts and referring to journals devoted to development and designing of analytical procedure and to critical techniques. Such a literature survey may lead to the compilation of a list of best possible procedures, and the ultimate selection must then be made in the light of the criteria with special consideration being given to questions of possible interpretations and to the equipment available.

1.8.1 Mechanical testing of coating

1.8.1.1 Standard test methods for measuring adhesion by tape test

The high performance of durability depends on the adherence of the coating to its substrate. Surface preparation (or lack of it) has a drastic effect on the adhesion of coatings; therefore there is an urgent need to develop such a method by which we can evaluate adhesion of a coating to different substrates or surface treatments. In the present time, these techniques is of considerable usefulness in the industry for quality control and product development purpose.

Coating adhesion is the minimum adhesion required for a coating to stick to the substrate. It governs the coating durability in many applications, and determines its performance. Good coating adhesion depends on many attributes of the interface region:

- Atomic bonding structure
- Elastic moduli and state of stress
- Thickness, purity and fracture toughness

Coating adhesion is strongly affected by the interactions between the coating and the substrate. A coating durability depends on two basic features:

- Cohesion—The inner strength of a material, which is determined by the strength of the molecular forces in the bulk.
- Adhesion—The strength of the bonds formed between one material and the other.

Lacking these two properties can create failures:

- Cohesive failures such as abrasion and cracking usually occur in the paint film due to aging and dissolving in solvent.
- Adhesive failures, which can be in the form of blister at the interface, lifting of the coating film or any other situation, those results from low adhesion at the interface.

Poor adhesion results in coating failure and determines the lifespan of the coating system, whereas good adhesion leads to durable coatings. Good adhesion will be evident when:

- Adsorption phenomenon forms interfacial bonds.
- Chemical bonds are formed at the interface between the coating and the substrate.
- Mechanical interlocking occurs when the paint dries.

All of these mechanisms do not have to be present all the time to form good adhesion. Different mechanisms may be at work depending on the paint system, but a good wetting or adsorption is always required for good adhesion.

For better performance coating must adhere to the substrates on which they are applied. There are a number of recognized methods for evaluation of how well a coating is bonded to the substrate. Three tests are used to measure coating adhesion: cross-cut test, scrape adhesion and pull-off test. Commonly used measuring techniques are performed with a knife or with a pull-off adhesion tester.

After any test, it is important to record if the bond failure was adhesive (failure at the coating/substrate interface) or cohesive (failure within the coating film or the substrate).

1.8.1.2 Knife test

Knife test requires the use of a utility knife to pick at the coating. It shows whether adhesion of a coating to a metal substrate or to another coating (in multi-coat systems) is at a generally adequate level. Evaluation is based on the degree of difficulty to remove the coating from the substrate. ASTM D6677 is the standard test method for evaluating adhesion by knife. Two cuts are made into the coating with a 45-degree angle in such a way that they intersect to form an "X" mark as shown in Figure 1.27. Then make an attempt to lift up the coating from the substrate or from the coating with the help of a pointed knife. This is a highly subjective test and its value depends upon the performing analyst, therefore this test should be carried out by a skilled person.

The Performance Evaluation Scale is based on both the degree of difficulty to remove the coating from the substrate and the size of removed coating chip. The data is shown in Table 1.4.

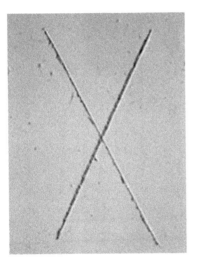 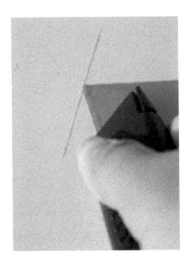

Figure 1.27: Measuring adhesion by knife test

Table 1.4: Rating performance of adhesion coating by knife test

Rating	Description
10	The coating is extremely difficult to remove; fragments no larger than approximately 0.8 by 0.8 mm removed with great difficulty.
8	The coating is difficult to remove; chips ranging from approximately 1.6 by 1.6 mm to 3.2 by 3.2 mm can be removed with some difficulty.
6	The coating is somewhat difficult to remove; chips ranging from approximately 3.2 by 3.2 mm to 6.3 by 6.3 mm can be removed with slight difficulty.
4	The coating is somewhat difficult to remove; chips in excess of 6.3 by 6.3 mm can be removed by exerting light pressure with the knife blade.
2	The coating is easily removed; once started with the knife blade, the coating can be grasped with one finger and easily peeled to a length of at least 6.3 mm.
0	The coating can be easily peeled from the substrate to a length greater than 6.3 mm.

1.8.1.3 Tape test

On metal substrates, an advanced version of the knife test is the tape test and is carried out by applying pressure sensitive tape over cut marks made on the coating. Standard test method for rating adhesion by tape test is done using ASTM D3359-17. In order for a coating to fulfill its function of protecting or decorating a substrate, the coating must remain adhered to the substrate. Since the substrate and its surface preparation (or lack thereof) has a drastic effect on the adhesion of coatings, a method to evaluate adhesion of a coating to different substrates, surface treatments or different coatings to the same substrate is of considerable usefulness in the industry. There are two variants of this test, the X-cut Tape Test (Test MethodA) and the Cross-Hatch Tape Test (Test MethodB). In Test Method A, the tape is placed on the center of the intersection of the cuts and then removed rapidly. The X-cut area is then examined for removal of coating from the interface of cut marks over coating substrate as shown in Figure 1.28.

Figure 1.28: Measuring adhesion of coating by Tape Test by Test Method A

The X-cut Tape Test is primarily intended for use at job sites meanwhile the Cross-Hatch Tape Test is applicable for use in the laboratory on coatings less than 125 microns dry film thickness. It uses a cross-hatch pattern rather than the X pattern, and incisions are made with a special cross-hatch cutter with multiple pre-set blades for obtaining proper spaced and parallel hatch. Pressure sensitive tape has been applied and pulled off; the intersection cross-cut area is then examined and rated. Inspect the X-cut area for removal of coating from the substrate or previous coating, and rate the adhesion in accordance with the following scale as given in Table 1.5:

Table 1.5: Rating of adhesion of coating by Tape Test Method A

5A	No peeling or removal
4A	Trace peeling or removal along incisions or at their intersection
3A	Jagged removal along incisions up to 1.6 mm ($^1/_{16}$ in.) on either side
2A	Jagged removal along most of incisions up to 3.2 mm ($^1/_8$ in.) on either side
1A	Removal from most of the area of the X under the tape
0A	Removal beyond the area of the X

1.8.1.4 Cross-cut adhesion test

Adhesion of the coating to mild steel surface was evaluated by cross-cut adhesion test. This test was carried out at room temperature as per ASTM D3359-09 (Test Method B). In order to carry out this test, a lattice pattern with either six or eleven cuts in each direction was made in the film to the substrate, the pressure-sensitive tape was applied over the lattice and then removed, and adhesion of coating on metal surface was evaluated by comparison with descriptions and illustrations as given in Figure 1.29.

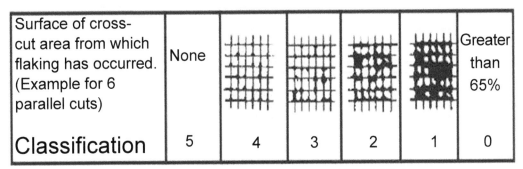

Surface of cross-cut area from which flaking has occurred. (Example for 6 parallel cuts)	None					Greater than 65%
Classification	5	4	3	2	1	0

Figure 1.29: Cross-cut adhesion test grading as per ASTM D3359-09

Grade 5: The edges of the cuts are completely smooth; none of the squares of the lattice is detached. Grade 4: Small flakes of the coating are detached at intersections; less than 5% of the area is affected. Grade 3: Small flakes of the coating are detached along edges and at intersections of cuts. The area affected is 5 to 15% of the lattice. Grade 2: The coating has flaked along the edges and on parts of the squares. The area affected is 15 to 35% of the lattice. Grade 1: The coating has flaked along the edges of cuts in large ribbons and whole squares have detached, the area affected is 35 to 65% of the lattice. Grade 0: Flaking and detachment are worse than Grade 1.

1.8.1.5 Pull-Off test

Pull-Off test is an advanced technique for evaluation of coating adhesion. The pull-off strength of a coating is a performance property that may be referenced in specifications. This test method serves as a means for uniformly preparing and testing coated surfaces, and evaluating and reporting the results. This test is carried out by a loading fixture, commonly called a dolly/stub, and is affixed by an adhesive to a coated specimen. By use of a pull-off adhesion tester, a uniform load is applied to the surface in increasing order until the dolly is pulled off. The force required to pull off the dolly yields the tensile strength in pounds per square inch (psi) or mega Pascals (MPa). The test failure will occur along the weakest plane within the system comprised of the dolly, adhesive, coating system and substrate. Standard test method for pull-off strength of coatings is done using ASTM D4541-17. Adhesion Tester, commercially available using fixed alignment pull-off tester is shown in Figure 1.30.

1.8.1.6 Mandrel bend test

Coating performance on the metal surface was evaluated using Mandrel Bend Test as per ASTM D522M/D522-93A, using a Mandrel test apparatus (see Figure 1.31). Coatings attached to substrates are elongated when the substrates are bent during the manufacture of articles or when the articles are abused in service. These test methods have been useful in rating attached coatings for their ability to resist cracking when elongated. They have been useful in evaluating the flexibility of coatings on flexible substrates. These test methods cover the determination of the resistance to cracking (flexibility)

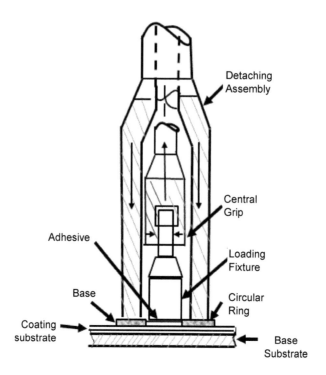

Figure 1.30: Fixed alignment pull-off tester

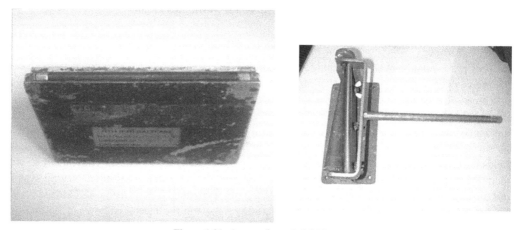

Figure 1.31: Image of mandrel 6.25 mm

of attached organic coatings on substrates of sheet metal or rubber-type materials. This test was carried out by placing the coated test samples over a mandrel (size 6.4 mm) with the uncoated side in contact, and with at least 50 mm overhang on either side using a steady pressure of the fingers, which bend the panel approximately 180 degrees around the mandrel at a uniform velocity in a time 1.0 second. Remove and examine the panel for cracking that are visible to the unaided eye.

1.8.1.7 *Taber abrasion resistance test*

This test method covers the determination of the resistance of organic coatings to abrasion produced by the Taber Abraseron coatings applied to a plane, rigid surface such as a metal panel. Abrasion resistance can be expressed in one or more of the following terms:

Wear index—1,000 times the loss in weight in milligrams per cycle.

Weight loss—The loss in weight in milligrams determined at specified number of cycles.

Wear Cycles per Mil—The number of cycles of abrasion required to wear a film through the substrate per mil of film thickness.

The coating on substrates can be damaged by abrasion during the manufacturing process, transporting and service. This test method has been useful in evaluating the abrasion resistance of attached coatings.

The image of the Taber AbraserTest Apparatus is shown in Figure 1.32. Taber Abraser consists of three main parts:

(i) *Abrasive Wheels*—Resilient calibrase wheels No. CS-10 or CS-17. The CS-17 wheels produce a harsher abrasion than the CS-10 wheels, therefore generally used for evaluation of abrasion resistance of ceramic tiles and wood specimen.

(ii) *Resurfacing Medium*—an S-11 abrasive disk, used for resurfacing the abrasion wheels.

(iii) *Vacuum Pick-Up Assembly*—consisting of a vacuum unit, a variable transformer suction regulator, a nozzle and a connecting hose with an adaptor.

Apply a uniform coating of the material to be tested to a plane, rigid panel. Specimens shall be a disk of diameter 4 inches (100 mm) or plate square with rounded corners and with a 1/4-inch (6.3-mm) hole centrally located on each panel. Prepare a minimum of two coated panels for the specimen. Mount the selected abrasive wheels on their respective flange holders, taking care not to handle them by their abrasive surfaces. Adjust the load on the wheels to 1000 g. Mount the resurfacing medium (S-11 abrasive disk) on the turntable. Lower the abrading heads carefully until the wheels rest squarely on the abrasive disk. Place the vacuum pick-up nozzle in position and adjust it to a distance of 1/32 in. (1 mm) above the abrasive disk.

Abrasion Resistance test of coated panels was carried out using Taber Abrasion apparatus (see Figure 1.32) as per ASTM D 4060-95. This test was carried out on the rigid panels (dimension

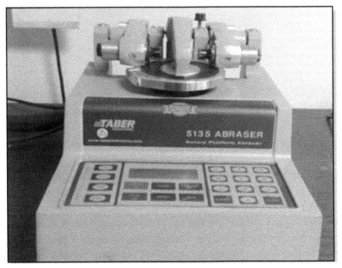

Figure 1.32: Image of Taber Abrasion Apparatus

10 cm²) having a uniform coating. The specimens were a disk of 100 mm in diameter or a plate (100 mm) square with rounded corners, and with a 6.3-mm hole centrally located on each panel. The test was performed at 25°C and 50% relative humidity for 24 hours. The organic coating was applied at uniform thickness to a plane. After curing, the surface was abraded by rotating the panel under 1.0 kg weighted abrasive CS-10 wheels. A test of 1,000 cycles was done to observe the weight loss in coatings. Start the vacuum pick-up and then the turntable of the abrader. Subject the test specimen to abrasion for the specified number of cycles or until wear through of the coating is observed. Abrasion resistance is calculated as a loss in weight at a specified number of abrasion cycles, as the loss in weight per cycle or as number of cycles required to remove a unit amount of coating thickness.

1.8.1.8 Scratch resistance test

Scratch resistance tests are performed on specimens to evaluate the scratch resistance of a particular material, to rank the relative scratch resistance of different materials or to determine the scratch coefficient of friction of materials. Tin panel coated with polymeric materials are loaded in Scratch Test Apparatus to evaluate the scratch resistance ability of coating. In this study, the scratch length shall be set at 150 mm. The test performs using a linearly increased normal load of 1.0 kg at a speed of 100 mm/s, and the stainless steel ball tip of 1.0 mm diameter is used to introduce the scratch on the coated metal surface. Examination of scratch on the metal surface is carried out by the unaided eye. The test equipment for scratch resistance is shown in Figure 1.33.

Figure 1.33: Image of Scratch Hardness Tester

1.9 New aged smart surface coating

New technologies and materials for corrosion protection by coatings are coming into the coatings science from other areas. The possibility of providing corrosion protection by incorporating the use of conductive polymers such as polyaniline, polypyrrole and their copolymers and nanocomposites polyaniline is being actively followed by researchers. There are several studies using conducting polymers deposited onto metals/alloys using various procedures like electro-polymerization, spray and dip-coating, water-dispersive formulations, solvent-casting, powder coating, etc., conducting polymers are new aged smart coatings for protection of metals from corrosion in various corrosive conditions. Conducting polymers can function by an anodic protection mechanism for ferrous alloys, and therefore can potentially replace carcinogenic and environmentally hazardous corrosion-inhibiting materials like Cr(VI) and cadmium (Cd) (Blisiak and Kowalik 2000; Vignati Davide et al. 2010; Praveen et al. 2012).

Designing of smart self-healing coatings of conducting polymers, which can be used for preventing corrosion of iron under hostile environmental conditions, is the aim for future research. The aim of the future work should be concentrated on designing new conducting polymer composites by incorporating specific filler materials, and suitably selecting a medium for polymerization so that the resultant epoxy coatings can be used for prevention of corrosion in saline water conditions. The main aspect should be focussed on:

* Designing of conjugated polymers which shows smart action and have self-healing ability. Self-healing mechanism (due to the redox property of the conducting polymer) leading to pin-hole and scratch free passivation.
* Environmental friendly/based on green technology (free from heavy metal ions and hazardous chromates), long service life and economic feasibility.

Our studies have indicated that PANI-SiO$_2$ composites prepared by in situ polymerization in the presence of per-fluoro octaonic acid (PFOA) as shown in Figure 1.34. This acts not only as surfactant but also act as a dopant, and wherein interaction of PFOA with SiO$_2$ particles leads to the formation of micelles, which acts as a soft template with the formation of emulsion; shows self-healing behaviour when these composite coating with epoxy are subjected to Salt Spray Test under marine environment (Dhawan et al. 2015).

Salt Spray Test of the PANI composite samples was done to evaluate the performance of the coating. It was observed that PANI composite with 6% loading (see Figure 1.35d) does not show any corrosion on the scribed X mark, whereas blank epoxy coated steel panel (Figure 1.35a), epoxy

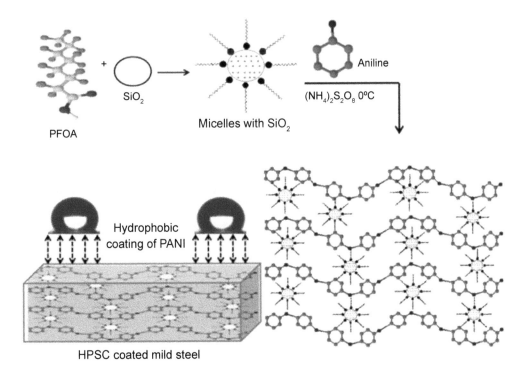

Figure 1.34: Self-healing conducting polyaniline composite (American Journal of Polymer Science, DOI: 10.5923/s. ajps.201501.02)

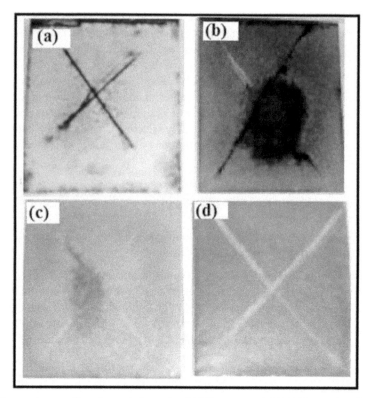

Figure 1.35: Salt spray images of powder coated sample of (a) Epoxy coated (b) PANI coated at 6% loading, (c) and (d) polyaniline composite coated at 1.5 and 6% loading level (American Journal of Polymer Science, DOI: 10.5923/s. ajps.201501.02)

coated MS panel with only polyaniline (Figure 1.35b), and epoxy encapsulated with 1.5% of modified PANI composite shows corrosion at the scribed mark.

Even self-healing Poly fibroblast formulations were prepared by Jason J. Benkoski, from The John Hopkins University, Applied Research Laboratory. In this study, silane-based hydrophobic primer additive like Octadimethyltrimethoxy silane (OTS), Isocyanoatopropyltrimethoxy silane (ITS) and Aminopropyltrimethoxy silane (APS) had shown good adhesion and even improved sealing across the scratch. During a laboratory experiment, new additive demonstrated its ability to prevent rusting for 6 weeks inside a chamber filled with salt fog. According to Frank Furman, Marine Captain and Research Program Manager of Department of Navy, this technology could cut maintenance cost and increase the time vehicles out in the field with the marines.

Development of self-healing coatings for corrosion protection on metallic structures has been nicely reviewed by Stankiewicz et al. (2016). According to their evaluation, the breakdown of research into self-healing coatings on a global basis concerning self-healing coatings for protection against corrosion, describe the use of polymers and composites contained within the coatings. Generally, self-healing protective coatings can be divided into two types: self-healing polymeric coatings and self-healing non-polymeric coatings (inorganic coatings). However, it is also reasonable to classify them according to the coatings functionality. Both types of coatings, for example, can prevent corrosion by releasing an active substance from capsules inserted into them. The addition of micro- or nanocapsules, loaded with corrosion inhibitors, into coatings is another approach to achieving a protective coating with self-healing attributes. Capsules are typically constructed from polymers, inorganic materials like silica, titania and calcium carbonate. Also, halloysite aluminosilicate nanotubes are found as an entrapment system for storage of anti-corrosion agents. An excellent study on Polyvinyl butyral with the nano-TiO_2-polyaniline system was explored by Radhakrishnan et al. (2009). In this system, three self-healing mechanisms are recognized. Barrier properties, redox behavior of polyaniline and formation of p–n junctions preventing charge transport after coating damage.

Another protective system, described as scratch and pin-hole tolerant/self-healing consists of a primer containing unique conducting polymer nano-dispersion, a top layer, and an optional interlayer. This coating technology system is called AnCatt, and is employed as an effective barrier for protecting metals against corrosion without using any heavy-metal pigments such as chromate, lead or zinc. Compared with current anti-corrosion coating products on the market, the AnCatt system offers 3–6x extension in the longevity of the coating. This coating system was tested in accordance with ASTM B117 for 13,000 h in a salt fog with no rusting or blistering of the substrate.

1.9.1 Conducting polymer-based coatings

In the past few decades, a lot of research has been carried out on conducting polymer-based coatings. Conducting polymers such as polypyrrole, polyaniline and polythiophene, etc., have been used for anti-corrosion application either by electrodepositing it on the metal surface or by direct addition to the corrosive medium as an inhibitor. Among the various conducting polymers, polyaniline and its derivative exhibit excellent anti-corrosive properties. This conducting polymer shows a dual protection mechanism by providing anodic protection to underlying metal, and shifting its potential to the passive region. Secondly, it acts as a barrier against highly corrosive conditions. Conducting polymer coatings on various metal substrates has been used in various industries. They are economically cheap to produce and can withstand against the ambient corrosive condition. The use of conducting polymers for the corrosion protection of metals has attracted excessive interest over the last thirty years. DeBerry in 1985 demonstrated that the polyaniline induced stabilization of the passive state for 400 series stainless steels in sulfuric acid solutions. The author observed that the conducting polymer-based organic coating would be able to maintain the surface potential of the metal substrate into the passive state, where a protective oxide film is formed on the substrates. Conducting polymer-based coating can efficiently bear the pin-holes and defects in a manner similar to that of the chromate-based coatings. This is described by the fact that oxygen reduction within the conducting polymer layer replenishes the conducting polymer charge consumed by metal oxidation. Along with the changing of the conducting polymers into the oxidation state, the potential turns into the passive state of the metal and inhibits the corrosion process. Even though the conducting polymers have been widely used as a potential material for protection of metal from corrosion, but still due to few limitations due to which these coatings are not able to fulfill physico-electrochemical and mechanical requirements of high-performance, anti-corrosive coatings under severe corrosion conditions such as in the presence of chlorides ions. Limitations of conducting polymers as anti-corrosive coatings include: (i) Irreversible consumption of the charge stored in the conducting polymers, which is capable of oxidizing the metal and resulting in the formation of a passive oxide layer, (ii) Anion-exchange properties, (iii) Porous structure and weak barrier properties and (iv) Poor adhesion to the metal substrate. Incorporation of conducting polymer-based composites, i.e., conducting polymer/inorganic fillers in to the coating system is one of the most efficient approaches to overcome the limitations of conducting polymers. Coating based on conducting polymer composites have presented better mechanical and physico-chemical properties improving the self-healing property, barrier effect, adhesion as well as water repellant behavior resulting the coating with better protection of metals against corrosion. Moreover, the development of conducting polymer-based coating system with commercial feasibility is also expected to advance by applying nanotechnology.

Numerous mechanisms of interaction are possible for the performance of conducting polymers on a variety of metal substrates, which generally depends on the type of metal substrate, use of specific conducting polymer and its quantity, its chemical composition, physical properties, type of dopant and its relative ratio with other active ingredients in the coating system.

Use of conducting polymers as anti-corrosive coating can be employed in several ways like a primer alone or as a primer coating with conventional top coat. It can also be blended with conventional

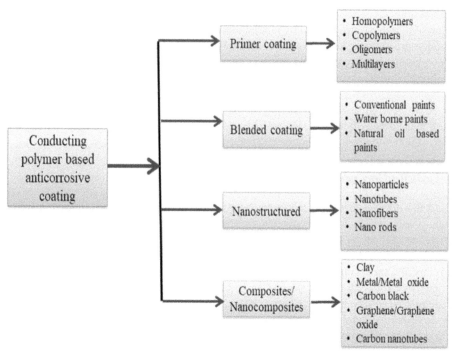

Figure 1.36: Various forms of conducting polymer-based anti-corrosive coating on metal substrates

polymer coatings (i.e., epoxy) or can be used as an anti-corrosive additive in the coating formulation, nanostructured coating and composites and nanocomposites based coating. All these methods are explored in Figure 1.36, and the following discussion under the heading of various conducting polymers.

1.9.2 Polyaniline and its derivatives

Polyaniline and its derivatives have been used for protection of metals in an aggressive environment. Ahmad and MacDiarmid (1996) observed that the electrochemically synthesized polyaniline was able to protect metals like iron and steel from corrosion. Polyaniline is found to be a good candidate for an anti-corrosive coating to replace highly hazardous chromium based coatings, which have an adverse effect on health and cause environmental concerns.

Among all conducting polymers, it is found to be the most promising candidate due to its good electrical conductivity, low-cost monomer, ease of synthesis, better environmental stability and its other various applications. Until the present, two forms of polyaniline have been most thoroughly explored for anti-corrosive coatings: these are the non-conducting emeraldine base (EB) and the conducting emeraldine salt (ES) form as shown in Figure 1.37.

On the basis of literature, different results of protection of metals by polyaniline were obtained and different mechanisms were suggested such as nature of dopants used, the different mode of synthesis of polyaniline, use of copolymers of aniline, different monomers and use of polyaniline/inorganic-based composites. Table 1.6 showed the kinds of literature including anti-corrosive performance of polyaniline and its derivatives, types of metal substrate used and different corrosive medium used during measurement.

Many researchers reported the corrosion inhibition behavior of polyaniline in acidic medium. Jeyaprabha et al. (2006) demonstrated the corrosion inhibition performance of water-soluble polyaniline for iron in 0.5 M H_2SO_4 solution. In order to improve the solubility of polyaniline, the authors had used dodecyl benzene sulfonic acid as a dopant. In this study, it was found that the

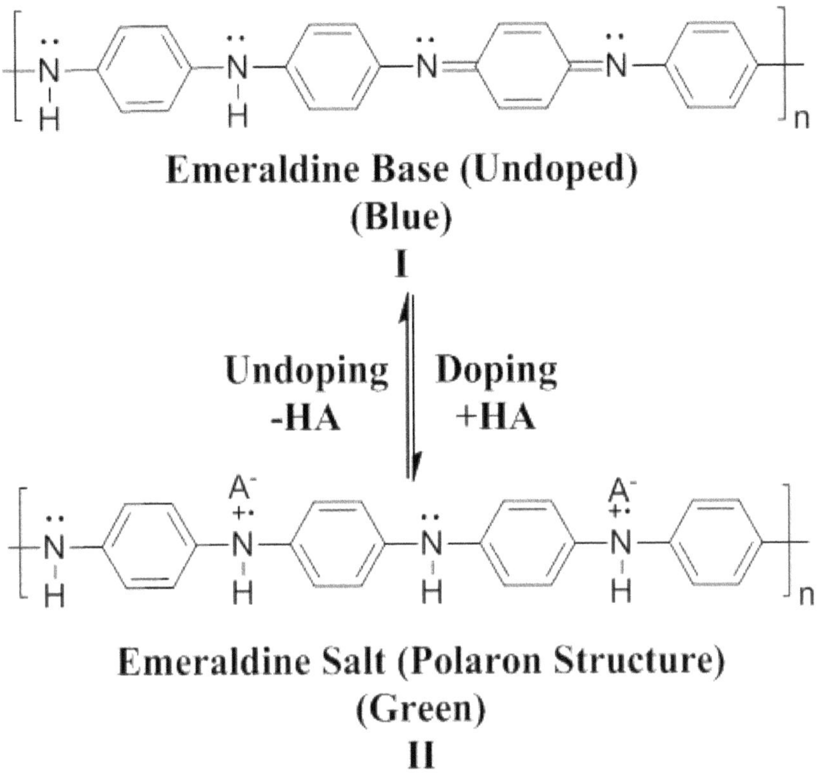

Emeraldine Base (Undoped)
(Blue)

I

Undoping | Doping
-HA | +HA

Emeraldine Salt (Polaron Structure)
(Green)

II

Figure 1.37: Undoped and doped forms of polyaniline used for corrosion protection

maximum corrosion inhibition efficiency of 84% had been obtained at a concentration of 100 ppm. Yi et al. (2013) demonstrated the corrosion inhibition property of sulfonated chitosan doped polyaniline for mild steel in acidic environment. The results revealed that sulfonated chitosan doped polyaniline had excellent corrosion inhibition for mild steel in 0.5 M hydrochloric acid media and it showed 92.3% corrosion inhibition efficiency at 40 ppm concentration. However, polyaniline had limited solubility in organic solvents and poor processability, due to the stiffness of its backbone (Ghosh et al. 2001). To overcome these drawbacks of polyaniline, extensive modifications have been made to this material (Pyshkina et al. 2008; Bhandari et al. 2009, 2011). Hence, several rings substituted anilines and copolymerization of aniline was found to be a successful modification method (Lie et al. 1997). Sazou (2001) carried out the electrodeposition of ring-substituted polyanilines such as o-toluidine, m-toluidine, o-anisidine and o-chloroaniline on iron surfaces from aqueous oxalic acid solutions, and evaluated their corrosion protection performance. Shinde and Patil (2010) reported that the electrochemical synthesis of poly(o-ethylaniline) on copper could remarkably reduce the rate of corrosion of copper in 3.5% NaCl solution.

Bhandari et al. (2008) showed that homopolymer and copolymers of aniline and ortho substituted aniline such as poly(o-isopropyl aniline) and poly(aniline-co-o-isopropyl aniline) that had superior solubility in organic solvents such as methanol and ethanol. Hence, these homopolymers and copolymers were found to be better corrosion inhibitors as compared to polyaniline. Similarly, Fuhua et al. (2011) evaluated the corrosion inhibitive effect of a polyaniline copolymer, namely, poly(aniline-co-o-anthranilic acid) of carbon steel in HCl medium. The effect of the different concentrations of the corrosion inhibitor was investigated and the results revealed that polyaniline copolymer solution acts as an effective corrosion inhibitor for carbon steel in 1.0 M HCl solution. The mechanism of corrosion inhibition had been explained on the basis of adsorption behavior of conducting copolymers. Copolymers of aniline showed better surface coverage on the metal substrate, this can be ascribed to

Table 1.6: Detail of polyaniline and its derivative, their protection efficiency, types of metal substrate used and different corrosive environment

Anticorrosive Materials	Type of metal used	Corrosive medium	Protection efficiency	Improved features	References
DBSA doped PANI	Iron	0.5M H_2SO_4	84%	Corrosion inhibition	Jeyaprabha et al. (2006)
Sulphonated chitosan doped PANI	Mild steel	0.5 M HCl	92.3% (at 40 ppm)	Corrosion inhibition	Yi et al. (2013)
Poly(Aniline-co-o-isopropylaniline)	Mild steel	1.0 M HCl	76% (80 ppm)	Corrosion inhibition	Bhandari et al. (2008)
Self-doped poly(aniline-co-4-amino-3-hydroxy-naphthalene-1-sulfonic acid)	Mild steel	1.0 M HCl	90% (70 ppm)	Corrosion inhibition	Bhandari et al. (2008)
Polyaniline-polyacrylic acid composites	316 stainless steel	0.5 M HCl	91.68% (200 ppm)	Corrosion inhibition	Syed et al. (2015)
Epoxy/polyaniline	Steel	3.5% NaCl meduim	Not measured	Anti-corrosive and barrier properties of coating improved	Saravanan et al. (2007)
Polyaniline/palm oil blend	Mild steel	NaCl	Not measured	Anti-corrosive and barrier properties of coating improved	Rashid et al. (2014)
Polyaniline-ZnO nanorod hybrid nanocomposites	Mild steel	3.5% NaCl	Not measured	Anti-corrosive and barrier properties of coating improved	Mostafaei and Nasirpouri (2014)
Polyaniline-clay nanocomposite	Cold rolled steel	5.0% NaCl	Not measured	Gas barrier property and mechanical strength improved	Yeh et al. (2001)
Polyaniline/ZnO nanocomposite	Aluminum and steel	1.0 M HCl	Not measured	Barrier effect and anti-corrosion protection improved	Alvi et al. (2016)
Epoxy/Cerium nitrate loaded polyaniline nanofibers	AA2024-T3	0.3 wt.% NaCl	Not measured	Superior corrosion protection performance and self-healing ability	Tavandashti et al. (2016)
Epoxy/PANI-nanofibers-CeO_2 grafted graphene oxide nanosheets	Mild steel	3.5% NaCl	Not measured	Barrier effect and anti-corrosion protection of GO nanosheet improved	Ramezanzadeh et al. (2018).
PANI-MWCNTs in epoxy	Steel	3.5 wt% NaCl	Protection for 120 h of immersion	Lower corrosion rate that of epoxy-coated steel	Deshpande et al. (2013)

the fact that more availability of p-electrons in the aromatic ring facilitates a flat orientation of the polymer on the metal surface. Bhandari et al. (2011) studied the corrosion inhibition performance of self-dopedpoly(aniline-co-4-amino-3-hydroxy-naphthalene-1-sulfonic acid) for the protection of iron in acidic medium. Results showed that the copolymers had remarkable corrosion inhibition performance as compared to the polyaniline. The corrosion inhibition efficiency was observed to be 90% by using 70 ppm concentration of copolymer in an acidic medium. Corrosion inhibition of iron in an acidic environment by copolymer had been explained on the basis of molecular adsorption. The electrochemical impedance studies showed that these copolymers protect the iron from corrosion in an aggressive medium through adsorption mechanism. The availability of lone pair of electrons on the hetro-atoms of co-monomer molecules facilitates it to show effective corrosion inhibition and greater coverage on the metal surface in acidic medium. Moreover, different functional groups present in naphthylamine ring such as $-NH_2$, $-OH$ and $-SO_3H$ are also responsible for enhancing the corrosion inhibition performance of iron in acidic medium. In acidic solution, amines act as a cation and adsorb onto the metal surface through electrostatic interaction between the positively charged anilinium cations, hence adsorbed chloride ions. The inhibitive action of ortho-methoxy substituted polyaniline, a new class of conducting polymer, on the corrosion of iron in acidic chloride solution has been evaluated by Electrochemical Impedance Spectroscopy (EIS), Linear Polarization Resistance (LPR), Weight Loss (WL) and by the Logarithmic Polarization Technique (LPT). Inhibition efficiencies of nearly 80–88% have been observed even at 25 ppm concentration as observed by Sathiyanaryanan et al. (1994). Double-layer capacitance studies indicate strong adsorption of polymer following the Temkin adsorption isotherm is largely responsible for its inhibitive action. Copolymer consists of aniline and substituted aniline act as a protonated species in acidic media, which adsorbs on the metal surface to form a monomolecular barrier. As with the barrier coatings, the adsorbed molecules restrict the anodic or cathodic corrosion reaction by reducing the rate of corrosion. Presence of delocalized pi-electrons on the copolymer chain form coordination type of bonds with the metal surface. Moreover, on the basis of literature, it was also assumed that the bulkiness of substitution in aromatic ring form a cluster network and attach more firmly on to the metal surface resulting in a good coverage of surface and enhance the inhibition efficiency. Better solubility exhibited by the copolymers as compared to polyaniline in common organic solvents such as alcohol, also plays an important role to enhance the corrosion inhibition performance of metals. The corrosion inhibition of iron in acid chloride solution, offered by a new class of polymer-*ortho* substituted poly ethoxy aniline, a conducting polymer, has also been studied by Sathiyanarayanan et al. (1992). Its corrosion inhibitive action was examined by the Tafel extrapolation method, linear polarization resistance method, impedance method and direct weight loss method. All these methods confirmed the effectiveness of the inhibitor, indicating also the possibility of monitoring its effectiveness by electrochemical techniques.

The copolymer based on aniline and anisidine has also been demonstrated by Ozyılmaz et al. (2005) as corrosion protective coating in 3.5 wt.% NaCl medium for copper. Hence, according to the literature, polyaniline has two different methods to protect the metal from corrosion. Firstly, it acts as a physical barrier and secondly, it passivates the electrode surface either if scratch takes place on the coating or if it gets ruptured. Syed et al. (2015) investigated the corrosion inhibition behavior of water-soluble polyaniline-polyacrylic acid composites for 316 stainless steels. The copolymer showed 91.68% corrosion inhibition efficiency in 0.5 M HCl medium at 200 ppm concentration.

Li et al. (1997) had shown that the anti-corrosion performance of polyaniline was further enhanced by the presence of a topcoat to increase the diffusional resistance against the corrosive species. Camalat et al. (2000) stated that electrochemically deposited polyaniline coating in the presence of oxalic acid showed good anti-corrosive properties in acidic medium. Recently Zhang et al. (2017) studied the corrosion protection performance of epoxy based different polyaniline coatings for mild steel substrate. The authors investigated the corrosion protection behavior of polyaniline synthesized in two different dopants such as hydrofluoric acid and camphor sulfonic acid. The results revealed that the coating based on camphor sulfonic acid doped polyaniline showed better protection of metal against corrosion as compared to hydrofluoric acid doped polyaniline. Kinlen et al. (2002) studied the corrosion protection performance of mild steel using sulfonic and phosphonic acid-doped polyanilines. The result was galvanic activity in pin-holes being eliminated when the phosphonic acid dopants were used in contrast to sulfonic acid dopants, which continued to allow galvanic activity in pin-holes. Dominis et al. (2003) demonstrated the comparison of polyaniline primers with different dopants for corrosion protection of steel. The results showed that the type of dopant and topcoat used had a significant influence on the corrosion protection as the dopant affects polymer conductivity and permeability. In addition, it was also observed that the dopant was released during galvanic coupling of the polymer to the metal. Dominis et al. (1998) also observed that polyaniline coatings without topcoats performed in a different way when different dopants were used in atmospheric corrosion testing.

Saravanan et al. (2007) investigated the anti-corrosive performance epoxy primer having polyaniline for steel substrate. In this study, it was observed that the metal repassivates due to the presence of polyaniline in the coating as a result of which epoxy formulated polyaniline coating showed a better anti-corrosive performance against the corrosive environment. Rashid et al. (2014) studied the polyaniline/palm oil blend for protection of mild steel, which showed significant corrosion protection to metal when exposed to the saline environment of 3.5 wt.% NaCl. Mostafaei and Nasirpouri (2014) developed the anti-corrosive epoxy based coating containing polyaniline-ZnO nanorod hybrid nanocomposites. The authors had observed that the epoxy coating having polyaniline-ZnO nanocomposites displayed higher corrosion resistance and provided better barrier properties in 3.5 wt.% NaCl medium as compared to unmodified epoxy and polyaniline modified epoxy coatings. The purpose of including ZnO in polyaniline matrix was to improve the mechanical integrity as well as the anti-corrosive performance of the coating. Recently, Ramezanzadeh et al. (2017) investigated the effects of highly crystalline and conductive polyaniline/graphene oxide composites on the corrosion protection performance of a zinc-rich epoxy coating. The results showed that the inclusion of 0.1 wt.% graphene oxide and GO and polyaniline/graphene oxide composites in zinc rich epoxy coating sample significantly improved its corrosion protection performance. Yeh et al. (2001) demonstrated the anti-corrosive performance of polyaniline-clay nanocomposite based coating on cold rolled steel in 5 wt.% NaCl medium. It was found that coatings showed better anti-corrosion performance than conventional polyaniline coating. In addition, the gas barrier property and mechanical strength were also investigated of the polyaniline-clay nanocomposite based coating and the results showed that the O_2 gas barrier of materials in the form of freestanding film had 400% reduction in permeability compared to conventional polyaniline. Incorporation of nanolayers of MMT clay in polyaniline matrix resulted in a reduction in mechanical strength (in the form of free-standing films).

Alvi et al. (2016) investigated corrosion inhibition behavior of zinc oxide-polyaniline nanocomposite for aluminum and steel in 0.1 M HCl and 1 MNaCl medium. The ZnO-polyaniline was found to be an excellent corrosion inhibitor for steel and aluminum due to its chain conformation and electronic properties. Their studies had shown that the ZnO-polyaniline nanocomposite based coatings on steel and aluminum inhibited corrosion due to the internal sacrificial electrode formation as well as the barrier effect at the ZnO-polyaniline and metal interface. A detailed review on conducting polyaniline nanocomposite-based paints for corrosion protection of steel has been given by Sazou and Deshpande (2017). The present review explains applications of PANI nanocomposites in steel anti-corrosion technology. The incorporation of inorganic fillers of different nature and size into conducting polyaniline (PANI)-based paint formulation extends the possibility of developing protective coatings with a self-healing capability and improved corrosion protection performance. The resulting PANI-based coatings are characterized as nanocomposite systems if the filler has nano-size dimensions. Nanofillers such as metal and metal oxide nanoparticles, clay, carbon nanotubes, graphene and other inorganic pigments combined with PANI give rise to a variety of PANI nanocomposites with interesting properties and potential applications.

1.9.3 Polypyrrole and its derivatives

Polypyrrole (PPy) is one of the potential materials among conducting polymers due to its better conductivity, environmental stability, ease of synthesis, anti-corrosive ability and eco-friendly nature (Redondo and Breslin 2007; Mollahosseini and Noroozian 2009). Two forms of polypyrrole which are generally used in corrosion protection are shown in Figure 1.38.

There is a lot of literature related to the corrosion protection of copper and its alloys by electrodeposited polypyrrole films (Fenelon and Breslin 2002; Fenelon and Breslin 2003; Prissanaroon et al. 2004; Redondo and Breslin 2007). El-Shazly and Wazzan (2012) investigated the corrosion resistance of buried steel by coating it with polypyrrole (PPy) layer. The results showed that polypyrrole

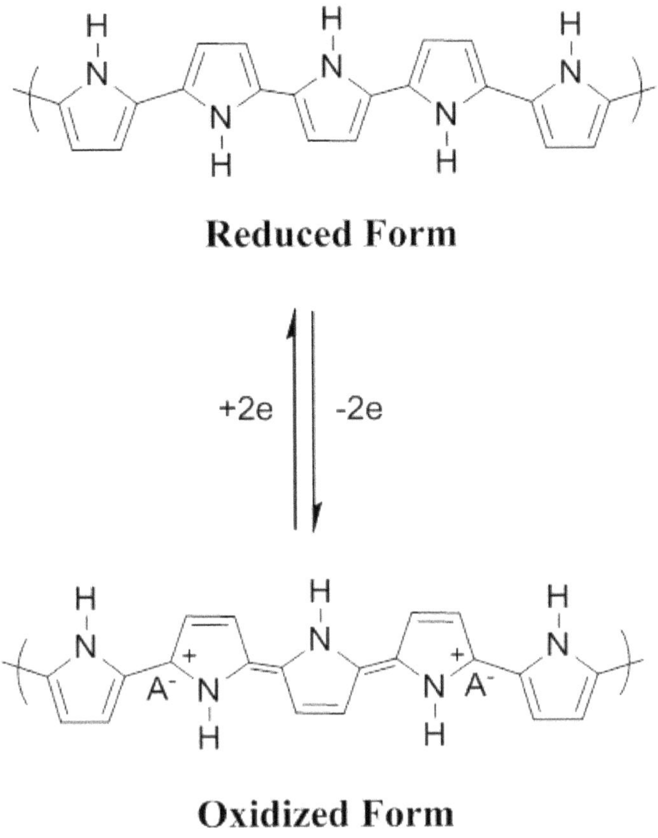

Reduced Form

+2e || -2e

Oxidized Form

Figure 1.38: Reduced and oxidized forms of polypyrrole

coated steel had remarkable corrosion protection when buried in the sand having various contaminants such as NaCl, H_2SO_4 and water.

Beck and Michaelis (1992) and Beck et al. (1994) studied the electropolymerization of pyrrole on iron from the oxalic acid medium. Results showed the poor adherence ability of polypyrrole on metal substrates. Le et al. (2001) performed the electrodeposition of polypyrrole on iron in potassium tetraoxalate and pyrrole solution. This procedure yielded strongly adherent and smooth polymer layers. Moreover, the synthesized polypyrrole film showed better corrosion protection in 3.0 wt.% NaCl medium. Su and Iroh (2000) demonstrated that the adherence ability of polypyrrole to the metal substrate increased with decreasing pH of oxalic acid. The authors proposed that –N-H groups present in polypyrrole ring which bonded to the steel surface are responsible for better adherence in acidic medium.

Kowalski et al. (2007) investigated the performance of bilayered polypyrrole coating for corrosion protection of steel. Their results revealed that the bilayered coating kept the steel in a passive state and prevented corrosion of the steels for a longer time period.

Martins et al. (2004) demonstrated the corrosion protection behavior of zinc-coated steel electrodes, which was electrochemically modified by polypyrrole films. The corrosion studies were carried out in NaCl, HCl and H_2SO_4 solutions. They found that the polypyrrole coatings increased the corrosion potential and reduced the corrosion current. Recently, Nautiyal et al. (2018) developed the high performance polypyrrole coating for corrosion protection and biocidal applications. Polypyrrole (PPy) coating was electrochemically synthesized on carbon steel using sulfonic acids as dopants. The coating was found to be an excellent anti-corrosive as well as an antimicrobial material.

On the whole, there is clear evidence that PPy shows good anti-corrosive properties, but it was found that maintenance of corrosion protective ability of polypyrrole alone is limited. During the synthesis of PPy, counter anions can be incorporated into the PPy matrix by neutralizing the positive charges on the polymer backbone. Counter anions play an important role in the change in physical properties and morphology. Small anions can be released from the polymer coating and provide some additional protection when the coating is reduced (Iroh and Su 2000; Tohumcu et al. 2014). Moreover, several efforts have also been made to increase the mechanical strength of polypyrrole coating on the metal surface. On the basis of the literature, it was found that polypyrrole and its nanocomposites are the most used polymer for corrosion protection of different metals. In addition, the coating based on these nanocomposites had better mechanical strength and adhesion on the metal substrates.

Babazadeh et al. (2012) investigated the physical properties of polypyrrole/TiO_2 nanocomposites prepared through a one-step *"in situ"* polymerization method. The obtained results revealed that polypyrrole containing TiO_2 nanoparticles had a strong effect on the morphology of polypyrrole-TiO_2 nanocomposites.

Hammache et al. (2003) demonstrated that corrosion protection behavior of iron using polypyrrole modified by copper microparticles, and the results showed enhanced corrosion protection ability by these composites.

Hussein et al. (2016) investigated the corrosion protection performance of nanocomposites based on PPy-sCNTs and PPy-NiLa nanocomposites for aluminum in a saline medium. The results showed that PPy-NiLa nanocomposite provided better protection than the PPy-sCNTs nanocomposite acidic and chloride solution.

Ates et al. (2015) investigated the corrosion protection performance of conducting polymer nanocomposites such as polyaniline-TiO_2 and polypyrrole-TiO_2 coating on Al1050 electrode in a saline medium. The results of these studies revealed that the corrosion protection efficiency of the PANI/TiO_2 (PE = 97.2%), PPy/TiO_2 (PE = 97.4%) nanocomposites coated on Al1050 electrode was higher than that of polyaniline (PE = 96.4%), polypyrrole (PE = 94.9%) and uncoated Al1050 electrodes. Authors concluded that the corrosion protection behavior of conducting polymers was enhanced by incorporation of nanofillers such as TiO_2 nanoparticles in the polymer matrix.

Recently Vu et al. (2018) evaluated the corrosion inhibiting ability of epoxy coating containing Silica/Polypyrrole-oxalate nanocomposite on carbon steel. Polymerization of pyrrole/silica was carried out in the presence of sodium oxalate solution. It was observed that the presence of oxalate anion enhanced the corrosion resistance ability of metal in 3.0% NaCl medium. The authors also concluded that such composites can be effectively used as a promising inhibitor for organic coating which can remarkably delay the degradation of metals in saline and humid conditions. Table 1.7 shows the details of polypyrrole based nanocomposites for protection of metals in different corrosive environments.

Also, Moynot et al. (2008) have designed self-healing coatings by making use of epoxy-amine microcapsules. The objective was to investigate the self-healing performance of encapsulated Mg^{2+} as inorganic film-formers, which should be automatically delivered when a crack propagates through the coating filled with capsules, ruptures the microcapsules and releases the agents in the vicinity of the bare metal under cathodic polarization. In the defect area, the production of hydroxyls ions should favor the formation of an inorganic deposit by precipitation of $Mg(OH)_2$.

A conducting polymer-based nanocapsule system for redox-triggered release of self-healing agents, which comprises a redox-sensitive polyaniline (PANI) shell and a self-healing agent encapsulated in the core synthesized by the mini-emulsion technique by Vimalanandan et al. (2013). Conducting polymers can store active anions as counter-charge to the oxidized polymer backbone, which can be released upon the onset of corrosion and subsequent reduction of conducting polymers. In the study, 3-Nitrosalicylic acid (3-NisA) was selected as a very efficient zinc corrosion inhibitor and was encapsulated in the PANI capsules. The release of 3-NisA was investigated by measuring the absorption at 225 nm with UV spectrophotometry under three different electrochemical potential conditions that simulate different situations during corrosion of a metal. When no potential is applied, the PANI shells act as an effective barrier preventing the release of 3-NisA from the core (areas with

Table 1.7: Details of the anti-corrosive performance of polypyrrole based nanocomposites in different corrosive environments

Anti-corrosive materials	Type of metal used	Corrosive medium	Observations and results	References
Epoxy/PPy-ZnO	Carbon steel	3.0 wt.% NaCl	Coating showed enhanced adherent and anti-corrosive properties	Valenca et al. (2015)
PVA/Molybdate doped PPy montmorillonite nanocomposites	Mild steel	3.0% NaCl	Protection effect was found to be doubled with montmorillonite and dopants anions	Trung et al. (2013)
Polypyrrole/multi-walled carbon nanotubes and PPY-MWCNT-COO⁻	60Cu–40Zn brass alloy	3.0% NaCl	PPy/MWCNT-COO− functionalized nanocomposite provided higher corrosion resistance coating than PPy/MWCNT alone	Davood et al. (2015)
Polyvinyl butyral/polypyrrole-carbon black	Carbon steel	Saline medium	The composites showed self-healing ability and improved anti-corrosion properties	Niratiwongkorn et al. (2016)
PPy doped with the $[PMo_{12}O_{40}]^{3-}$	Iron	0.1 M NaCl	Coating showed self-healing ability	Porebska et al. (2005)
DBSA doped PPy-montmorillonite nanocomposies	Cold rolled Steel	5.0% NaCl meduim	Anti-corrosive and barrier properties of coating improved	Yeh et al. (2003)

white background). The reduction of PANI (gray-shaded areas) causes an increase of the permeability of the shell and a release of inhibitor is observed.

1.10 Conclusion

Corrosion is an omnipresent and omnipotent phenomenon that causes severe deterioration to the metallic surfaces. We encounter corrosion everywhere, be it in the presence of air, water or soil. Needless to say, corrosion has its roots in the surrounding environment. It is a powerful detrimental force that turns tons of metals into scrap every year. The estimation of direct and indirect losses due to corrosion is very difficult to predict. In some instances, it causes huge monetary loss whereas in other instances, it is responsible for the loss of precious human lives. The aim of the present chapter is to give detailed description of the protection of metallic surfaces from corrosion. The chapter starts with a general discussion about the basics of corrosion process. The cost of corrosion in various sectors of different countries is mentioned in this chapter. The initial sections of this chapter mention a brief discussion about the various types of corrosion processes occurring on the metal surface. The electrochemistry of corrosion is discussed in terms of various anodic and cathodic reactions taking place on the metal surface. The EMF series and galvanic series are listed for various metals. The chapter mentions the thermodynamic aspects of corrosion by presenting the Pourbaix diagram, whereas the kinetic aspects of the corrosion process is discussed in terms of polarization technique and Electrochemical Impedance Spectroscopy (EIS). A detailed discussion of the methods of corrosion protection is given in this chapter.

Application of protective coatings on metal surfaces is one of the efficient methods to curb corrosion of metals. Various types of protective coatings are listed in the chapter. Apart from the conventional protective paints/coatings, the chapter gives a detailed discussion on the new age smart coatings based on conducting polymers like polyaniline, polypyrrole, etc. An elaborate discussion on the corrosion resistant behavior of the conducting polymer-based coatings is carried out in the chapter. Apart from this, the corrosion resistant properties of various derivatives of these conducting polymeric coatings are also discussed in the chapter. The corrosion protection mechanisms of conducting polymers and their derivatives are highlighted to exhibit physical barrier effect, anodic protection and self-healing mechanism. The later sections of this chapter provides a detailed discussion about the standard test methods of mechanical testing of coatings including tape test, cross-cut adhesion test, pull-off test, Mandrel Bend Test, Taber Abrasion Resistance Test, etc. Details of the testing instruments and the relevant calculations are given in this chapter. Smart coatings for corrosion protection are the main aims which should be considered while designing a coating on the metal surface. Various groups have done excellent work in this direction and the whole idea is to minimize the loss due to corrosion protection.

References

Ahmad, N. and MacDiarmid, A. G. (1996). Inhibition of corrosion of steels with the exploitation of conducting polymers. Synthetic Metals, 78: 103–110.

Akpolat, S. and Bilgic, S. (2014). The protective effect of polypyrrole coating on the corrosion of steel electrode in acidic media. Protection of Metals and Physical Chemistry of Surfaces, 50: 266–272.

Aoki, I. (2001). Ac-impedance and Raman spectroscopy study of the electrochemical behavior of pure aluminum in citric acid media. Electrochimica Acta, 46(12): 1871–1878.

Alfred, T. (2005). Fats and Fatty Oils. Ullmann's Encyclopedia of Industrial Chemistry. Weinheim: Wiley-VCH. doi:10.1002/14356007.a10_173.

Alsebrook, W. E. (1955). The coating of magnesium alloys, Corrosion Technology, 2. 113 (April 1955).

Alvi1, F., Aslam, N. and Shauka, S. F. (2015). Corrosion inhibition study of zinc oxide-polyaniline nanocomposite for aluminum and steel. American Journal of Applied Chemistry, 3(2): 57–64.

Ates, M., Kalender, O., Topkaya, E. and Kamer, L. (2015). Polyaniline and polypyrrole/TiO_2 nanocomposite coatings on Al1050: electrosynthesis, characterization and their corrosion protection ability in saltwater media. Iranian Polymer Journal, 24(7): 607–619.

Babazadeh, M., Gohari, F. R. and Olad, A. (2012). Characterization and physical properties investigation of conducting polypyrrole/TiO_2 nanocomposites prepared through a one-step "*in situ*" polymerization method. Journal of Applied Polymer Science, 123: 1922–1927.

Bagal, N. S., Vaibhav, S. K. and Deshpande, P. P. (2018). Nano-TiO_2 phosphate conversion coatings a chemical approach. https://doi.org/10.1515/eetech-2018-0006, Electrochemical Energy Technology, 4: 47–54.

Beck, F. and Michaelis, R. (1992). Strongly adherent, smooth coatings of polypyrrole oxalate on iron. Journal of Coating Technology, 64: 59–67.

Beck, F., Michaelis, R. and Schloten, F. (1994). Film forming electropolymerization of pyrrole on iron in aqueous oxalic acid. Electrochimica Acta, 39: 229–234.

Bellucci, F. and Nicodemo, L. (1993). Water transport in organic coatings. Corrosion, 49(3): 235–247.

Bhandari, H., Sathiyanarayana, S., Choudhary, V. and Dhawan, S. K. (2008). Synthesis and characterizations of proceesible polyaniline derivatives for corrosion inhibition. Journal of Applied Polymer Science, 111: 2328–2339.

Bhandari, H., Bansal, V., Choudhary, V. and Dhawan, S. K. (2009). Influence of reaction conditions on the formation of nanotubes/nanoparticles of polyaniline in the presence of 1-amino-2-naphthol-4-sulphonic acid and its applications as electrostatic charge dissipation material. Polymer International, 58: 489.

Bhandari, H., Choudhary, V. and Dhawan, S. K. (2011). Influence of self doped Poly(aniline-co-1-amino-2-naphthol-4-sulphonic acid) on corrosion inhibition behavior of iron in acidic medium. Synthetic Metals 161: 753–762.

Bhandari, H., Choudhary, V. and Dhawan, S. K. (2011). Influence of self-doped poly(aniline-co-4-amino-3-hydroxy-naphthalene-1-sulfonic acid) on corrosion inhibition behavior of iron in acidic medium. Synthetic Metals 161: 753–762.

Blisiak, J. and Kowalik, J. A. (2000). Comparison of the *in vitro* genotoxicity of tri- and hexavalent chromium. Mutation Research/Genetic Toxicology and Environmental Mutagenesis, 469: 135–45.

Bockris, J. O. M. and Reddy, A. K. M. (1970). Modern Electrochemistry: An Introduction to an Interdisciplinary area. Plennum Press, New York.

Bockris, J. O. M. and Reddy, A. K. M. (1977). Modern Electrochemistry. Vol. 1, Plenum Press, New York.

Bokris, J. O. M. and Yang, B. (1991). Mechanism of corrosion inhibition of iron in acid solution by acetylenic alcohols. Journal of Electrochemical Society, 138: 2237–2252.

Bonora, P., Deflorian, F. and Fedrizzi, L. (1996). Electrochemical impedance spectroscopy as a tool for investigating underpaint corrosion. Electrochimica Acta, 41(7): 1073–1082.

Boocock, S. K. (1994). A Report on SSPC programs to research performance evaluation methods. Journal of Protective Coatings and Linings, 11(10): 51–58.

Boocock, S. K. (1995). Meeting industry needs for improved tests. Journal of Protective Coatings and Linings, 12(9): 70–76.

Brockes, A. (1964). Der Zusammenhang von Farbstaerke und Teilchengroesse von Buntpigmentennach der Mie-Theorie. Optik, 21(10): 550.

Budakian, R., Weninger, K., Hiller, R. A. and Putterman, S. J. (1998). Letters to nature: Picosecond discharges and slip-stick friction at a moving meniscus of mercury on glass. Nature, 391: 266–268.

Cáceres, L., Vargas, T. and Herrera, L. (2007). Determination of electrochemical parameters and corrosion rate for carbon steel in un-buffered sodium chloride solutions using a superposition model. Corrosion Science, 49(8): 3168–3184.

Camalet, J. L., Lacroix, J. C. and Nguyen, T. D. (2000). Aniline electropolymerization on platinum and mild steel from neutral aqueous media. Journal of Electroanalytical Chemistry, 485: 13–20.

Carr, W. (1971). Effect of pigment dispersion on the appearance and properties of paint films. Journal of the Oil & Colour Chemists Association, 54: 1093.

Chromey, F.C. (1960). Evaluation of mie equations for colored spheres. Journal of the Optical Society of America, 50(7): 730.

Connell, George A. (inventor) (1958). Methods and compositions for controlling fires. Published November 4, 1958.

Dahl, P. F. (1997). Flash of the Cathode Rays: A History of J.J. Thomson's Electron. CRC Press, pp. 49–52.

Davy, H. (1812). Elements of Chemical Philosophy. p. 85.ISBN 0-217-88947-6. This is the likely origin of the term "arc".

Davide, V. A. L., Beye Dominik, J. and Pettine, Mamadou, L. (2010). Chromium (VI) is more toxic than chromium (iii) to freshwater algae: a paradigm to revise. Ecotoxicology and Environmental Safety, 73(5): 743–9.

Davoodi, A., Honarbakhsh, S. and Farz, G. A. (2015). Evaluation of corrosion resistance of polypyrrole/functionalized multi-walled carbon nanotubes composite coatings on 60Cu–40Zn brass alloy. Progress in Organic Coatings, 88: 106–115.

DeBerry, D. W. (1985). Modification of the electrochemical and corrosion behavior of stainless steels with an electroactive coating. Journal of Electrochemical Society, 132(5): 1022–1026.

Deflorian, F., Fedrizzi, L. and Bonora, P. (1994). Determination of the reactive area of organic coated metals using the breakpoint method. Corrosion, 50(2): 113–119.

Deshpande, P. P., Vathare, S. S., Vagge, S. T., Tomšík, E. and Stejskal, J. (2013). Conducting polyaniline/multi-wall carbon nanotubes composite paints on low carbon steel for corrosion protection: electrochemical investigations, Chemical Papers, 67(8): 1072–1078.

Dominis, A. J., Spinks, G. M. and Wallace, G. G. (2003). Comparison of polyaniline primers prepared with different dopants for corrosion protection of steel. Progress in Organic Coatings, 48(1): 43–49.

Dominis, A., Spinks, G. M. and Wallace, G. G. (1998). Conducting polymer coatings on steel: adhesion and corrosion protection. pp. 1229–1233. *In*: Proceedings of the 56th Annual Technical Conference, Society of Plastic Engineering.

Doubleday, D. and Barkman, A. (1950). Reading the Hegman Grind Gauge. Paint, Oil and Chemical Review June 22, 113: 34–39.

Donahoe, D., Zhao, K., Murray, S. and Ray, R. M. (2008). Accelerated life testing. Encyclopedia of Quantitative Risk Analysis and Assessment.

Dybwad, G. L. and Mandeville, C. E. (1967). Generation of light by the relative motion of contiguous surfaces of mercury and glass. Physical Review, 161: 527–532.

El-Shazly, A. H. and Wazzan, A. A. (2012). Using polypyrrole coating for improving the corrosion resistance of steel buried in corrosive mediums. International Journal of Electrochemical Science, 7: 1946–1957.

Evans, U. R. (1976). The Corrosion and Oxidation of Metals. 2nd Supp. Vol., Arnold, London, page no. 82.

Eric, K. (1999). Dispersions: Characterization, Testing, and Measurement, p. 243. Marcel Dekker, Inc.

Farstad, D. K. (1961). A new water-soluble linseed oil. Amer. Paint J., 45. 44.

Flexner, Bob (1993). Understanding Wood Finishing. Pan Macmillan. p.77. ISBN0875965660.

Fenelon, A. M. and Breslin, C. B. (2002). The electrochemical synthesis of polypyrrole at a copper electrode: corrosion protection properties. Electrochimica Acta, 47: 4467–4476.

Fenelon, A. M. and Breslin, C. B. (2003). The electropolymerization of pyrrole at a CuNi electrode: corrosion protection properties. Corrosion Science, 45: 2837–2850.

Feser, R. M. S. (1990). Steel Research, 61: 482–489.

Francis, Hauksbee. (1705). Several experiments on the mercurial phosphorus, made before the Royal Society, at Gresham-College. Philosophical Transactions of the Royal Society of London, 24: 2129–2135.

Fuhua, S., Xiutong, W., Jianqiang, Y. and Baorong, H. (2011). Corrosion inhibition by polyaniline copolymer of mild steel in hydrochloric acid solution. Anti-Corrosion Methods and Materials, 58/3: 111–115.

Ghosh, P., Siddhanta, S. K., Haque, S. R. and Chakrabarti, A. (2001). Stable polyaniline dispersions prepared in nonaqueous medium: synthesis and characterization. Synthetic Metals, 123(1): 83–9.

Gibson, W.F. and Pikas, J.L. (1993). Conductive polymeric cable anodes for pipelines with deteriorating coatings. Materials Performance, 32(3): 24–26.

Gluck, H. (1964). The impermanence's of painting in relation to artists' materials. Journal of the Royal Society of Arts, Volume CXII.

Grundmeier, G., Schmidt, W. and Stratmann, M. (2000). Corrosion protection by organic coatings: electrochemical mechanism and novel methods of investigation. Electrochimica Acta, 45(15): 2515–2533.

Gullichsen, J. and Hannu, P. (2000). Chemical Pulping. Papermaking Science and Technology. 6B. Finland.pp. B378–B388. ISBN 952-5216-06-3.

Hussein, M. A., Al-Juaid, S. S., Abu-Zied, B. M. and A-Elhagag, A. (2016). Hermas electrodeposition and corrosion protection performance of polypyrrole composites on aluminum. International Journal of Electrochemical Science, 11: 3938–3951.

https://ancatt.com/anti-corrosion-coatings/10-07-2015.

Iroh, J. O. and Su, W. (2000). Corrosion performance of polypyrrole coating applied to low carbon steel by an electrochemical process. Electrochimica Acta, 46(1): 15–24.

Jane Alexander. (2017). Understand the Danger: Pitting Corrosion. In Efficient Planet.

Jeyaprabha, C., Sathiyanarayanan, S. and Venkatachari, G. (2006). Polyaniline as corrosion inhibitor for iron in acid solutions. Journal of Applied Polymer Science, 101: 2144 –2153.

Jones, Frank, N. (2003). Alkyd Resins. doi:10.1002/14356007.a01_409.

Kamaraj, K., Devarapalli, R., Siva, T. and Sathiyanarayanan, S. (2015). Self-healing electrosynthesied polyaniline film as primer coat for AA, 2024-T3. Materials Chemistry and Physics., 153: 256–265.

Kendig, M. and Scully, J. (1990). Basic aspects of electrochemical impedance application for the life prediction of organic coatings on metals. Corrosion, 46(1): 22–29.

Kinlen, P. J., Ding, Y. and Silverman, D. C. (2002). Corrosion protection of mild steel using sulfonic and phosphonic acid-doped polyanilines. Corrosion Science Section, 58(6): 490–497.

Kowalski, D., Ueda, M. and Ohtsuka, T. (2007). Corrosion protection of steel by bi-layered polypyrrole doped with molybdophosphate and naphthalene disulfonate anions. Corrosion Science, 49: 1635–1644.

Koul, S., Dhawan, S. K. and Chandra, R. (2001). Compensated sulphonated polyaniline—correlation of processibility and crystalline structure. Synthetic Metals, 124: 295–299.

Kumar, A. S., Bhandari, H. and Dhawan, S. K. (2013). Polymer International, 62: 1192–1201.

Le, N., Huan, Garcia, B and Deslouis, Claude and Xuân, Lê. (2001). Corrosion protection and conducting polymers: polypyrrole films on iron. Electrochimica Acta, 46: 4259–4272.

Leigraf, C. T. E. (2000). Graedel, Atmospheric corrosion. Electrochemical Society Series, John Wiley, New York, pp. 149–157.

Li, P., Tan, T. C. and Lee, J. Y. (1997). Corrosion protection of mild steel by electroactive polyaniline coatings. Synthetic Metals, 88: 237–242.

Mansfeld, F., Jeanjaquet, S. and Kendig, M. (1986). An electrochemical impedance spectroscopy study of reactions at the metal/coating interface. Corrosion Science, 26(9): 735–742.

Martins, J. I., Reis, T. C., Bazzaoui, M., Bazzaoui, E. A. and Martins, L. (2004). Polypyrrole coatings as a treatment for zinc-coated steel surfaces against corrosion. Corrosion Science, 46: 2361–2381.

Mohebbi, H. and Li, C. Q. (2011). Experimental investigation on corrosion of cast iron pipes. International Journal of Corrosion Volume 2011, Article ID 506501, 17 page.

Mohd, R., Sabir, S., Afidah, R. and Waware, A. (2014). Umesh polyaniline/palm oil blend for anticorrosion of mild steel in saline environment. Journal of Applied Chemistry Volume 2014, Article ID 973653, 1–6.

Mollahosseini, A. and Noroozian, E. (2009). Electrodeposition of a highly adherent and thermally stable polypyrrole coating on steel from aqueous polyphosphate solution. Synthetic Metals, 159: 1247–1254.

Moutarlier, V., Gigandet, M.P., Normand, B. and Pagettia, J. (2005). EIS characterisation of anodic films formed on 2024 aluminium alloy, in sulphuric acid containing molybdate or permanganate species. Corrosion Science, 47(4): 937–951.

Mostafaei, A. and Nasirpouri, F. (2014). Epoxy/polyaniline–ZnO nanorods hybrid nanocomposite coatings: Synthesis, characterization and corrosion protection performance of conducting paints. Progress in Organic Coatings, 77(1): 146–159.

Nair, R. R., Geim, A. K., Wong, S. L., Waters, J., Kravets, V. G. and Su, Y. (2014). Impermeable barrier films and protective coatings based on reduced graphene oxide, The University of Manchester.

Nanan, K. (2017). In How Metallic Coatings Protect Metals from Corrosion, Corrosion Pedia.

Nautiyal, A., Qiao, M., Cook, J. E., Zhang, X. and Huang, T. -S. (2018). High performance polypyrrole coating for corrosion protection and biocidal applications. Applied Surface Science, Volume 427, Part A, 1 January 2018, Pages 922–930.

Niratiwongkorn, T., Luckachan, G. and Mittal, V. (2016). Self-healing protective coatings of polyvinyl butyral/polypyrrole-carbon black composite on carbon steel. RSC Advances, 6(49): 1039/C6RA01619G.

Ozyılmaz, A. T., Colak, N., Sangun, M. K. and Yazıcı, B. (2005). The electrochemical synthesis of poly(aniline-co-o-anisidine) on copper and their corrosion performances. Progress in Organic Coatings, 54: 353–359.

Pebere, N., Picaud, Th., Duprat, M. and Dabosi, F. (1989). Evaluation of corrosion performance of coated steel by the impedance technique. Corrosion Science, 29(9): 1073–1086.

Paliwoda-Porebska, G., Stratmann, M., Rohwerder, M., Potje-Kamloth, K., Lu, Y., Pich, A. Z. and Adler, H. -J. (2005). On the development of polypyrrole coatings with self-healing properties for iron corrosion protection. Corrosion Science, 47: 3216–3233. 10.1016/j.corsci.2005.05.057.

Petitjean, J., Aeiyach, S., Lacroix, J. C. and Lacaze, P.C. (1999). Ultra-fast electropolymerization of pyrrole in aqueous media on oxidizable metals in a one-step process. Journal of Electroanalytical Chemistry, 478: 92–100.

Pyshkina, O. A., Kim, B., Korovin, A. N., Zezin, A., Sergeyev, V. G. and Levon, K. (2008). Interpolymer complexation of water-soluble self-doped polyaniline. Synthetic Metals, 158: 999–1003.

Pourbaix, M. (1966). Atlas of Electrochemical Equilibrium in Aqueous Solutions, Pergemon Press, NewYork.

Praveen, C. V., Pradeep, Kiran, J. A. and Bhaskar, M. (2012). Cadmium toxicity—a health hazard and a serious environmental problem—an overview. International Journal of Pharma and Bio Sciences, 2(4): 235–46.

Radhakrishnan, S., Siju, C. R., Mahanta, D., Patil, S. and Madras, G. (2009). Conducting polyaniline-nano-TiO_2 composites for smart corrosion resistant coatings. Electrochimica Acta, 54: 1249–54.

Ramezanzadeh, B., Mohamadzadeh Moghadam, M. H., Shohani, N. and Mahdavian, M. (2017). Effects of highly crystalline and conductive polyaniline/graphene oxide composites on the corrosion protection performance of a zinc-rich epoxy coating. Chemical Engineering Journal, 320(15): 363–375.

Ramezanzadeh, B., Bahlakeh, G. and Ramezanzadeh, M. (2018). Polyaniline-cerium oxide (PAni-CeO_2) coated graphene oxide for enhancement of epoxy coating corrosion protection performance on mild steel. Corrosion Science, 137: 111–126.

Redondo, M. I. and Breslin, C. B. (2007). Polypyrrole electrodeposited on copper from an aqueous phosphate solution: corrosion protection properties. Corrosion Science, 49: 1765–1776.

Ruhi, G., Bhandari, H. and Dhawan, S. K. (2014). Progress in Organic Coating, 77(9): 1484.

Sabin, A. H. (1916). Technology of Paint and Varnish, 2d ed. John Wiley and Sons, New York, N.Y., 10016.

Sambyal, P., Ruhi, G., Bhandari, H. and Dhawan, S. K. (2015). Surface and Coating Technology, 272,129.

Santos, J. R., Jr., Mattoso, L. H. C., Motheo, A. J. 1998. Electrochimica Acta, 43: 309.

Saravanan, K., Sathiyanarayanan, S., Muralidharan, S., Azim, S. S. and Venkatachari, G. (2007). Performance evaluation of polyaniline pigmented epoxy coating for corrosion protection of steel in concrete environment. Progress in Organic Coating, 59(2): 160–167.

Sathiyanarayanan, S., Dhawan, S. K., Trivedi, D. C. and Balakrishna, K. (1992). Soluble conducting poly ethoxy aniline as an inhibitor for iron in HCl. Corrosion Science, 33(12): 1831–1841.

Sathiyanarayanan, S., Azim, S. S. and Venkatachari, G. (2007). Preparation of polyaniline–TiO_2 composite and its comparative corrosion protection performance with polyaniline. Synthetic Metals, 157: 205–213.

Sazou, D. (2001). Electrodeposition of ring-substituted polyanilines on Fe surfaces from aqueous oxalic acid solutions and corrosion protection of Fe. Synthetic Metals, 118: 133–147.

Sazou, D. and Deshpande, P. P. (2017). Conducting polyaniline nanocomposite-based paints for corrosion protection of steel. Chemical Papers, 71(2): 459–487.

Shinde, V. and Patil, P. P. (2010). Evaluation of corrosion protection performance of poly(o-ethyl aniline) coated copper by electrochemical impedance spectroscopy. Materials Science and Engineering: B, 168: 142–150.

Singh, I. (2003). Corrosion and sulphate ion reduction studies on Ni and Pt surfaces in with and without V_2O_5 in (Li, Na, K) 2 SO_4 melt. Corrosion Science, 45(10): 2285–2292.

Souto, R. M., Laz, M. a. M. and Reis, R. L. (2003). Degradation characteristics of hydroxyapatite coatings on orthopaedic TiAlV in simulated physiological media investigated by electrochemical impedance spectroscopy. Biomaterials, 24(23): 4213–4221.

Stern, M. and Geary, A. L. (1957). Journal of the Electrochemical Society, 104(1): 56–63.

Stankiewicz, A. and Barker, M. B. (2016). Development of self-healing coatings for corrosion protection on metallic structure. Smart Materials and Structures, 25084013.

Su, W. and Iroh, J. O. (2000). Electrodeposition mechanism, adhesion and corrosion performance of polypyrrole and poly(N-methylpyrrole) coatings on steel substrates. Synthetic Metals, 114: 225–234.

Syed, J. A., Tang, S., Lu, H. and Meng, X. (2015). Water-soluble polyaniline–polyacrylic acid composites as efficient corrosion inhibitors for 316SS. Industrial & Engineering Chemistry Research, 54(11): 2950–2959.

Tavandashti, N. P., Ghorbani, M., Shojaei, A., Gonzalez-Garcia, Y., Terryn, H. and Mol, J. M. C. (2016). pH responsive Ce(III) loaded polyaniline nanofibers for self-healing corrosion protection of AA2024-T3. Progress in Organic Coating, 99: 197–209.

Thomas, N. L. (1989). Journal of Protective Coatings & Linings, 6(12): 63.

Tohumcu, C., Tas, R. and Can, M. (2014). Increasing the crystallite and conductivity of polypyrrole with dopant used. Ionics, 20(12): 1687–1692.

Trung, V. Q., Hoan, P. V., Phung, D. Q., Duc, L. M. and Le Thi Thu Hang. (2014). Double corrosion protection mechanism of molybdate-doped polypyrrole/montmorillonite nanocomposites. Journal of Experimental Nanoscience, 9(3): 282–292.

Tsai, C. and Mansfeld, F. (1993). Determination of coating deterioration with EIS: Part II. Development of a method for field testing of protective coatings. Corrosion, 49(9): 726–737.

Uhlig, H. (1978). History of Passivity, Experiments and Theories, eds. R. Frankenthal and J. Kruger. The Electrochemical Society of America. New Jersey.

Valençaa, D. P., Alvesa, K. G. B. Melob, C.P- de and Bouchonneau, N. (2015). Study of the efficiency of polypyrrole/ ZnO nanocomposites as additives in anticorrosion coatings. Materials Research, 18(Suppl 2): 273–278.

Vignati, Davide A. L., Dominik, J., Beye, M. L., Pettine, M. and Ferrari, B. J. (2010). Chromium (VI) is more toxic than chromium (iii) to freshwater algae: a paradigm to revise. Ecotoxicology and Environmental Safety, 73(5): 743–9.

Vimalanandan, A., LvLi-Ping, Tran, T. H., Landfester, K., Crespy, D. and Rohwerder, M. (2013). Redox-responsive self-healing for corrosion protection. Advanced Materials, 25: 6980–6984.

Vu, Van T. H., Dinh, Thanh T. M., Pham, Nam T., Nguyen, Thom. T., Nguyen, Phuong T. and To Hang, T. X. (2018). Evaluation of the corrosion inhibiting capacity of silica/polypyrrole-oxalate nanocomposite in epoxy coatings. International Journal of Corrosion. 1–10. 10.1155/2018/6395803.

Wagner, C. (1965). Corrosion Science, 5: 751.

Walter, G. (1991). The application of impedance spectroscopy to study the uptake of sodium chloride solution in painted metals. Corrosion Science, 32(10): 1041–1058.

Wang, J., Duh, J. and Shih, H. (1996). Corrosion characteristics of coloured films on stainless steel formed by chemical, INCO and ac processes. Surface and Coatings Technology, 78(1): 248–254.

Wessling, B. (1996). Corrosion prevention with an organic metal (polyaniline): surface ennobling, passivation, corrosion test results. Materials & Corrosion, 47: 439.

Whitney, W. R. (1903). Journal of Chemical Society. The Corrosion of Iron, 25: 395.

William Walker. (1913). The corrosion of iron and steel. Industrial & Engineering Chemistry, 5(6): 444–445.

Yagan, N. O. and Pekmez, A. Yildiz. (2006). Progress in Organic Coatings, 57: 314–318.

Yamashita, M., Miyuki, H., Matsuda, Y., Nagano, H. and Misawa, T. (1994). Corrosion Science, 36: 283–299.

Yeh, J. -M., Liou, S. -J., Lai, C. -Y. and Wu, P. -C. (2001). Enhancement of corrosion protection effect in polyaniline via the formation of polyaniline-clay nanocomposite materials. Chemistry of Materials, 13(3): 1131–1136.

Yeh, J. -M., Chin, C. -P. and Chang, S. (2003). Enhanced corrosion protection coatings prepared from soluble electronically conductive polypyrrole-clay nanocomposite materials. Journal of Applied Polymer Science, 88: 3264–3272.

Yi, Y., Liu, G., Jin, Z. and Feng, D. (2013). The use of conducting polyaniline as corrosion inhibitor for mild steel in hydrochloric acid. International Journal of Electrochemical Science, 8: 3540–3550.

Yamashita, M. and Uchida, H. (2002). Recent research and development in solving atmospheric corrosion problems of steel industries in Japan. Hyperfine Interactions, 139/140: 153–166.

Zayed, A. M. and Sagues, A. A. (1990). Corrosion at surface damage on epoxy-coated reinforcing steel. Corrosion Science, 30(10): 1025–1044.

Zhang, Y., Shao, Y., Liu, X., Shi, C., Wang, Y., Meng, G., Zeng, X. and Yang, Y. (2017). A study on corrosion protection of different polyaniline coatings for mild steel. Progress in Organic Coatings, 111: 240–247.

2

Conducting Polymers

2.1 Conducting polymers

About four decades ago, before the discovery of polyacytylene (PA), no one would have ever thought that organic polymers could be electrical conductors as good as metals. It is because polymers have always been considered as insulators. In 1970s, a polymer chemist Hideki Shirikawa, an inorganic chemist Alan G. MacDiarmid and a physicist Alan J. Heegar had changed this view that the electrical conductivity of conjugated organic polymers can be altered through doping (Chiang et al. 1977; Shirakawa et al. 1977, 2001, 2003; Heeger 2001; MacDiarmid 2001). They observed that when a thin film of PA oxidized by the iodine vapors, the electrical conductivity of PA increased by a billion times. These three eminent scientists and co-workers published this work in the Journal of Chemical Society, Chemical Communications as "Synthesis of electrically conducting organic polymers: Halogen derivatives of PA $(CH)_n$," in the year 1977 (Shirakawa et al. 1977). For this fascinating discovery and development of conducting polymers, they received the Nobel Prize in Chemistry in the year 2000.

So, organic polymers that possess electrical and optical properties of metal while retaining the mechanical properties and processability, etc., associated with conventional polymers are knowns as Intrinsically Conducting Polymers (ICPs) or synthetic metals. This class of ICPs differs from those conducting polymers which are simply a mixture of insulating polymers with an electrically conducting filler material such as metal nanoparticles, graphene, CNTs, carbon fibers and graphite. These electrically conducting fillers form a conductive network by distributing throughout the polymer matrix.

Conducting polymers were considered as a futuristic electronic materials with promising properties and have attracted the attention of many material scientists because of their boundless applications like energy storage, EMI shielding, corrosion inhibition, sensors, batteries, anti-static, supercapacitors, organic solar cells, light emitting diodes, field emission transitions, electrochromic devices, etc., that can be attributed to their tailorable electrical conductivity, ease of synthesis, low processing cost, corrosion resistance, light weight and good environmental stability.

Although initally, ICPs were neither processable nor air-stable, new generations of these materials are now processable into powders, films and fibers and are also air stable. The typical electronic feature of undoped conducting polymers is the π-conjugated system, which is formed by the overlap of carbon p_z orbitals and alternating carbon-carbon bond lengths.

After the discovery of PA much research has been started in the field of conducting polymers and many new conducting polymers have been synthesized. Examples are Polyphenylene (PP), Polyphenylene vinylene (PPV), Polypyrroles (PPy), Polythiophene (PTh), Poly(3, 4-ethylene dioxythiophene) (PEDOT), Polyaniline (PANI) which have been nicely provided in the book by Skotheim (2007). However, their processability to make either film or coated layers is rather difficult because it is infusible and insoluble in common solvents, and has poor mechanical properties. These limitations can be overcome by means of using bulky dopants as studied by Song et al. (2004) and Han et al. (1989), insertion of substitutional groups either on the ring or in the nitrogen atom Zotti et al. (1992), blending with other polymers as studied by Anand et al. (1998), Pud et al. (2003) or by preparing composites (Wang et al. 2009; Haba et al. 2000; Zhang et al. 2003; Cochet et al. 2001; Tung et al. 2011).

2.2 Structure of conducting polymers

The unique characteristic of conducting polymers is the delocalized π-conjugated system (presence of alternate single and double bonds) in the polymer backbone. The conjugation of π-electrons in the polymer chain generates the highest occupied molecular orbital (HOMO) and lowest unoccupied molecular orbital (LUMO) leading to a system that can be readily oxidized or reduced. Hideki Shirakawa showed that PA, which is the simplest polyconjugated system, can be made conductive by reaction with bromine or iodine vapors. Spectroscopic studies reveal that the redox reaction takes place and the polymer chain transformed from neutral polymer chains into polycarbocations with simultaneous insertion of the equivalent amount of Br_3^- or I_3^- anions between the polymer chains, in order to neutralize the positive charge generated on the polymer chain. This exciting discovery initiated extensive and systematic research in the field of conducting polymers for various aspects of the chemistry and physics in their neutral (undoped) and charged (doped) states.

In many of the cases, polymers are insulators in their neutral undoped form and they show conducting when treated with electron acceptors/donors (dopants) by a process known as 'doping'. The electrical conductivity of a conducting polymer can be tailored by chemical manipulation such as the nature of the dopant, the degree of doping and blending with other polymers (McCullough et al. 1993). Figure 2.1 shows the chemical structure of some conjugated polymers in their neutral insulating state. Undoped conjugated polymers are semiconductors with band gaps ranging from 1 to 4 eV (Bredas et al. 1984), therefore their room temperature conductivities are very low (Kievlson et al. 1988), typically of the order of 10^{-8} S/cm or lower. However, doping can lead to an increase in conductivity of polymer by many orders of magnitude (Bredas et al. 1985).

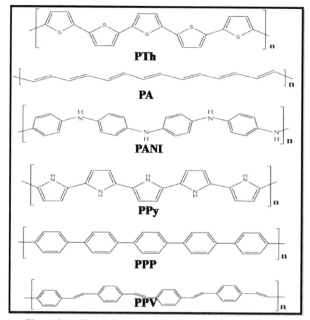

Figure 2.1: Chemical structures of some conjugated polymers

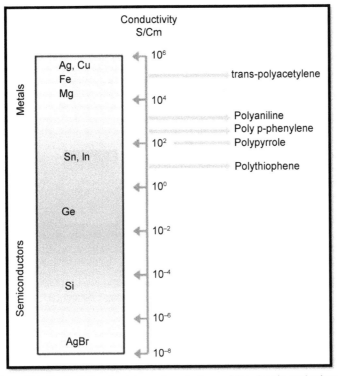

Figure 2.2: Conductivity of metals, semi-conductors and doped conjugated polymers

During the doping process, an undoped polymer having low conductivity, typically in the range of 10^{-10} to 10^{-5} S/cm, is converted to doped polymer, which is maximum conductivity in the range of $1–10^4$ S/cm. The highest value reported till date has been obtained for iodine-doped PA ($> 10^5$ S/cm). Much of the combined research efforts of industrial, academic and R&D research institutions have been directed toward developing materials that are stable (mechanically and electronically) for their use in nanoscience and technology, which are easily processible and can be produced by a simple process at low cost. Recent advances in the field of ICPs have led to a variety of materials with great potential for commercial applications such as rechargeable batteries done by Novak et al. (1997), Wang et al. (2002), Park et al. (2007), MacDiarmid et al. (1987), light emitting diodes (LEDs) as explored by Gustafsson et al. (1992), Burroughes et al. (1990), Kim et al. (2002) and Kim et al. (2016), field-effect transistors (FETs) studied by Lei et al. (2016), Park et al. (2016), solar cells Zhang et al. (2016), Wang et al. (2008) and Gunes et al. (2007), EMI shielding Wang et al. (2007), Joo et al. (2000), Lakshmi et al. (2009), Agnihotri et al. (2015), Koul et al. (2000), electrostatic charge dissipation Saini et al. (2016), electrochromic devices Beaujuge et al. (2010), Mortimer et al. (2006), Somani et al. (2003), supercapacitors as studied by Frackowiak et al. (2006), Mastragostino et al. (2003), Snook et al. (2011), artificial muscles explored by Otero et al. (1995 and 1998), Baughman (1996), corrosion control studied by Tallman et al. (1999), Wrobleski et al. (1993), Michalik et al. (2005) and sensors as studied by Miasik et al. (1986) and Janata et al. (2003), etc.

2.3 Methods of doping

To make conjugated polymer electrically conducting, it is necessary to introduce mobile carriers into the conjugated system by doping. The electrical conductivity has resulted from the existence of charge carriers and the ability of charge carriers to move along the π-bonded "highway". Doped conjugated polymers are good conductors for two reasons:

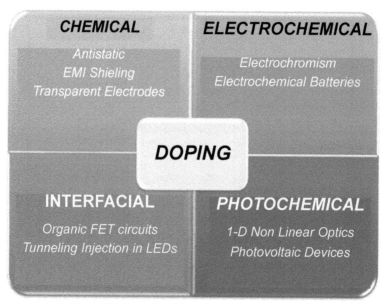

Figure 2.3: Types of doping and their related application

- Doping introduces charge carriers into the electronic structure.
- The attraction of an electron in one repeat unit to the nuclei in the neighboring units leads to carrier delocalization along the polymer chain and to charge carrier mobility, which is extended into three dimensions through interchange electron transfer.

The term "doping" used in the field of conducting polymers is somewhat different from the term used in the field of conventional inorganic semiconductors. Dopants interact with polymers by oxidizing or reducing them, and do not participate in the charge transport mechanism directly. Charge injection onto conjugated chains leads to a wide variety of interesting and important phenomena that defines the field of conducting polymer. As summarized in Figure 2.3, reversible charge injection by "doping" can be accomplished in a number of ways.

2.3.1 Chemical doping by charge transfer

The initial discovery of the ability to dope conjugated polymers involved charge-transfer redox chemistry, oxidation (positive/p-type doping) or reduction (negative/n-type doping) are illustrated by the following examples:

a) p-type doping

$$(\pi\text{-polymer})_n + 3/2\ ny(I_2) \rightarrow [(\pi\text{-polymer})^{+y}(I_3^-)y]_n$$

b) n-type doping

$$(\pi\text{-polymer})_n + [Na + (naphthalide)^{\cdot}]_y \rightarrow [(Na+)_y((\pi\text{-polymer})^{-y}] + (naphthalene)^{\circ}$$

When the doping level is sufficiently high, the electronic structure evolves toward that of a metal.

2.3.2 Electrochemical doping

Although chemical doping is an efficient and straight forward process, it is typically difficult to control. Complete doping to the highest concentrations yields reasonably high quality materials. However, attempts to obtain intermediate doping levels often result in inhomogeneous doping. Electrochemical doping was adopted to circumvent this problem as explored by Kawai et al. (1991). In electrochemical doping, the electrode supplies the redox charge to the conducting polymer, while ions diffuse into (or out of) the polymer structure from the nearby electrolyte to compensate the electronic charge. The doping at any level can be achieved by setting the electrochemical cell at a fixed applied voltage. Electrochemical doping is illustrated by the following examples:

a) p-type doping

$(\pi\text{-polymer})_n + [Li^+(BF_4^-)]sol^n \rightarrow [(\pi\text{-polymer}) + BF_4^-] + Li^o$

b) n-type doping

$(\pi\text{-polymer})_n + Li(electrode) \rightarrow [(Li^+)_y(\pi\text{-polymer})^-_y]_n + Li^+(BF_4^-)sol^n$

It should be pointed out that n-doped conjugated polymers are much more reactive than the p-type. Usually, their conductivity drops drastically upon exposure to ambient atmosphere, even for a short duration due to its reaction with aerial oxygen. As indicated in Figure 2.3, each of the methods of charge-injection doping leads to unique and important phenomena. In the case of chemical and/or electrochemical doping, the induced electrical conductivity is permanent until the carriers are chemically compensated or until the carriers are purposely removed by "undoping".

2.3.3 *Mechanism of conductivity*

To explain the electronic conduction phenomena in organic conducting polymers, new concepts including solitons, polarons and bipolarons have been proposed by solid-state physicists. These are as follows:

2.3.4 *Polymers with degenerate ground states*

In π-conjugated polymers with degenerate ground states, solitons are important and dominant charge storage species. PA, $[CH_x]$, is the only known polymer with a degenerate ground state due to its access to two possible configurations as shown in Figure 2.4. The two structures differ from each other by the exchange of the carbon-carbon single and double bonds. While PA can exist in two isomeric forms: cis and trans-PA, the trans-acetylene form is thermodynamically more stable and the cis-trans isomerization is irreversible.

A soliton can be viewed as an excitation of the radical from one potential well to another well of the same energy (Figure 2.5 degenerate PA). A neutral soliton occurs in pristine trans-PA when a chain contains an odd number of conjugated carbons, in which case there remains an unpaired π-electron, a radical, which corresponds to a soliton (Figure 2.5). In a long chain, the spin density in a neutral soliton (or charge density in a charged soliton) is not localized on one carbon but spread over several carbons which gives the soliton a width. Starting from one side of the soliton, the double bonds become gradually longer and the single bonds become shorter so that while arriving at the other side the alternation has completely reversed.

This implies that the bond lengths do equalize in the middle of a soliton. The presence of a soliton leads to the appearance of a localized electronic level at mid-gap, which is half-occupied in the case of a neutral soliton and empty (doubly occupied) in the case of a positively (negatively) charged soliton (Figure 2.5). Similarly, in n-type doping, neutral chains are either chemically or electrochemically reduced to polycarbonium anions, and simultaneously charge-compensating cations are inserted into the polymer matrix. In this case, negatively charged spinless solitons are the charge carriers.

2.3.5 *Polymers with non-degenerate ground states*

Consider the structure of other conductive polymers such as poly-p-phenylene (PPP), polypyrrole (PPy), poly (3, 4-ethylenedioxythiophene) (PEDOT) and polyaniline (PANI). These polymers do

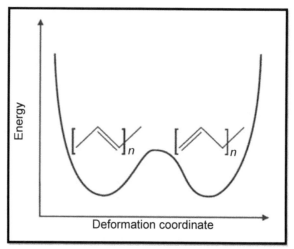

Figure 2.4: Energetically equivalent forms of degenerated polyacetylene

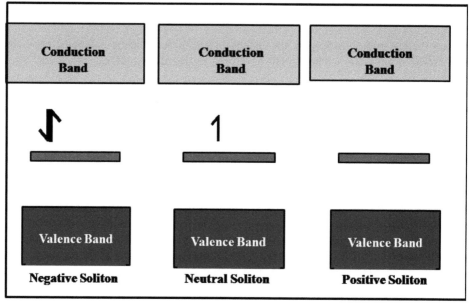

Figure 2.5: Band structure of a neutral, positive and negative solitons in trans-PA

not support soliton-like defects because the ground state energy of the quinoid form is substantially higher than the aromatic benzenoid structure (Figure 2.6). As a result, the charge defects on these polymers are different as explained by Bredas et al. (1981). As an example, consider the oxidation of PPy (Figure 2.7). The removal of one electron from the π-conjugated system of PPy results in the formation of a radical cation. In solid-state physics, a radical cation that is partially delocalized over a segment of the polymer is called a polaron. It is stabilized through the polarization of the surrounding medium, hence the name. Since it is really a radical cation, a polaron has the spin 1/2. The radical and cation are coupled to each other via local resonance of the charge and the radical.

The presence of a polaron induces the creation of a domain of quinone-type bond sequence within the PPy chain, exhibiting an aromatic bond sequence. The lattice distortion produced by this has higher energy than the remaining portion of the chain. The creation and separation of these defects cost energy, which limits the number of quinoid-like rings that can link these two species,

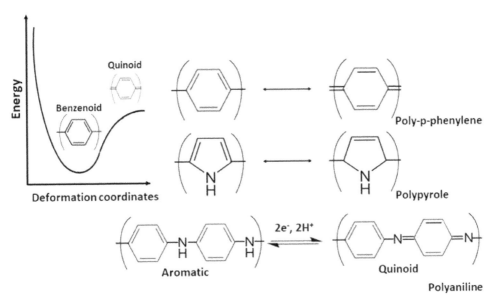

Figure 2.6: Resonance structures of conjugated polymers

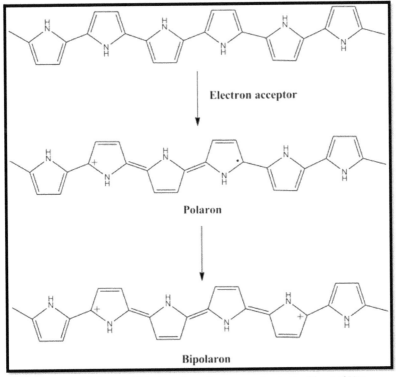

Figure 2.7: Polaron and bipolaron formation on π-conjugated backbone of PPy

i.e., radical and cation together. In the case of PPy, it is believed that the distortion extends over four pyrrole rings. Upon further oxidation, the subsequent loss of another electron can result in two possibilities: the electron can come from either a different segment of the polymer chain thus creating another independent polaron, or from a polaron level (removal of an unpaired electron) to create a dication separating the domain of quinone bonds from the sequence of aromatic-type bonds in the polymer chain referred to as a bipolaron. This is of lower energy than the creation of two distinct polarons; therefore, at higher doping levels it becomes possible for two polarons to combine to form a bipolaron, thereby replacing polarons with bipolarons. Bipolarons also extend over four pyrrole rings. Calculations on PPy and PTh indicate that two polarons in close proximity are unstable with respect to the formation of a bipolaron. The two free radicals combine, leaving behind two cations separated by a quinoidal section of the polymer chain. Figure 2.7 shows that the two cations have some freedom to separate. However, higher energy of the quinoid section between them binds them together resulting in a correlated motion. The net effect is the formation of a doubly charged defect acting like a single entity and delocalized over several rings (3–5), i.e., a bipolaron. The formation of bipolarons implies a net free energy gain in forming a closed shell defect from two open shell structures. The quinoid form has higher energy than the aromatic benzenoid form but its electron affinity is higher and the ionization potential lower as studied by Bredas et al. (1985). This leads to the formation of two localized states in the band gap. As the doping is increased, additional states are created in the gap and they finally evolve into two narrow bands.

The theoretically expected evolution of the electronic structure with doping is shown schematically in Figure 2.8. At low doping levels the defects are polarons, which tends to combine at higher doping levels to form bipolarons. The two bands inside the band gap are empty in the case of bipolarons, while the lower polaron band is half-filled. Figure 2.8 also shows the optical transitions characteristic of the charge defects.

The bandgap transition W_0 increases in energy with doping level because the interband states are derived from states at the band edges. The doping results in the appearance of mid-gap transitions. Three in gap transitions are characteristics of polaron states, while bipolaron states are noted for

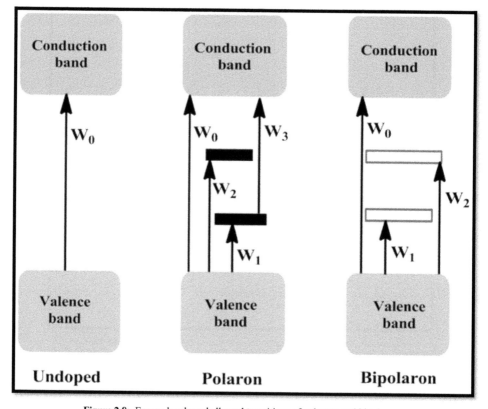

Figure 2.8: Energy levels and allowed transitions of polarons and bipolarons

the absence of the transition W_3. The evolution of the optical spectra with doping level provides experimental evidence for the picture presented in Figure 2.7 for PPy and other polymers as interpreted by Kaufman et al. (1985). At low doping levels, PPy shows the three interband transitions characteristic of polaron states. As the doping level increases, the transition W_3 decreases in intensity and finally disappears.

This behavior provides evidence for the formation of bipolarons in the PPy chain. As an internal check for the assignment of transitions, note that the sum (in energy values) of W_1 and W_2 approximately add up to the interband transition W_0. Let us now consider some peculiarities of the PANI system. In the conventional conductive polymers, e.g., PPy, PTh, PA and PPP, oxidative doping results in the removal of electrons from the bonding π-system. In PANI, in contrast, the initial removal of electrons is from the non-bonding nitrogen lone pairs. Unlike other conducting polymers, the quinoid form is not just simply an alternative resonance form. Its formation requires reduction and deprotonation so that it actually differs in chemical composition from the benzenoid form (Figure 2.9).

This peculiarity of the PANI structure makes doping by an acid-base reaction possible (see Figure 2.9) possible. In addition, the constituent parts of both the polaron and the bipolaron are very tightly bound owing to valence restrictions. The radical and cation of the polaron are confined to a single aniline residue. The bipolaron is confined to and identical with a (doubly protonated) quinone-diimine unit. This narrow confinement may destabilize bipolarons with respect to polarons owing to the coulomb repulsion between the cations. As indicated above, the doping of PANI can be achieved via two routes. Doping by oxidation of the leucoemeraldine form results in the formation of radical cations, which may then convert to bipolarons. Alternatively, protonation of the emeraldine base form leads to the initial formation of bipolarons that may rearrange with neutral amine units to form radical cations.

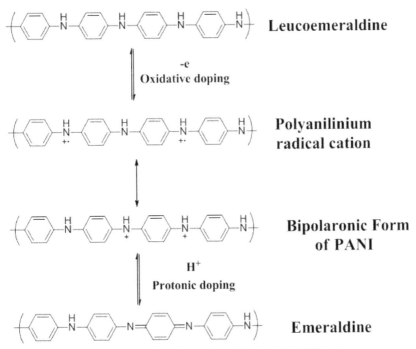

Figure 2.9: Oxidative and protonic acid doping in the PANI system

2.4 Poly(3, 4-ethylene dioxythiophene) (PEDOT)

After the discovery of polyacetylene, various conducting polymers were developed like polyaniline (PANI), polypyrrole (PPy), polythiophene (PTh), poly-phenylene vinylene (PPV) and poly-p-phenylene (PPP). In these conducting polymers, PANI, had shown some great potential due to its ability to form processable conductive forms at low-cost but the presence of benzidine moieties (Bremer et al. 1997; Kinlen et al. 1998) in the polymer chain yielded toxic products on degradation. Whereas PPy, PTh, PPV had disadvantages like being insoluble and infusible (Jonas et al. 1991). To circumvent these problems, numerous substituents were attached to these polymers to form derivatives which gave control to the physical and electronic properties but the actual accessible electronic properties of the parent polymers were degraded. In the 1980's Bayer's AG, in Germany, developed a new polythiophene derivative poly(3, 4-ethylene dioxythiophene) (PEDOT). Bayer's commercialized PEDOT under the trade name of "BAYTRON P", which was prepared by chemical or electrochemical polymerization methods (Lefebvre et al. 1999); the polymer produced was insoluble in nature yet possessed exciting properties. The insolubility problem of the polymer could be stepped aside by using polyelectrolyte and surfactants like polystyrene sulfonic acid (PSSA) (Louwet et al. 2003) and dodecyl benzene sulfonic acid (DBSA) (De et al. 2009). The resultant PEDOT/PSS and PEDOT/DBSA possess very good conductivity (Kim et al. 2011), optical transparency in doped form (Sotzing et al. 2002), good film-forming properties (Jönsson et al. 2002), high visible light transparency (Zhang et al. 2013) and excellent environmental stability (Nardes et al. 2009). PEDOT/PSS finds various technological applications which include anti-static coating for photographic films (Kirchmeyer et al. 2005) and hole transport layer in organic photovoltaic's and light emitting diodes (Wakizaka et al. 2004; Jong et al. 2000). The conducting layer of PEDOT/PSS is also applied to electroluminescent, organic field effect transistors. During its initial age of development, PEDOT was found to be an insoluble polymer but still possessed some interesting properties. To step aside the insolubility problem of PEDOT, Bayer AG (now Covestro) polymerized 3, 4-ethylene dioxythiophene (EDOT) monomer in an aqueous polyelectrolyte (polystyrene sulfonic acid, PSSA) solution using the sodium persulfate as the oxidizing agent. The chemical structure of PEDOT/PSS is shown in Figure 2.10.

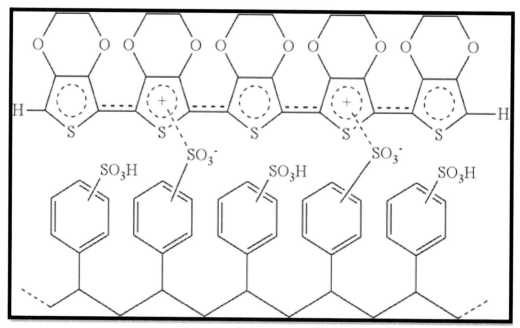

Figure 2.10: Chemical structure of PEDOT/PSS

PEDOT/PSS is an aqueous dispersion of positively charged conjugated PEDOT and negatively charged saturated polystyrene sulfonate (PSS). PSS is a polymeric surfactant that also acts as dopant in the polymer chain, which helps to disperse and stabilize PEDOT in water and other solvents. The dark blue aqueous dispersion of PEDOT/PSS was first commercialized by Bayer AG under the name of "Baytron P". It easily forms a thin-film on flexible and stiff surfaces by different solution processing techniques like spray technique, spin coating, screen printing, ink jet printing, etc. Its electrical conductivity ranges from 10^{-2} to 10^3 S cm^{-1}. The superior electrical conductivity and environmental stability are few properties that make PEDOT a suitable candidate for applications in energy and storage fields.

2.4.1 Synthesis of PEDOT/PSS

PEDOT/PSS can be synthesized by the method given in previous literature by Louwet et al. (2003), in which 3, 4-ethyelene dioxythiophene (75 mmol) is mixed with 6 wt% of PSS solution and ammonium persulfate (104 mmol) in water (~ 2 L). After stirring for 30 minutes at room temperature, Fe_2SO_4 is added to the mixture and then stirred for 24 hours. Figure 2.10.1 shows the synthesis of PEDOT/PSS. After 24 hours of vigorous stirring, a dark-blue colored aqueous suspension of PEDOT/PSS mixture is obtained via purifying it by an ion exchange method.

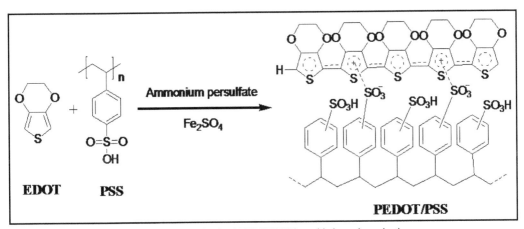

Figure 2.10.1: Synthesis of PEDOT/PSS by oxidative polymerization

2.4.2 Mechanism of EDOT to PEDOT/PSS polymerization

Figure 2.10.2 describes the general mechanism of polymerization of EDOT to PEDOT/PSS polymer. The reaction follows four-step mechanism: (1) In the first step, the removal of an electron from the monomer takes place; (2) Then, the resonance stabilization of the radical cation takes place after oxidation of monomer formed; (3) In the next step dimerization of the radical cation takes place; (4) In this step, deprotonation of the dimeric cation takes place to form a stable dimer; (5) And the final step is the chain propagation in which dimer again undergoes these previous given steps to form the polymer.

2.4.2.1 Synthesis of PEDOT in DBSA medium

Synthesis of PEDOT doped with DBSA was carried out using the emulsion polymerization. The DBSA functions both as a surfactant and dopant in the polymerization. In this method, 0.1 M of EDOT and 0.3 M of DBSA was homogenized in an aqueous medium at room temperature for 60 minutes to form EDOT–DBSA micelles. The polymerization was initiated by the dropwise addition

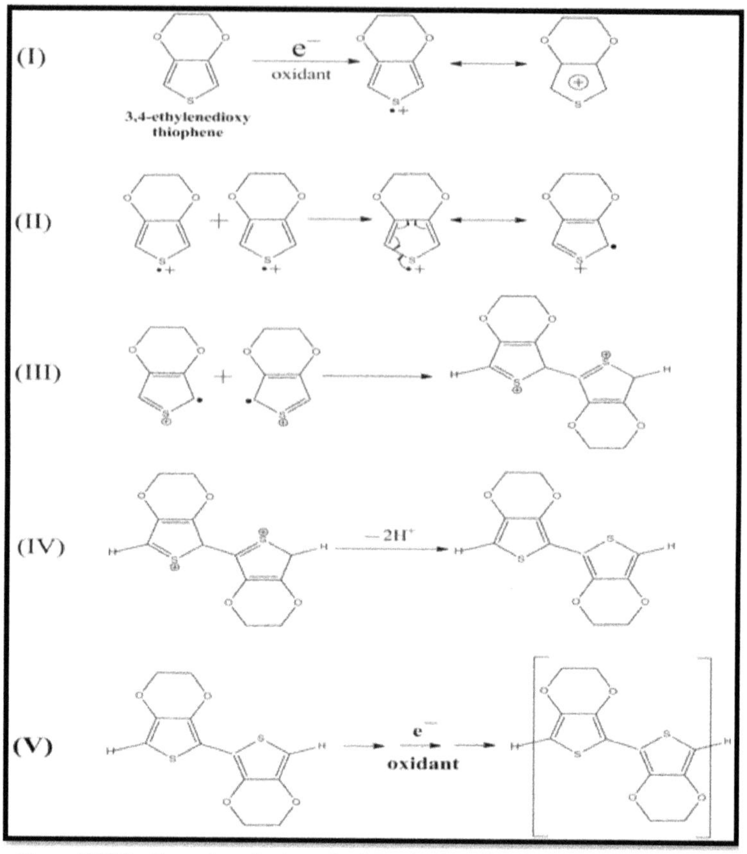

Figure 2.10.2: General mechanism of polymerization of EDOT to PEDOT polymer

of oxidant, 0.1 M aqueous APS solution, at $-2°C$ with constant stirring for 8 hours. The mechanism of polymerization of EDOT monomer to PEDOT polymer is shown in Figure 2.10.2. A bluish-green suspension containing precipitates of PEDOT was obtained thereafter. The resulting suspension was demulsified using isopropyl alcohol due to the formation of stable micro-emulsions of PEDOT, water and DBSA and then filtered in Buchner funnel. The filtered precipitate obtained was then washed with distilled water and dried at $60°C$ in a vacuum oven.

2.4.2.2 *Synthesis of PEDOT/MWCNT composites*

The *in situ* emulsion polymerization of EDOT was carried out in the presence of MWCNTs in an aqueous medium as shown in Figure 2.10.3. In a typical synthesis, 0.3 M DBSA (surfactant) was homogenized in water for an hour to form an aqueous emulsion. To this aqueous emulsion, a calculated amount of MWCNTs was added (5.0, 10 and 15 wt%) and then homogenized by using the ART-Miccra D-8 (No.-10956) homogenizer at 12000 rpm for 2 h. The wt% of filler was taken with respect to the monomer (EDOT) weight. To this, 0.1 M of EDOT was added and stirred for 8 hours to form an emulsion. The oxidant APS (0.1 M) was added dropwise with vigorous stirring while keeping the temperature of the reactor at $-2°C$. The bluish/black precipitate of polymer composite so obtained was treated with isopropyl alcohol (de-emulsifier) with vigorous stirring for 2 hours. The resulting precipitate was filtered, washed thoroughly with distilled water and then dried at $60°C$ in a vacuum oven. The size obtained of PEDOT/MWCNT nanocomposites was 80–90 nm.

The morphology of the PEDOT and PEDOT/MWCNT composites were investigated using Scanning Electron Microscopy and HRTEM. Figure 2.10.4(a) and (b) presents the SEM micrographs

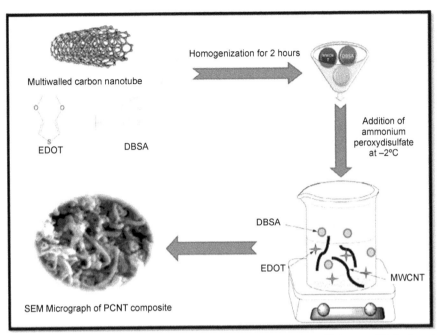

Figure 2.10.3: Synthesis of PEDOT/MWCNT composite (Farukh et al., Composites of Science and Technology, https://doi.org/10.1016/j.compscitech.2015.04.004)

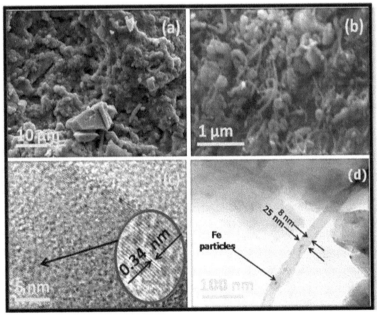

Figure 2.10.4: (a) SEM of PEDOT, (b) SEM of PEDOT/MWCNT10 composite (c) HRTEM of PEDOT/MWCNT (d) TEM micrograph presents MWCNTs diameter of approximately 25 nm (Farukh et al., Composites of Science & Technology, doi.org/10.1016/ j.compscitech. 2015.04.004)

of PEDOT and PEDOT/MECNT composites wherein a SEM micrograph of PEDOT shows regular granular, whereas SEM of composite shows the presence of fibrilar MWCNT in PEDOT matrix. Here, the MWCNTs act as a template for the polymerization of EDOT monomer, resulting in the formation of PEDOT coated MWCNTs. Figure 2.10.4(c) shows MWCNT HRTEM image. Inset highlighted image confirms the crystalline phase of MWCNTs having inter-planar distance 0.34 nm embedded in amorphous PEDOT matrix, whereas in Figure 2.10.4(d) the deposition of PEDOT over MWCNT can be seen. The work revealed that PEDOT/MWCNT composite increases the shielding performance in X-band and Ku-band. EMI shielding result data showed increasing the MWCNT's incorporation also increases the shielding effectiveness of PEDOT/MWCNT composites. The maximum shielding effectiveness of 58 dB was observed in PEDOT/MWCNT composite. The results showed that these composites could be used as a promising shielding material. From all the above the results of PEDOT/MWCNT composites suggest that these could be used as effective and light-weight wave absorbing materials for commercial applications. Even these materials can be tested for suitable corrosion preventive material.

2.4.2.3 PEDOT as corrosion inhibitor

The corrosion behavior of steel samples coated with the epoxy paint containing PEDOT/PSS paint was investigated in seawater by Hou et al. (2012). For this purpose, Electrochemical Impedance Spectroscopy was utilized, and surface morphology of coatings after corrosion was observed using a scanning electron microscope. It was found that the addition of small PEDOT/PSS to the epoxy resin increased its corrosion protection efficiency. Meanwhile, the possible mechanism was discussed.

Poly(3,4-ethylenedioxythiophene) graphene oxide composite coatings for controlling magnesium implant corrosion was investigated by Catt et al. (2016). They observed that PEDOT/GO coating significantly reduced the rate of corrosion as evidenced by lower Mg ion concentration and pH of the corrosion media. Aquirre et al. (2015) studied Poly (3,4-ethylenedioxythiophene), as a coating on stainless steel AISI 470 as corrosion inhibition, found that the conductivity of PEDOT facilitates the charge delocalization, which hinders the formation of localized anodic or cathodic regions. Therefore, the stable surface inhibits the corrosion reaction, which requires the local aggregation of the charge.

Epoxy based coating with the addition of zinc oxide-poly (3,4-ethylenedioxy thiophene) doped with poly(styrene sulphonate) hybrid nanocomposite additive was introduced as paint/metal surface coating by Nur Ain et al. (2018). The main finding showed that the addition of hybrid nanocomposite had increased corrosion protection yet enhanced corrosion process when excess additives were loaded into the epoxy coating. Addition of 2 wt.% ZnO-PEDOT:PSS was found significantly to improve corrosion behaviour and provided optimum corrosion protection to stainless steel 316L as the corrosion rate for 0 day, 15 days and 30 days of immersion duration is 0.0022 mm/yr, 0.0004 mm/yr and 0.0015 mm/yr respectively.

2.5 Polyaniline

Out of all conducting polymers, polyaniline (PANI) in particular is being studied more and more and in recent years has been the center of considerable scientific interest; although polyaniline is not a new material and its existence has been known for a long time. Fritzche undertook the first attempts to analyze 'aniline blacks'. This was followed by studies by Nietski who tried to use the process for tinting cotton. Several authors studied the products of chemical and electrochemical oxidation of aniline and Jozefowicz et al. (1989) led to a more precise understanding of PANI.

In the 1980s, MacDiarmid et al. (1985) and Geneis et al. (1985) had answered many of the queries related to these conducting polymers. The interest of researchers in polyaniline could possibly be linked to numerous applications that exist for electronic conducting polymers and, on the other hand, to the fact that aniline is a cheap product and also a very stable material. Industrialists are also catching up, and the leading world chemical industries now have research programs based on PANI. The concern is related to the polymerization mechanism, the structure of the material in its oxidation states, the redox mechanisms, the electronic and ionic conduction mechanisms, the role of the doping ions, protons, etc. Related to the electrochemical behavior, the interpretation of capacitive effects and electrostatic interactions are very important. Still, scientists have not completely understood all the

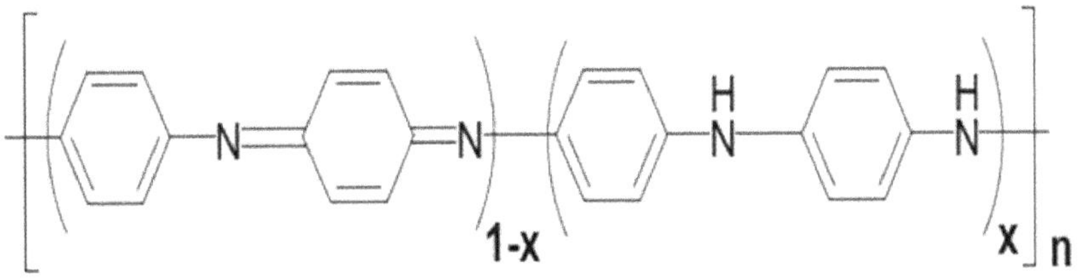

Figure 2.11: General formula for polyaniline base

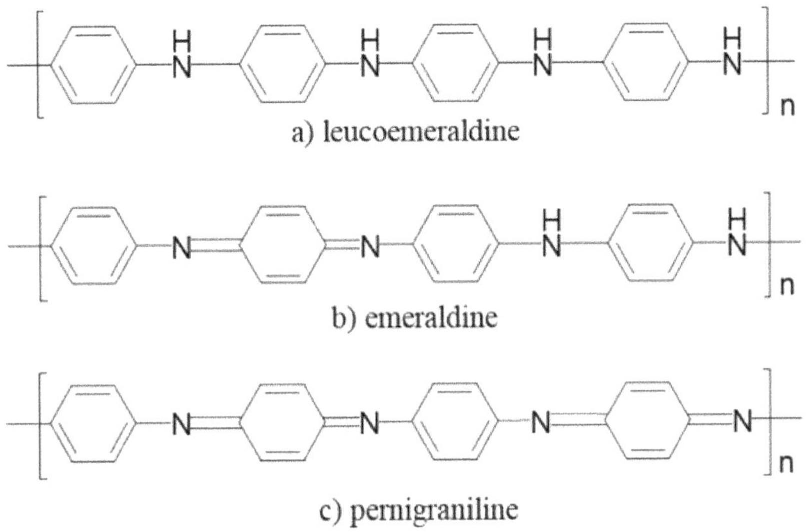

a) leucoemeraldine

b) emeraldine

c) pernigraniline

Figure 2.12: Different forms of polyaniline

observed phenomena and the existing models need to be elaborated to explain the experimental results. Polyaniline (PANI) ranks highest among electrically conducting polymers. Its high conductivity and chemical variability make it suitable for a number of applications. In the course of polymerization, PANI has the ability to create thin conducting films with very good adhesion on various base materials. In reality, "polyaniline" is a name for the whole family of polymers, which can be described by the formula presented in (Figure 2.11).

The correctness of the formula presented above was proved, among others, by spectroscopic studies by Wudl and his co-workers (1987). As mentioned earlier, three principal forms of polyaniline (Figure 2.12) can be distinguished. All these forms exhibit interesting spectroscopic properties, but two of them—fully reduced (a) and fully oxidized (c) are environmentally unstable. Leucoemeraldine (a) (white powder) is a strong reducing agent that easily reacts with air (oxygen) giving emeraldine (b) as the product Pernigraniline (c) (red-purple, partially crystalline powder) composed of oxidized units, easily undergoes hydrolytic type degradation via chain scission. Emeraldine base (dark blue powder with a metallic gloss)—a semi-oxidized form of PANI is stable in air and can be stored for a long time without undergoing chemical changes. Emeraldine is the most extensively studied form of polyaniline. Thus, in the majority of publications, names "polyaniline" and "emeraldine" are used interchangeably. In the case of PANI—contrary to other conjugated systems—contribution to the conjugation is given not only pi electrons of aromatic rings interaction but also by interactions between the lone electron pairs of nitrogen atoms and electrons (π-interactions). Additionally, in emeraldine base relatively strong interactions between amine and imine groups in neighbor chains via hydrogen bonds are present. These phenomena are responsible for very difficult processibility of polyanilines in the base form. A partial dissolution of PANI base is possible only when solvent—polymer hydrogen bond interactions replace interchain interactions. Few solvents dissolve emeraldine base, for example:

NMP (N-methyl pyrrolidinone), DMSO (di-methyl sulphoxide), DMA (N,N-dimethylacetamide). In concentrated PANI solutions, gelation process is sometimes observed. To inhibit this process a mixture of solvents can be applied (combination of electron donors and acceptors that interact with amine and imine groups respectively). Conjugation as well as the presence of hydrogen bonds cause not only insolubility in the majority of common solvents but is also the reason why PANI is infusible. At high temperatures (above 400°C), polymer gradually decomposes without melting. Polyaniline in its base form exhibits the electrical conductivity lower than 10^{-6} S/cm. The poor conductivity is limited by the band gap between HOMO and LUMO levels, i.e., 3.8 eV. It is possible to convert PANI into conducting form.

2.5.1 Synthetic routes to polyaniline

The polymerization process of aniline and its derivatives have been investigated for more than a century and several synthetic routes have been proposed. Due to the fact that only the emeraldine base is stable in ambient conditions, the most popular polymerization procedures are aimed at the preparation of this form. When pernigraniline or leucoemeraldine are required, the oxidation or the reduction of emeraldine usually obtains them.

2.5.2 Chemical oxidative polymerization

The oldest and still the most popular way for the preparation of polyaniline is the chemical oxidative polymerization. Optimization of the process was widely investigated during the past two decades. Several papers devoted to the aniline polymerization in organic solvents have been published but the reaction in aqueous solutions at low pH has still been the most extensively studied. Different inorganic oxidants such as KIO_3, $KMnO_4$, $FeCl_3$, K_2CrO_4, $KBrO_3$, $KClO_3$, $(NH_4)_2S_2O_8$ were tested. Reactions were carried out in different acids (i.e., HCl, H_2SO_4, $HClO_4$, para-toluene sulfonic acid (p-TSA), DBSA, H_3PO_4). The effect of the temperature of the reaction medium and the acid concentration were also investigated as variables of the process. It turned out that the temperature had a specially pronounced influence on the properties of the obtained product, particularly on its molecular weight. According to Adams and his co-workers, polyaniline obtained at 18°C had Mw = 4200 Da, Mw = 3000 Da, the polymer synthesized at 0°C had Mw = 20400 Da, Mw = 122000 Da, PANI obtained at –25°C had Mw = 43500 Da, Mw = 209000 and for polyaniline obtained at –35°C Mw = 19100 Da, Mw = 166000 Da.

Thus, there exists no need for the reaction temperature to decrease below –25°C since upon further temperature lowering the macromolecular parameters of the polymer worsens. Another important parameter influencing the properties of the resulting polymer is the monomer to the oxidant molar ratio. This problem was investigated by Pron and his co-workers. Due to the fact that different oxidants are capable to accept a different number of electrons, these authors proposed a unified coefficient k. It was found that coefficient k had to be $\gg 1$ (large excess of aniline) when a good quality polymer was required. Aniline was first polymerized to pernigraniline, which was then reduced to emeraldine by the reaction with the excess of the monomer. The use of additional reducing agents such as $FeCl_2$ facilitated the reduction of pernigraniline; the values of k could be lower in this case but still higher than 1. Chemical oxidative polymerization of aniline using HCl and $(NH_4)_2S_2O_8$ can be described by the following chemical equation:

$$4x\ (C_6H_7N.HCl) + 5x\ (NH_4)_2S_2O_8 \rightarrow (C_{24}H_{18}N_4.2HCl)_x + 5x(NH_4)_2SO_4 + 2xHCl + 5xH_2SO_4$$

It is clear from the above equation that in acidic media emeraldine is obtained in its protonated state. If needed, it is usually transformed into the base form by deprotonation in 0.1 M ammonia aqueous solution. Several interesting modifications of the oxidative polymerization of aniline have been developed. In the emulsion polymerization, the organic phase consists of the solvent, aniline and the protonating agent, usually a functionalized sulfonic acid. The water phase contains the oxidant (usually ammonium persulfate). The polymer is formed in the organic phase and does not precipitate since it is protonated with acids containing solubilizing substituents. As a result, soluble processible polyaniline with a relatively high molecular weight can be obtained in a one-step process. A new approach of the oxidative polymerization of aniline is the enzymatic polymerization. Horseradish peroxydase/H_2O_2 system was used to oxidize the monomer molecules. An important element of the reaction environment was sulfonated polystyrene, which was used as a template to favor "para" coupling of aniline units. Polyaniline/sulfonated PS complex was soluble in water. The conductivity of the reaction product was 0.1 S/cm. The main disadvantage of this method is the difficulty in separation of both polymers.

2.5.3 *Mechanism of oxidative polymerization of aniline*

Mechanism of oxidative polymerization was investigated by many authors independently on the proposed mechanism; it was believed that the polymerization was initiated by the oxidation of the monomer to a radical cation (stabilized by resonance) as shown in Figure 2.13.

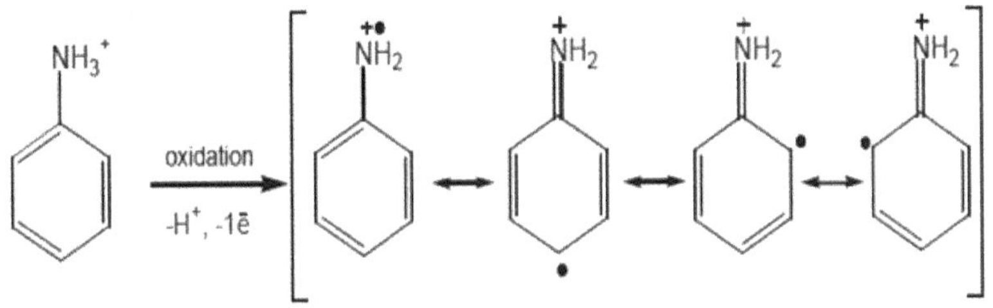

Figure 2.13: The formation of aniline radical cation

The electrophilic substitution of the formed radical cation with another monomer of aniline leads to formation and coupling of two radical cations as shown in Figure 2.14.

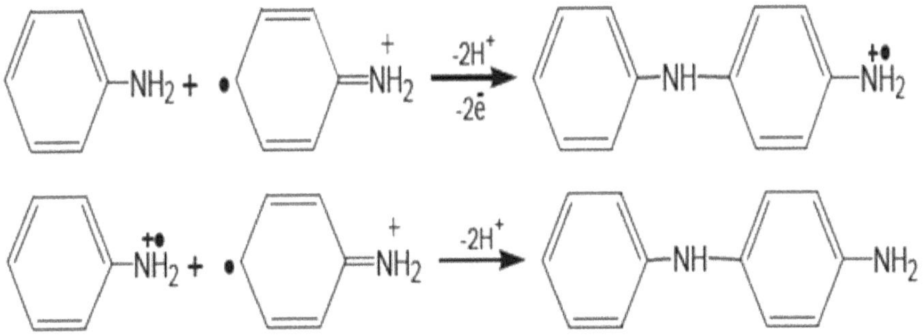

Figure 2.14: The coupling of two radical cations

In the propagation step, the dimer is being oxidized to a radical cation and then it can couple with the radical cation, formed by the oxidation of the monomer (Figure 2.15) or with another dimer-type radical cation.

The polymer chain obtained via the above-mentioned coupling is formally in the most reduced state (leucoemeraldine form). In reality, during the propagation step, it undergoes further oxidation to pernigraniline, as schematically depicted in Figure 2.16. The radical cations of aniline formed in this polymer reduction process either initiate the growth of a new chain or alternatively can participate in the propagation of already growing chain.

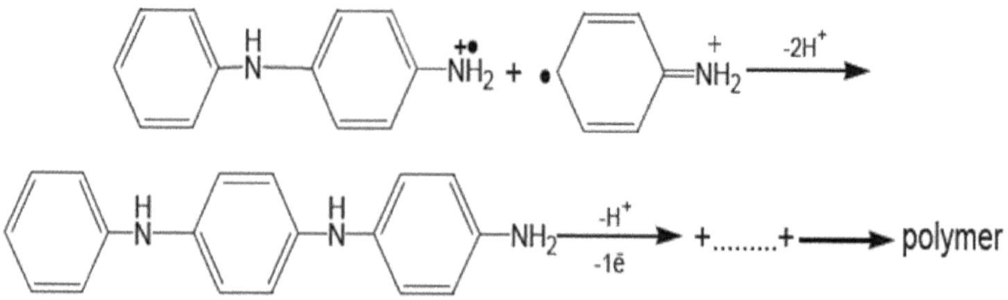

Figure 2.15: Propagation of the polyaniline chain

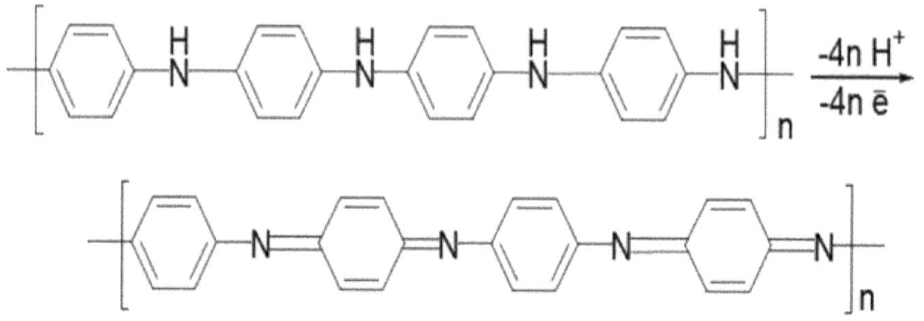

Figure 2.16: Oxidation of leucoemeraldine to pernigraniline

In the next step, the totally oxidized polymer is being reduced to the semi-oxidized state of emeraldine in red-ox reaction with the monomer (Figure 2.17).

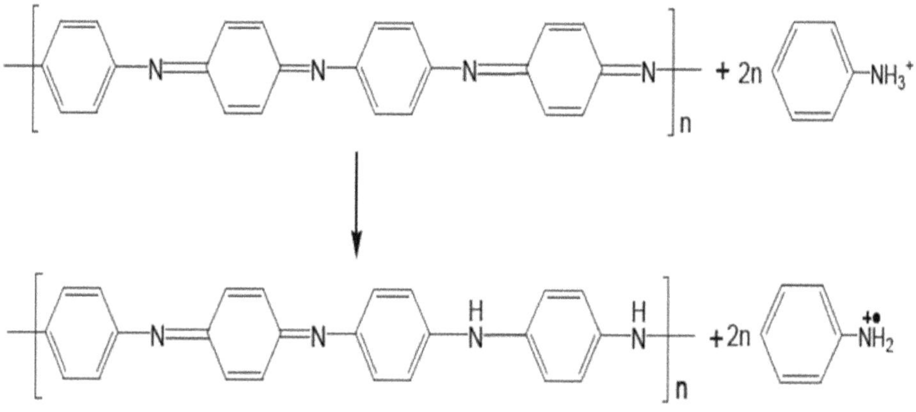

Figure 2.17: Reduction of fully oxidized form to partially oxidized form

The polymer as depicted in Figure 2.17 is in its undoped form. In reality, in a highly acidic reaction medium, it undergoes protonation of the imine nitrogens (Figure 2.18).

Polyaniline is highly conducting, and readily converts between its various oxidation states. These oxidation states differ from each other by the number of quinoid rings, which range from zero to two in the elementary unit of four rings with the other rings being benzenoid. The interconversion between the emeraldine base, with three benzenoid rings, the leucoemeraldine base (LB), with four benzenoid rings, and the inter-conversion between the base and conducting salt can be seen in Figure 2.19. In all the cases, electroneutrality of the polymer is maintained by the presence of counter anions.

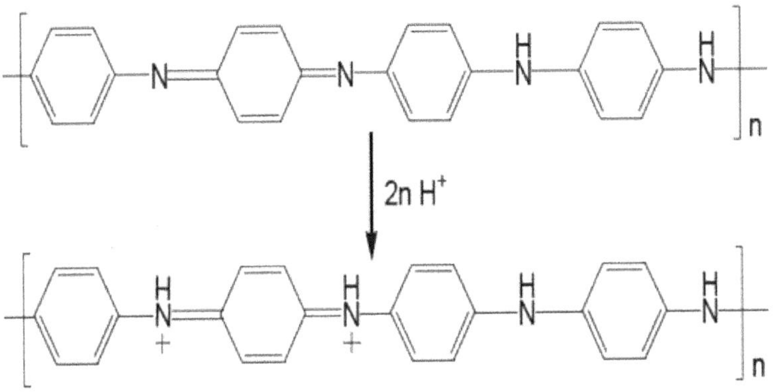

Figure 2.18: Protonation of the emeraldine base to doped PANI form

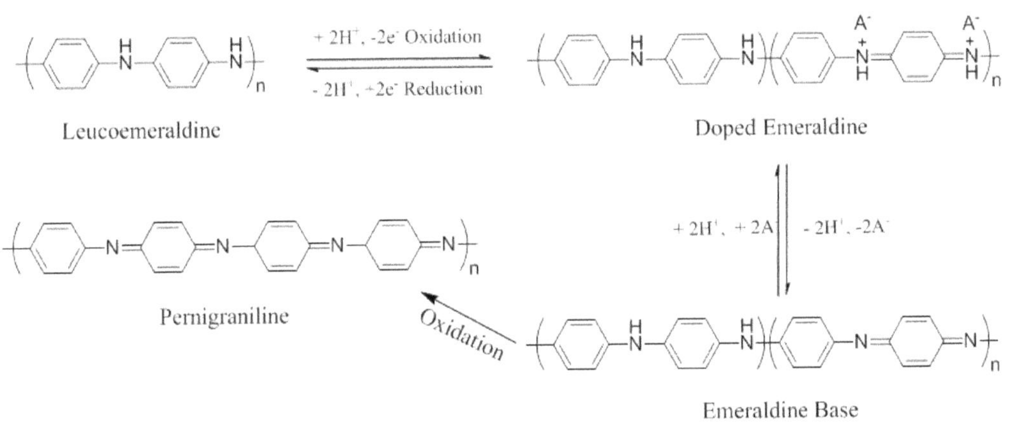

Figure 2.19: Scheme showing the inter-conversion of polyaniline

2.5.4 *Electrochemical polymerization of aniline*

The electrochemical polymerization of aniline is carried out in acidic medium by keeping the concentration of monomer to the protonic electrolyte from 0.1 to 0.5 to 1.0. Prior to electrochemical polymerization, the solution is deoxygenated by passing argon gas for 30 minutes. The polymerization is conventionally carried out at 0.75 V to +0.8 V versus saturated calomel electrode (SCE) (charge passed Q = 0.32 C/cm^2) on platinum, indium-tin-oxide (ITO)-coated glass plate (resistivity 20–25~ cm) or stainless-steel electrodes. The polymer film growth was also studied by sweeping the potential between –0.2 to +0.8 V at a scan rate of 20–50 mV/s. Dhawan et al. (1993) have done extensive studies on the electrochemical polymerization of aniline. On electrochemical polymerization of aniline in sulphosalicylic acid (SSA) medium using the cyclic voltammetry by switching the potential from –0.2 to +0.8 V (SCE) (Figure 2.19.1(a)) and –0.2 to 1.4 V (SCE) (Figure 2.19.1(b)) at a scan rate of 50 mV/s.

The peak appearing at 0.97 V versus SCE in the first cycle corresponds to oxidation of aniline, whereas the corresponding peak in H$_2$SO$_4$ medium appears at 0.8 V versus SCE. This suggests that the generation of anilinum radical cations (essential for polymerization of aniline) occurs at a higher potential in SSA medium. In subsequent cycles, new oxidation peaks (II and III) and appears indicating that these radical cations undergo further coupling and the peak current increases continuously with successive potential scans, indicating the build up of electroactive polyaniline on the electrode surface. Though, the peak potential of the peak observed in the first cycle is at 0.97 V (very broad; formation beginning at around 0.74 V), this suggest that even by keeping the potential around this value can

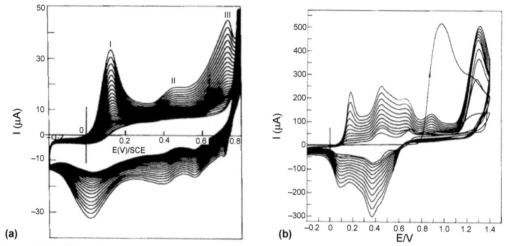

Figure 2.19.1: Electrochemical polymerization of aniline in 1.0 M sulfosalicylic acid solution by cyclic voltammetry by sweeping the potential between (a) –0.2 and 0.8 V, and (b) –0.2 and 1.4 V on Pt electrode versus SCE at a scan rate of 50 mV/s (Dhawan et al. (1993). Synthetic Metals, 58: 309–324).

lead to formation of polymer at a slow growth rate, which is beneficial to obtaining a more-ordered thin polymer film useful for electrochromic displays. Figure 2.19.2 shows the cyclic voltammogram of polyaniline film obtained by potential sweeping in a blank 1.0 M SSA medium (not containing the monomer).

It shows two main redox couples at 0.140 V (peak I) and 0.72 V (peak III). In order to explain the electrochemical behavior of polyaniline, the formation of radical cations near peak I, which are subsequently oxidized into imines near peak III has been suggested (Glarum et al. 1987) and can be represented in the Figure 2.19.3.

Numerous kinds of literature are available on the corrosion protection of conducting polymer-based coatings. Xiaogang Yang et al. 2010 reported the preparation of nanostructured polyaniline by

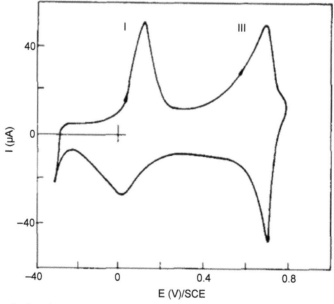

Figure 2.19.2: Cyclic voltammogram of Polyaniline (Dhawan et al. (1993). Synthetic Metals, 58: 309–324)

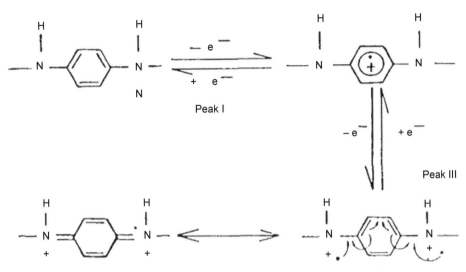

Figure 2.19.3: Inter-conversion of amine moieties to imine moieties in PANI

conventional polymerization, interfacial polymerization and direct mixed reaction respectively. The polyaniline nanofibers, which were synthesized by direct mixed reaction, also have more excellent corrosion protection than conventional aggregated polyaniline for mild steel. Araujo et al. 2001 studied the anti-corrosive properties of undoped polyaniline coated onto mild steel and galvanized steels by a classical methodology used for organic coatings. Thin PANI films are porous, and the adhesion onto mild and galvanized steel is very poor. The lack of adhesion leads to bad performance, even when PANI is top coated with high performance paint as an epoxy one. Such behavior does not depend on the composition of the solution employed for the immersion tests.

The corrosion inhibition of iron in acid chloride solution offered by a new class of polymer-*ortho* substituted poly ethoxy aniline, a conducting polymer studied by Sathiyanarayanan et al. (1992). Polyaniline and substituted polyaniline are of utmost interest because the electronic properties of these polymers can be modified through variations of either the number of protons or the number of electrons or both. Upon protonation at –N = sites, the conductivity of polymer increases. The protonated form in polyaniline or substituted polyaniline, which is stable below a pH of 4, becomes adsorbed on the metal surface, forming an effective barrier for diffusion of aggressive ions responsible for corrosion. Various studies using monomers like ethoxy aniline, aniline and poly ethoxy aniline soluble in CH_3 OH indicate delocalization of π-electrons on the polymer chain with the formation of quaternary ammonium nitrogen, which is responsible for the increased adsorption on the iron surface. The delocalization of π-electrons is ruled out in the case of monomers, and hence they provide less inhibition. The suggestion that the quaternary ammonium nitrogen formation finds relevant support in the fact that quaternary ammonium salts have been extensively used as corrosion inhibitors. Their study indicated that corrosion prevention efficiency is much larger at a lower concentration than quaternary ammonium salts or monomers like ethoxy aniline. This is traced to the greater availability of π-electron in the aromatic ring whose orientation on a metal surface is usually coplanar. It is envisaged that the quaternary ammonium nitrogen and electrons on the aromatic ring in the polymer help in promoting both strong adsorption and uniform surface coverage by the polymers. Impedance measurements of the polymer doped with sulphamic acid (Figure 2.19.4) indicates charge transfer resistance (Rct) increases with an increase in polymer concentration.

The capacitance calculations show a surface coverage to the extent of 0.75–0.85 (surface coverage $0 = 1 – C_{dli}/C_{dl}$, where C_{dli} is the double layer capacitance with inhibitor and C_{dl} is the double layer capacitance without inhibitor. Samui et al. (2005) reported the corrosion protection of mild steel by

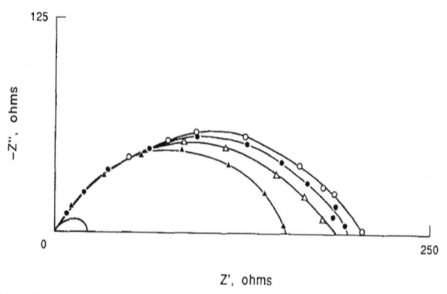

Figure 2.19.4: Nyquist plots for the polymerin 1.0 N HCl medium with different concentrations of polyethoxy aniline doped with sulphamic acid (Sathiyanarayanan et al., Corrosion Science, doi.org/10.1016/0010-938X(92)90187-8)

coating it with paint containing polyaniline-hydrochloride as pigment. The barrier property contributes appreciably towards corrosion resistance of the PANI-HCl containing paint films. The paint system has shown appreciable corrosion resistance without any top barrier coat. Radhakrishnan et al. described the coatings prepared from polyaniline-nano-TiO_2 particles synthesized by *in situ* polymerization and found to exhibit excellent corrosion resistance much superior to polyaniline in aggressive environments. These coatings have good gloss and shiny surface, which is not easily obtained in a conventional coating, prepared with commercial micron size particle additives. Such systems can be used as a primer coat or even as a single coat on steel where color is not a very important criterion.

Elaine Armelin et al. carried out the corrosion protection imparted by epoxy paint, modified by the addition of polyaniline emeraldine base. For this purpose, accelerated assays through corrosion cycles, which simulate the marine conditions, have been performed using home-made robotized equipment. They concluded that inorganic corrosion inhibitors, which may have detrimental effects on both the environment and human health, can be replaced by a small concentration of environmental friendly organic polymers.

Patil reported the preparation of hybrid composites containing zinc oxide and polyaniline as nano-additive dispersions with poly (vinyl acetate) as the major matrix. The results are explained on the basis of enhancement in barrier properties due to nano-particulate additives in PVAc-ZnO-PANI film together with the redox behavior of PANI and protective oxide layer formed near the substrate. The chemical synthesis of water-soluble, self-doped conducting polymer to get higher solubility and corrosion efficiency had been studied by Shukla et al. (2008). It was observed that the inhibition efficiency of polyanthranilic acid increases with an increase in the inhibitor concentration.

Yeh et al. (2001) reported the preparation of a series of nanocomposite materials that consisted of emeraldine base of polyaniline and layered montmorllonite (MMT) clay by effectively dispersing the inorganic nanolayers of MMT clay in organic polyaniline matrix via *in situ* polymerization. The corrosion protection effect of polyaniline–clay nanocomposite (PCN) materials at low clay loading compared to conventional polyaniline was demonstrated by a series of electrochemical measurements of corrosion current on CRS in 5 wt.% aqueous NaCl electrolyte.

Even studies on copolymers of aniline with -amino-naphthol-sulphonic acid were studied by Bhandari et al. (2010). It was observed that the copolymer showed a larger degree of surface coverage onto the iron surface, reflecting a higher protection for corrosion of the iron in acidic medium. In

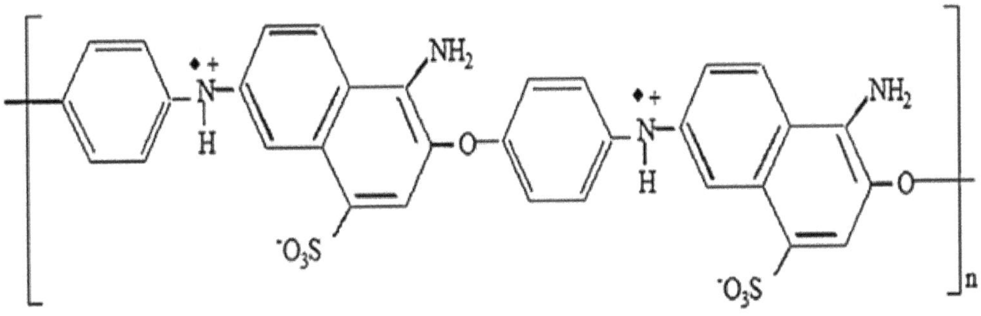

Figure 2.19.5: Self-doped copolymer of aniline and ANSA

addition, the film constitutes physical as well as chemical barrier layer due to the presence of –OH and –NH groups in ANSA unit, which provides passivity protection in polymer coatings. The mechanism of corrosion protection of iron by these copolymers was investigated by surface morphology and EIS techniques. Bhandari et al. (2011) even synthesized self-doped copolymers of aniline with 4-amino-3-hydroxy-naphthalene-1-sulfonic acid (ANSA), which were used to study the corrosion inhibition of iron in a highly corrosive medium. Moreover, they observed that by copolymerization, it was possible to obtain a processable and soluble conducting polymer. Inhibition efficiency of copolymer also increased with an increase in the copolymer concentration. Inhibition efficiency of the copolymer was studied by Tafel polarization curves as well as Electrochemical Impedance Spectroscopy. The structure of the copolymer of aniline with ANSA is shown in Figure 2.19.5.

Nyquist plot of the copolymer (Figure 2.19.6) in 90:10 and 80:20 molar ratios showed inhibition efficiency of 70% and 85% respectively, whereas 50:50 molar ratios showed maximum inhibition efficiency of 90%.

EIS and AFM studies of these copolymers revealed that they inhibit corrosion of iron through adsorption mechanism. These studies clearly revealed that the copolymers have excellent corrosion inhibition properties.

Sathiyanarayanan et al. (2005) reported polyaniline coatings which are found to protect steel from corrosion. The corrosion protection by emeraldine salt of polyaniline has been ascribed to substrate ennoblement due to the redox property of polyaniline. Martyak Nicholas et al. (2002) found that poor adhesion of polyaniline (PANI) is observed if the steel is not properly passivated. They also

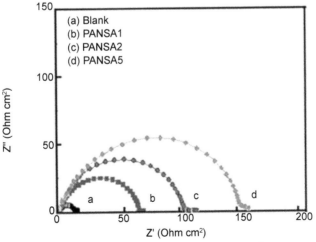

Figure 2.19.6: Nyquist plot of copolymer of aniline and ANSA in 1.0 M HCl in the frequency range from 0.1 Hz to 10^5 Hz (Bhandari et al., Synthetic Metals doi:10.1016/j.synthmet.2011.01.026)

found that during PANI deposition, the Fe (III) layer is reduced to Fe (II) on the reverse scan but the adhesion of the PANI coating is excellent.

Moraes Sandra et al. (2003) found that polyaniline forms an excellent adherent coating on stainless steel in aniline containing phosphate buffer solutions. Inhibition effect depends on the oxidation state of the polymer as well as the characteristics of the passive layer formed on the metallic substrate. Arindam Adhikari et al. (2008) reported conducting polyaniline prepared in the presence of methane sulfonic acid as a dopant by chemical oxidative polymerization. The results from CV and polarization curves show that the conducting polymer coating induces a passive-like behavior and greatly increases the corrosion resistance of the PANI-MeSA/PVAc coated carbon steel in the aggressive NaCl solution.

Gozen Bereket et al. (2005) reported that steel and adherent polymer films of polyaniline and of poly (2-anisidine) were grown on 304-stainless steel by cyclic voltammetry. PANI formed by electrochemical polymerization behaved in a similar manner to polyaniline with regard to corrosion protection. Camalet et al. (1998) reported the electro-synthesis of polyaniline film on mild steel in aqueous p-toluene sulfonic acid solution. The general performance of the deposit (adhesion, resistance to deformation) remains to be improved by the formulation of appropriate electrochemical baths.

Sudeshna Chaudhari et al. (2007) reported that Poly(o-anisidine)-dodecylbenzene sulfonate coatings were synthesized on stainless steel from an aqueous solution containing o-anisidine and DBSA. This study reveals that the POA-DBSA coating has excellent corrosion protection properties and it can be considered as a potential coating material to protect stainless steel against both localized and general corrosion in aqueous 3% NaCl. Vandana Shinde et al. (2003) studied the electrochemical polymerization of 2, 5 dimethylaniline in an aqueous solution of oxalic acid on low carbon steel under galvanostatic condition. The ECP of MLA in oxalic acid solution results into the deposition of uniform and well-adherent conducting PDMLA coatings on LCS substrates. Nicholas M. Martyak reported the cyclic voltammetric study on steel to understand the effects of oxalic acid on the polymerization of aniline. Aniline polymerizes on this iron (III) layer, forming predominately the emeraldine salt. Reversing the scan reduces the iron (III) coating to the iron (II) species and is necessary to achieve good adhesion of PANI to steel. The kinetics of passivation is crucial for obtaining good adhesion of the PANI coatings.

An extensive research has been done, and many protective coating systems have been developed and tried so far all over the world. Metallic coating like galvanizing, nickel, etc., non-metallic coating like fusion bonded epoxy and cement-based coating are some of the examples of the coating systems, which are most widely used with advantages and limitations. In recent years, there has been an increasing use of galvanizing as a protective method against corrosion of reinforcing steel, particularly in the UK and USA. Many researchers have evaluated a few non-metallic coatings and have suggested that the fusion bonded epoxy coating could be considered as a protective coating for reinforcing steel. Cement-based coatings are also adopted for corrosion protection of rebars embedded in concrete. However, many adverse reports have been presented by various researchers for these coating systems. It is reported that most of the organic coatings have pin-holes, and accelerated corrosion of steel takes place through these pin-holes. In order to prevent the corrosion of steel in the pin-holes, coating system containing conducting polymer such as polyaniline can be highly useful as discussed already. Even though a lot of literature is available about the protective ability of polyaniline for different steel as presented earlier, no report is available so far on the usage of polyaniline-containing coating for the protection of steel reinforcements in concrete.

2.5.5 *Corrosion protection by conducting polymers*

Figure 2.20 shows three mechanisms by which PANI coatings may act as a corrosion inhibition system. In case of A, the open-circuit potential (OCP) of the metal substrate is raised to a level at which the formation of a protective oxide on the surface occurs (Abu et al. 2005; Luo et al. 2006). Another possible mechanism is case B, where once the PANI component of the blend is reduced; it releases its counter anion which complexes with metal ions released from the metal surface, forming

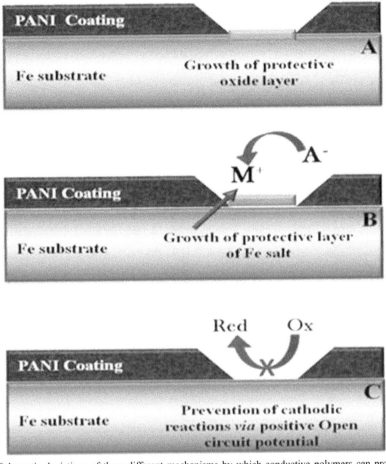

Figure 2.20: Schematic depictions of three different mechanisms by which conductive polymers can provide corrosion protection

an insoluble salt (Kinlen et al. 1999). In a third mechanism, the PANI component of the coating raises the OCP of the interface to values where cathodic reactions are kinetically hindered.

Numerous kinds of literature are available on the corrosion protection of conducting polymer-based coatings. Yang et al. (2010) reported the preparation of nanostructured polyaniline by conventional polymerization, interfacial polymerization and direct mixed reaction respectively. The polyaniline nanofibers, which were synthesized by direct mixed reaction, also have more excellent corrosion protection than conventional aggregated polyaniline for mild steel. Araujo et al. (2001) studied the anti-corrosive properties of undoped polyaniline coated onto mild steel and galvanized steels by a classical methodology used for organic coatings. Thin PANI films are porous, and the adhesion onto mild and galvanized steel is very poor. The lack of adhesion leads to a bad performance even when PANI is top coated with high performance paint as an epoxy one. Such behavior does not depend on the composition of the solution employed for the immersion tests. Samui et al. (2003) reported the corrosion protection of mild steel by coating it with paint containing polyaniline-hydrochloride as pigment. The barrier property contributes appreciably towards corrosion resistance of the PANI-HCl containing paint films. The paint system has shown appreciable corrosion resistance without any top barrier coat. Radhakrishnan et al. (2009) described the coatings prepared from polyaniline-nano-TiO$_2$ particles synthesized by *in situ* polymerization and found to exhibit excellent corrosion resistance much superior to polyaniline in aggressive environments. These coatings have good gloss and shiny surface, which is not easily obtained in the conventional coating that are prepared with commercial

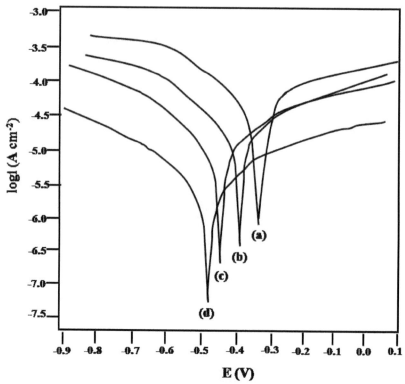

Figure 2.21: Tafel extrapolation graphs of PANI/SiO$_2$ composites with different loadings of the composite in epoxy (Anoop et al., Polymer International, DOI 10.1002/pi.4406)

micron size particle additives. Such systems can be used as a primer coat or even as a single coat on steel, where color is not a very important criterion.

A new smart coating of polyaniline–SiO$_2$ composite for protection of mild steel against corrosion in strongly acidic medium was studied by Anoop et al. (2011). The coating of the poyaniline/SiO$_2$ composite was carried out on mild steel surface by using a powder coating technique. The corrosion protection efficiency of the PANI composite was evaluated by electrochemical measurements of corrosion potential, polarization resistance, corrosion current and Tafel extrapolation methods in 1.0 N HCl medium. The Tafel polarization behavior of the PANI/SiO$_2$ composite epoxy coatings on mild steel surface in 1.0 N HCl medium is shown in Figure 2.21.

Corrosion current values (i$_{corr}$) were reduced from 132 µAcm^{-2} for uncoated mild steel samples to 107.6 µAcm^{-2} for epoxy-coated mild steel samples to 0.09 µAcm^{-2} for PANI/SiO$_2$ coated mild steel samples. The corrosion current values (i$_{corr}$) decreased with increasing concentration of PSC in an epoxy resin from 15.4 µAcm^{-2} at 1.5 wt% to 0.09 µAcm^{-2} at 6.0 wt% loading of conducting material in an epoxy resin. Meanwhile, PANI-coated mild steel showed i$_{corr}$ of the order of 10.9 µAcm^{-2} at 6.0% loading. The corrosion protection efficiency calculated from the Tafel parameter revealed that PANI-coated mild steel showed a 25% protection efficiency at 1.5 wt% loading while PSC-coated mild steel showed 88% protection efficiency at the same loading level. Up to 99.9% protection efficiency had been achieved by using 6.0 wt% PSC in epoxy resin. The coating performance of epoxy coated polymers was investigated in a corrosive atmosphere by exposing the coatings for 60 days. Corrosion rates (CR, mm year^{-1}) of PANI-SiO$_2$ composite coatings were also calculated by the weight loss method. It was observed that CR was highest for uncoated mild steel in HCl medium. After 60 days of immersion, the CR for uncoated mild steel was 7.25 mm year^{-1}. Epoxy- and PANI-coated samples showed CR values of 6.37 mm year^{-1} and 1.90 mm year^{-1} respectively, whereas in the case of the PANI/SiO$_2$ composite coated sample with 6% loading the CR value decreased to 0.73 mm year[-1]. Olad et al. (2010) reported the results from studies on nanocomposite of polyaniline (PANI) with natural clinoptilolite. The reversible electroactive behavior of PANI/Clino nanocomposite coated with 1 and 5% clinoptilolite content coated iron samples. Bagherzadeh et al. (2011) reported that a micropolyaniline and a dispersion of nano polyaniline particles can be used as an anti-corrosion additive in new water-based hardener. From salts pray and cross-cut adhesion tests, it was concluded that using nano PANI as an additive and in a small portion led to a considerable increase in anti-corrosion efficiency of water-based epoxy coating. Patil et al. (2006) reported the preparation of hybrid composite containing zinc oxide and polyaniline as nano-additive dispersions with poly (vinyl acetate) as the major matrix. The results are explained on the basis of enhancement in barrier properties due to nano-particulate additives in PVAc-ZnO-PANI film together with the redox behavior of PANI and protective oxide layer formed near the substrate.

The chemical synthesis of water-soluble self-doped conducting polymer to get higher solubility and corrosion efficiency has been studied by Shukla et al. (2008). It was observed that the inhibition efficiency of polyanthranilic acid increases with an increase in the inhibitor concentration. Yeh et al. (2001) reported the preparation of a series of nanocomposite materials that consisted of emeraldine base of polyaniline and layered montmorllonite (MMT)clay by effectively dispersing the inorganic nanolayers of MMT clay in organic polyaniline matrix via *in situ* polymerization. The corrosion protection effect of polyaniline clay nanocomposite (PCN) materials at low clay loading compared to conventional polyaniline was demonstrated by a series of electrochemical measurements of corrosion current on CRS in 5 wt% aqueous NaCl electrolyte. Huang et al. (2005) reported that a solvent-free mechanochemical route for the synthesis of polyaniline had been developed. Resistance measurements indicate that 1 h is sufficient for reactions.

Chen et al. (2011) reported that a novel approach for preparing waterborne corrosion protection by polyaniline (PANI) containing coatings. The waterborne PANI/P-PVA nanoparticles exhibited significant dispersibility in aqueous media. Marjanovic et al. (2009) reported that self-assembled conducting; paramagnetic polyaniline. Polyaniline nanotubes have been synthesized by the oxidative polymerization of aniline with ammonium peroxydisulfate in an aqueous medium. This novel composite which combines unique properties of 1D PANI nanostructures and mesoporous PANI/Zeolite hybrid materials can be applied as an electronic component, sensor, and catalyst. Ogurtsov et al. (2006) reported the protective properties of polyaniline coatings electrochemically deposited by galvanostatic mode from oxalic acid solutions onto the surface of low-carbon steel by measuring polarization curves. The polyaniline coatings at low current density exhibited higher corrosion retardation coefficients. Sathiyanarayanan et al. (2004) reported polyaniline coatings which were found to protect steel from corrosion. The corrosion protection by emeraldine salt of polyaniline had been ascribed to substrate ennoblement due to the redox property of polyaniline. Martyak et al. (2002) found that poor adhesion of polyaniline (PANI) was observed if the steel was not properly passivated. They also found that during PANI deposition, the Fe (III) layer was reduced to Fe (II) on the reverse scan but the adhesion of the PANI coating was excellent.

Moraes et al. (2003) found that polyaniline formed an excellent adherent coating on stainless steel in aniline containing phosphate buffer solutions. Inhibition effect depends on the oxidation state of the polymer as well as the characteristics of the passive layer formed on the metallic substrate. Adhikari et al. (2008) reported conducting polyaniline prepared in the presence of methane sulfonic acid as a dopant by chemical oxidative polymerization. The results from CV and polarization curves show that the conducting polymer coating induces a passive-like behavior and greatly increases the corrosion resistance of the PANI-MeSA/PVAc coated carbon steel in the aggressive NaCl solution.

Bereket et al. (2005) reported that steel and adherent polymer films of polyaniline and of poly (2-anisidine) were grown on 304-stainless steel by cyclic voltammetry. PANI formed by electrochemical polymerization behaved in a similar manner to polyaniline with regard to corrosion protection. Camalet et al. (1998) reported the electrosynthesis of polyaniline film on mild steel in aqueous p-toluene sulfonic acid solution. The general performance of the deposit (adhesion, resistance to deformation) remains to be improved by the formulation of appropriate electrochemical baths.

Raotole et al. (2006) reported that strongly adherent polyaniline coatings were electrochemically synthesized on mild steel from an aqueous salicylate medium. The results of these studies revealed that the corrosion resistance of the polyaniline-coated mild steel was significantly higher and the corrosion rate was considerably lower than that of uncoated steel. Martyak et al. (2002) reported that cyclic polarization measurements made on steel, iron oxalate-coated steel and polyaniline in a high pH solution, and in the absence of chloride in the caustic solution caused significant pitting in the bare steel and slight pitting in the oxalate coated steel. The oxalate layer was necessary to achieve good adhesion of PANI to steel. PANI was sufficient to prevent pitting corrosion of steel in high pH media containing chloride. CP results showed no pitting corrosion in the scratched area.

Ansari et al. (2009) reported that the corrosion inhibition properties of polished steel plates coated with a polyaniline blend with nylon 66 via cast method with formic acid as the solvent. One can create corrosion protection materials that are environmentally promising as potential substituents for highly toxic lead and chromium pigments. Rajagopalan et al. (2010) reported that uniform and adherent polyaniline-polypyrrole composite coatings were electrochemically polymerized on low carbon steel under aqueous conditions. The structure of the coatings was analyzed as the function of electrochemical deposition parameters. Adhesion studies showed that the composite coatings formed using an equimolar feed ratio of monomers were more adherent than the homopolymeric coatings. Sankar et al. (2011) reported that aniline was electropolymerized in phenyl phosphoric acid medium to yield a greenish black smooth adhesive film of polyaniline on Pt/mild steel electrodes with an electronic conductivity of 8 S/cm. Electropolymerisation of aniline in PPA medium acted as a good barrier against the corrosive medium. Narayanasamy et al. (2010) reported that polyaniline and poly (N-methyl aniline) have been electrodeposited on mild steel from oxalic acid bath using the cyclic voltammetric technique. Copolymer coatings and composite-bilayer coatings have exhibited better protection than the individual homopolymers. In the case of composite-bilayer coatings, metal-PANI-PNMA showed higher stability and better performance than metal-PNMA-PANI.

Ren et al. (2010) reported that polyaniline coating doped with DBSA anions is electrodeposited galvanostatically on 304 stainless steel used as bipolar plates of a proton-exchange membrane fuel cell. As a barrier to the inward permeation of corrosive species, and as an effective catalytic oxidizer maintaining the passive state of the substrate alloy. Yagan et al. (2007) reported that homopolymers and bilayers of polyaniline and polypyrrole coatings have been electropolymerized on mild steel by potentiodynamic synthesis technique in aqueous oxalic acid solutions. PANI/PPy and PPy coatings on cheap material such as mild steel may exhibit very attractive properties for technological applications such as the production of electrolytic hydrogen and enzyme electrode.

Ganash et al. (2011) reported that electrodeposited polyaniline films on stainless steel surfaces are made from two different acidic solutions containing the aniline monomer. In comparison with sulfate, counter ion hinders the autocatalytic oxidation of aniline monomer and hence decreases the electropolymerization. Dominis et al. (2003) reported that conducting polymers can provide corrosion protection to metals in aqueous environments, the processes involved are complex and the mechanisms are not fully understood. It is likely that the dopant used with the conducting polymer will affect the corrosion processes since the dopant affects the polymer conductivity and permeability, and the dopant is also released during the galvanic coupling of the polymer to the metal.

Le et al. (2009) reported the corrosion resistance of conducting polyaniline coatings deposited on 316L stainless steel at various cycles of cyclic voltammetry by electro-polymerization in a sulfuric acid solution containing fluoride. The increase in the number of cycles, increased the thickness and enhanced the performance of the PANI coating due to low porosity. Chang et al. (2007) reported that the DBSA-doped polyaniline (PANI) Na+-monomorillonite (MMT) clay nanocomposite (PCN) materials have been successfully prepared with DBSA as emulsifier and dopant by the emulsion polymerization of aniline. They further found that the PANI and clay materials doped with HCl had a higher electric conductivity than that doped with HNO_3 and H_2SO_4, which lead to a faster doping rate.

Grgur et al. (2006) reported the corrosion behavior, sorption characteristics and thermal stability of epoxy coatings electrodeposited on mild steel. Dissolved oxygen, oxides dedoped (leucoemeraldine)

PANI form (doping with chloride anions) increased the local corrosion potential and faster dissolution of the mild steel substrate. Popovic et al. (2005) reported the corrosion behavior of mild steel covered by electrochemically deposited polyaniline film on steel, epoxy coating and the PANI/epoxy coating system during cathodic deposition of epoxy coatings on PANI coated mild steel.

Camalet et al. (1998) reported the electrodeposition of protective polyaniline films on mild steel from aqueous oxalic acid. The electropolymerization of aniline occurred on a surface passivated by the precipitation of a Fe(II)-oxalate layer and led to strongly adherent films with a controlled thickness and with the same structure as the idealized emerldine base, but the deposits appeared to be less oxidized. These PANI coatings exhibited very promising properties against corrosion in acidic solution and appeared to be more effective than PPy coatings. Sazou et al. (1997) reported the electropolymerization of aniline on iron disc electrode in aqueous solutions, by using various inorganic and organic acids under potentiodynamic, potentiostatic and galvanostatic conditions. Smooth and well-adhered polyaniline coatings were obtained. The inhibitive properties of the PANI coatings seemed to be quite good. This behavior might be attributed to stabilization of the passive oxide film of iron in an acidic corrosive medium. Iroh et al. (2003) synthesized different kinds of conducting polymers, including polypyrrole–polyaniline composites. The processability and corrosion performance of PPy/PANI, composite coatings were significantly better than those for either PPy or PANI coatings. The corrosion rate and corrosion potentials of the bi-layer coatings were significantly better than those for the PPy/PANI modified low carbon steel.

Wang et al. (2011) reported that polyaniline was deposited onto mesoporous carbon-silica coated stainless steel using electropolymerization method. The mesoporous C-SiO$_2$-N films had a potential application as a protective coating of the bipolar plate material. Ozyilmaz et al. (2010) reported that Poly (o-anisidine) and polyaniline coatings were synthesized on a platinum surface. The superior corrosion protection properties of the PANI and POA-L films could be due to polymer resistance and the formation of oxide layers on the metal surface. Kumar et al. (2008) reported that the corrosion resistant behavior of polymer metal bilayer coatings, viz. polyaniline, polyaniline-nickel, nickel-polyaniline, polyaniline, polyaniline-zinc, zinc-polyaniline. The PANI layer was, however, porous in nature while the porosity of PANI+Ni and PANI+ Zn bilayer coatings were significantly reduced due to the ability of the metal clusters to fill the pores of the PANI layer.

Chaudhari et al. (2006) reported that Poly(o-anisidine)-dodecylbenzene sulfonate coatings were synthesized on stainless steel from an aqueous solution containing o-anisidine and DBSA. This study revealed that the POA-DBSA coating had excellent corrosion protection properties and it could be considered as a potential coating material to protect stainless steel against both localized and general corrosion in aqueous 3% NaCl. Salunkhe et al. (2016) had given an overall review on the utilization of conducting polymers for corrosion protection. In the review paper attempts had been made to summarize extensive studies of polyaniline, polypyrrole and polythiophene polymer and their anti-corrosive properties. Several researchers had reported diverse views about corrosion protection by CPs and hence various mechanisms had been suggested to explain their anti-corrosion properties. These included anodic protection, controlled inhibitor release as well as barrier protection mechanisms. Different approaches had been developed for the use of CPs in protective coatings (dopants, composites, blends). Shinde et al. (2003) studied the electrochemical polymerization of 2, 5 dimethylaniline in an aqueous solution of oxalic acid on low carbon steel under galvanostatic condition. The ECP of MLA in oxalic acid solution resulted into the deposition of uniform and well-adherent conducting PDMLA coatings on LCS substrates. Martyak et al. (2002) reported the cyclic voltammetric studies on steel to understand the effects of oxalic acid on the polymerization of aniline. Aniline polymerizes on this iron (III) layer formed predominately the emeraldine salt. Reversing the scan reduced the iron (III) coating to the iron (II) species and was necessary to achieve good adhesion of PANI to steel. The kinetics of passivation was crucial for obtaining good adhesion of the PANI coatings.

Beard et al. (1997) reported the electrochemical response of polyaniline in its undoped intermediate oxidation state coated on cold rolled steel or glass substrate. The spontaneous electrochemistry between steel and polyaniline was changed in the presence of oxygen. Ogurtsov et al. (2006) studied the protective properties of polyaniline coatings electrochemically deposited in the galvanostatic mode from oxalic acid solutions onto the surface of low-carbon steel by measuring polarization curves. The polyaniline coatings obtained at low current density exhibit higher corrosion retardation coefficients. This was apparently due to a decrease in the extent of excessive oxidation of polyaniline synthesized at lower potentials.

Zhu et al. (2008) reported the preparation method of a self-supporting doped-polyaniline film electrode and its open-circuit potential in $NaClO_4$ and Na_2SO_4 solutions with different pH value as well as cathodic polarization. PANI would act as a cathode when it was coupled with metals. Potentiodynamic cathodic polarization had shown that PANI film electrode was much more difficult to reach a steady state due to its large RC constant and unique redox behavior. Vera et al. (2007) reported that polyaniline and poly (ortho-methoxyaniline) were prepared by chemical synthesis and evaluated their corrosion protection properties on carbon steel and copper in an aggressive media such as sodium chloride. The formation of a homogeneous oxide film at the metal-polymer interface, which was due to the reduction of the polymer. On the other hand, the use of poly (ortho-methoxy aniline) resulted in a less compact and rougher film. There was also no evidence of the formation of an oxide film, likely explaining why it is less effective as a protective agent than polyaniline.

Shinde et al. (2003) had synthesized strongly adherent poly (2,5-dimethylaniline) coatings on low carbon steel substrates, with an objective of examining the possibility of using this polymer for corrosion protection of steel in a chloride environment. This study clearly revealed that the poly (2, 5-dimethylaniline) coating has an excellent corrosion protection of low carbon steel in aqueous 3% NaCl. Martina et al. (2011) reported that composite films of polyaniline and carbon nanotubes were prepared by electrochemical co-deposition from solutions of the corresponding monomer containing two different kinds of CNTs. As the PANI-CNTs/films showed better anti-corrosion properties than PANI, they could be suitable as protective coatings on electrode materials for new electronic devices.

Ye et al. (1995) reported the preparation of conducting poly (aniline-co-butyl aniline) copolymer and were evaluated by cyclic voltammetry and impedance spectroscopy methods. The comparative study indicated that PANI, PABA and PNBA films decreased in the order PANI < PABA < PNBA. The porosity of the three polymer films and the diffusion coefficients of the ions in these films decreased in the same order.

2.6 Polypyrrole

Various approaches for synthesis of polypyrrole have been reported. Electrochemical polymerization and chemical polymerization are very frequently used techniques for obtaining conducting polymers. In 1979, Diaz et al. prepared the polymer in the form of flexible films by electrolysis of an aqueous solution of pyrrole. This work gave a start to the extensive use of electrochemical synthesis of PPy and other conducting polymers. Electrochemical polymerization is performed in a one-compartment cell with three electrodes. In the electrochemical oxidation method, pyrrole and an electrolyte salt are dissolved in a suitable solvent and then the solution is subjected to oxidation, resulting in the deposition of a conducting PPy film on the inert anodic working electrode like Pt, Au, glassy carbon or stainless steel. The advantages of electrochemical deposition of polypyrrole are that films can be prepared simply with the one-step procedure and exact control of thickness. The properties of electronically conducting polymers strongly depend on their synthesis condition such as the film growth rate, oxidation potential, current density, the nature of solvent and the kind and concentration of dopant anion. Electrochemical polymerization is generally preferred as cleaner polymers are produced, and it provides a better film thickness and morphology control when compared to chemical oxidation as reviewed by Sadaki et al. (2000). Though, the bulk quantity of polypyrrole can be obtained by using chemical polymerization of the monomer by selecting transition metal ions in water or various other solvents (Machida et al. 1989). Chemical polymerization is a simple and fast process as compared to the electrochemical polymerization process as there is no need for any special instrument. Aqueous and anhydrous iron (III) chloride is the most widely used oxidant for the chemical oxidative polymerization (Armes 1987). The yield and conductivity of the chemically produced polypyrrole depend on various factors, among which are the choice of solvent and oxidant, monomer/oxidant ratio, time and temperature of the reaction. However, a number of other methods such as Interfacial and Supercritical Fluid Polymerization (Lu et al. 1998), Plasma Polymerization (Wang et al. 2004), Radiolysis Polymerization (Karim et al. 2007) and Copolymerization (Prasannan et al. 2009) have been reported in the literature.

The main strategies for all synthesis methods are to synthesize new and novel structures, increasing the order of the polymer backbone (and also the conductivity), good processing ability, easier synthesis, more defined three-dimensional structure, stability in certain solvents such as water and many other applications unique properties.

Electrochemical polymerization of pyrrole leads to the formation of polypyrrole films on the electrode. The polymerization of pyrrole to polypyrrole is shown in Figure 2.22:

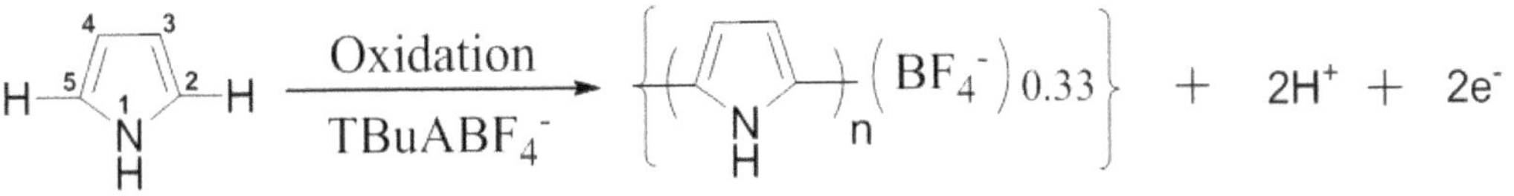

Figure 2.22: Oxidative polymerization of pyrrole

In the polypyrrole matrix, pyrrole units are positively charged, which are balanced by dopant anions. On applying a negative potential, dopants are detached from the polypyrrole leading to the formation of the undoped polypyrrole film. However, dopants can again be attached in the polymer matrix by applying an anodic potential. Doped and undoped forms of PPy is shown in Figure 2.23.

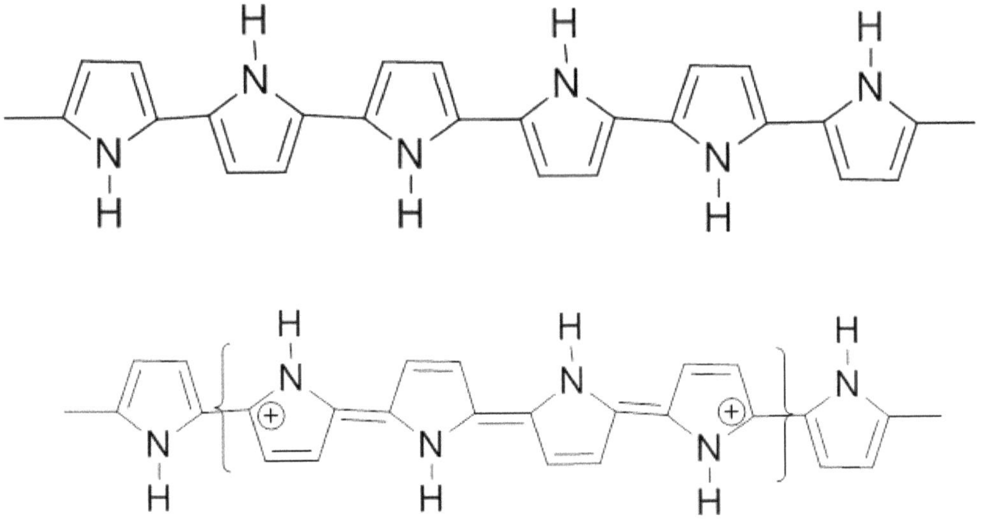

Figure 2.23: Undoped and doped form of polyprrole

In the undoped state, polypyrrole becomes less conducting and its color changes to yellowish green whereas, on dedoping, polymer becomes conducting due to the presence of charge carriers and film turning greenish blue.

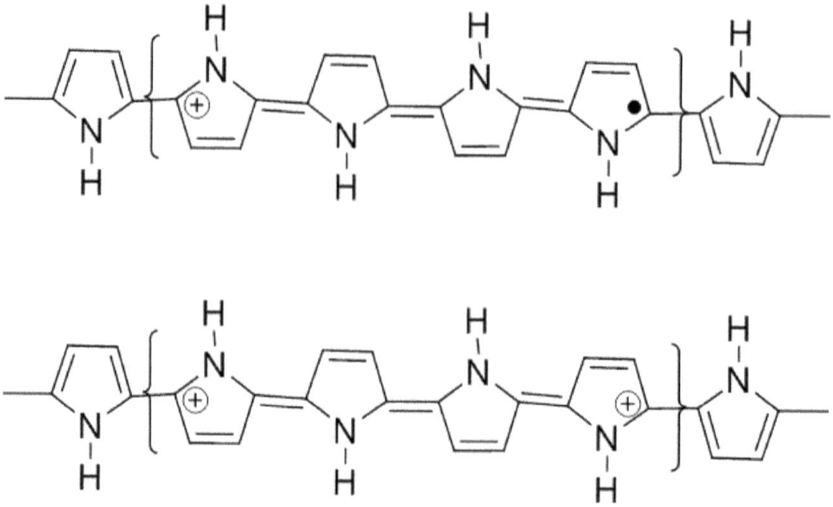

Figure 2.24: Polarons and bipolarons present in polypyrrole matrix

Low doping in polypyrrole leads to a generation of polarons or cation radicals and on increasing the doping concentration, bipolarons or dications are formed as shown in Figure 2.24.

Unlike polyacetylene, which has degenerate ground state and the two forms of polyacetylene are mirror images of each other, two forms of polypyrrole, benzenoid form and quinoid form are not identical which is shown in Figure 2.25.

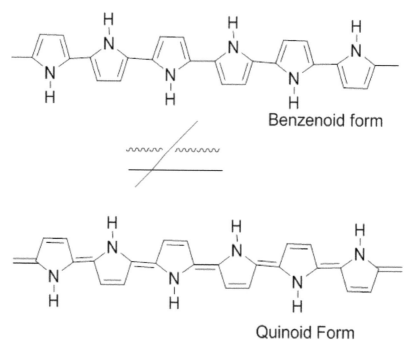

Benzenoid form

Quinoid Form

Figure 2.25: Benzenoid and quinoid forms of polypyrrole

2.6.1 *Polypyrrole as corrosion inhibitor*

The corrosion inhibition property of polypyrrole and the reinforcing ability of fly ash is utilized to design coatings with superior corrosion resistance for saline conditions. Ruhi et al. (2015) have done extensive studies on the designing of PPy/flyash composites for studying its corrosion behavior in 3.5% NaCl. Electrochemical parameters obtained from Tafel extrapolation in marine conditions show that a protection efficiency of 99% is achieved using PPY/flyash composite. The mechanism of prevention of corrosion of Fe substrate is shown in Figure 2.26.

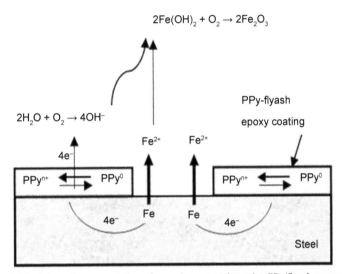

Figure 2.26: Schematic representation of corrosion prevention using PPy/flyash composite

It was observed that polypyrrole and flyash composite showed enhanced thermal stability because of the interactions of PPy with fly ash. Better corrosion behavior was explored from the data obtained from Tafel polarization, impedance analysis and salt spray tests. It was also observed that loading of 1% and 2% in epoxy showed better corrosion response compared to the epoxy coating on the Fe substrate. PPy-flyash composite acted as an effective barrier layer to passivate the metal surface and delayed the degradation of the coating under extremely corrosive saline conditions. Even polypyrrole-biocompatible polymer composites were designed by Ruhi et al. (2018). The reported work explains that the presence of Gum Acacia in the PPy matrix improves the corrosion inhibition response of the coatings under severe marine conditions. The polypyrrole-gum acacia composite coatings will in future emerge as a better corrosion inhibitive coating compared to conventional chromate-based and phosphate coatings.

The synergistic effect of the corrosion inhibition properties of polypyrrole and Gum Acacia is the basic reason for the superior corrosion resistance of the coatings. Polypyrrole shows redox property and intercepts the electron released from the metal and utilize them to the reduction of oxygen at coating/electrolyte interface. This reaction assists in the formation of a passive oxide layer at the polymer/metal interface, which shifts the corrosion potential of the mild steel to noble direction. Further, the Gum Acacia assists in the formation of the oxide layer on the metal surface. Gum Acacia is a branched complex polysaccharide having pronounced corrosion inhibition property. It simply adsorbs on the metal surface through their oxygen and nitrogen atoms and blocks the cathodic and anodic sites. In this way, the two constituents of the composite synergistically improves the corrosion resistance of the epoxy coating system. Salt spray test results of epoxy coating and epoxy loaded with different concentrations of PPy/Gum Acacia composite for 120 days shows that, whereas epoxy coated steel panel shows pronounced corrosion along the scribe mark (Figure 2.27). However, no

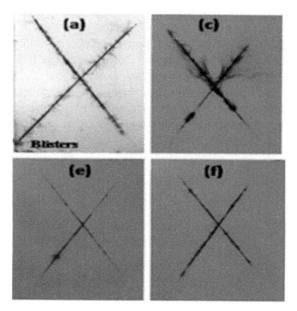

Figure 2.27: Photographs of (a) epoxy coated steel panel (EC), epoxy with (c) 1.0 wt% (e) 3.0 wt% and (f) 4.0 wt% loading of Polypyrrole/Gum Acacia composite after 120 days of exposure to 5.0 % NaCl in salt spray fog (Ruhi et al., Advanced Material Letters, DOI: 10.5185/amlett.2018.7007.

extended corrosion was observed at the scribe mark where composite was incorporated (Figure 2.26 c, e, f). PPY/GA composites present in the epoxy system act as reinforcing the material in the coating.

Their synergy towards corrosion inhibition enhances the overall corrosion resistance of the epoxy coating system. The composite effectively inhibits the progress of under-coating corrosion and formation of oxide scale at coating/metal interface.

Conducting polymer and their composites are subjects of intensive research because of their wide spread applications as smart coatings which can be used for preventing corrosion of iron under hostile environmental conditions. The aim of this chapter was to understand the role of conducting polymers, its synthesis and designing of conducting polymer composites by incorporating filler materials and suitably selecting a medium for polymerization so that the resultant epoxy coatings can be used for prevention of corrosion of iron in saline water conditions.

References

Abu, Y. M. and Aoki, K. (2005). Corrosion protection by polyaniline-coated latex microspheres. Journal of Electroanalytical Chemistry, 583(1): 133–139.

Adhikari, A., Claesson, P., Pan, J., Leygraf, C., Dedinaite, A. and Blomberg, E. (2008). Electrochemical behavior and anticorrosion properties of modified polyaniline dispersed in polyvinylacetate coating on carbon steel. Electrochimica Acta, 53: 4239–4247.

Agnihotri, N., Chakrabarti, K. and De, A. (2015). Highly efficient electromagnetic interference shielding using graphite nanoplatelet/poly (3, 4-ethylenedioxythiophene)–poly (styrenesulfonate) composites with enhanced thermal conductivity. RSC Advances, 5(54): 43765–43771.

Anand, J., Palaniappan, S. and Sathyanarayana, D. (1998). Conducting polyaniline blends and composites. Progress in Polymer Science, 23(6): 993–1018.

Armes, S. P. (1987). Optimum reaction conditions for the polymerization of pyrrole by iron(III) chloride in aqueous solution. Synthetic Metals, 20: 365–371.

Ansari, R. and Alikhani, A. H. (2009). Application of polyaniline/nylon composites coating for corrosion protection of steel. Journal of Coatings Technology and Research, 6: 221–227.

Arauio, W. S., Margarit, I. C. P., Ferreira, M., Mattos, O. R. and Neto, P. L. (2001). Undoped polyaniline anticorrosive properties. Electrochimica Acta, 46: 1307–1312.

Armelin, E., Aleman, C. and Irbarren, J. I. (2009). Anticorrosion performances of epoxy coatings modified with polyaniline: A comparison between the emeraldine base and salt forms. Progress in Organic Coatings, 65: 88–93.

Aguirre, J., Armijo, F., Walczak, M., Iglesia, R. de la, Pizarro, G. and Vargas, I. (2015). Poly (3,4-ethylenedioxythiophene) used as a coating on stainless steel AISI 470. Evaluation of corrosion inhibition performance. Interdisciplinary Project VRI -UC 2013-1 and Fondecyt project No110041.

Bagherzadeh, M. R., Ghasemi, M., Mahadevi, F. and Shariatpanahi, H. (2011). Investigation on anticorrosion performance of nano and micro polyaniline in new water-based epoxy coating. Progress in Organic Coatings, 72: 348–352.

Baughman, R. (1996). Conducting polymer artificial muscles. Synthetic metals, 78(3): 339–353.

Beard, B. C. and Spellane, P. (1997). XPS evidence of redox chemistry between cold rolled steel and polyaniline. Chemistry of Materials, 9(9): 1949–1953.

Beaujuge, P. M. and Reynolds, J. R. (2010). Color control in π-conjugated organic polymers for use in electrochromic devices. Chemical Reviews, 110(1): 268–320.

Bereket, G., Hur, E. and Sahin, Y. (2005). Electrochemical synthesis and anti-corrosive properties of polyaniline, poly(2-anisidine), and poly(aniline-*co*-2-anisidine) films on stainless steel. Progress in Organic Coatings, 54: 63–72.

Brédas, J., Chance, R. and Silbey, R. (1981). Theoretical studies of charged defect states in doped polyacetylene and polyparaphenylene. Molecular Crystals and Liquid Crystals, 77(1-4): 319–332.

Bredas, J., Chance, R. and Silbey, R. (1982). Comparative theoretical study of the doping of conjugated polymers: polarons in polyacetylene and polyparaphenylene. Physical Review B, 26(10): 5843.

Brédas, J., Thémans, B., Fripiat, J., André, J. and Chance, R. (1984). Highly conducting polyparaphenylene, polypyrrole, and polythiophene chains: An ab initio study of the geometry and electronic-structure modifications upon doping. Physical Review B, 29(12): 6761.

Bredas, J. L. and Street, G. B. (1985). Polarons, bipolarons, and solitons in conducting polymers. Accounts of Chemical Research, 18(10): 309–315.

Bremer, L., Verbong, M., Webers, M. and Van Doorn, M. (1997). Preparation of core-shell dispersions with a low tg polymer core and a polyaniline shell. Synthetic Metals, 84(1): 355–356.

Burroughes, J., Bradley, D., Brown, A., Marks, R., Mackay, K., Friend, R., Burns, P. and Holmes, A. (1990). Light-emitting diodes based on conjugated polymers. Nature, 347(6293): 539–541.

Catt, K., Li, H. and Tracy, Cu X. (2017). Poly (3,4-ethylenedioxythiophene) graphene oxide composite coatings for controlling magnesium implant corrosion. Acta Biomaterialia, 48: 530–540.

Camalet, J. L., Lacroix, J. C., Aejyach, S. and Lacaze, P. C. (1998). Characterization of polyaniline films electrodeposited on mild steel in aqueous *p*-toluenesulfonic acid solution. Journal of Electroanalytical Chemistry, 445: 117–124.

Chang, K. C., Jang, G. W., Peng, C. W., Lin, C. Y., Shieh, J. C., Yeh, J. M., Wang, J. C. and Li, W. T. (2007). Comparatively electrochemical studies at different operational temperatures for the effect of nanoclay platelets on the anticorrosion efficiency of DBSA-doped polyaniline/Na+–MMT clay nanocomposite coatings. Electrochimica Acta, 52: 5191–5200.

Chaudhary, S., Sanikar, S. R. and Patil, P. P. (2007). Poly(*o*-ethylaniline) coatings for stainless steel protection. Progress in Organic Coatings, 58: 54–63.

Chen, F. and Liu, P. (2011). Conducting polyaniline nanoparticles and their dispersion for waterborne corrosion protection coatings. ACS Applied Materials & Interfaces, 3: 2694–2702.

Chiang, C. K., Fincher, C., Jr., Park, Y. W., Heeger, A. J., Shirakawa, H., Louis, E. J., Gau, S. C. and MacDiarmid, A. G. (1977). Electrical conductivity in doped polyacetylene. Physical Review Letters, 39(17): 1098.

Chiang, C., Park, Y., Heeger, A., Shirakawa, H., Louis, E. and MacDiarmid, A. G. (1978). Conducting polymers: halogen doped polyacetylene. The Journal of Chemical Physics, 69(11): 5098–5104.

Cochet, M., Maser, W. K., Benito, A. M., Callejas, M. A., Martínez, M. T., Benoit, J. -M., Schreiber, J. and Chauvet, O. (2001). Synthesis of a new polyaniline/nanotube composite: "*in-situ*" polymerization and charge transfer through site-selective interaction. Chemical Communications, 2001(16): 1450–1451.

De Jong, M., Van Ijzendoorn, L. and De Voigt, M. (2000). Stability of the interface between indium-tin-oxide and poly (3, 4-ethylenedioxythiophene)/poly (styrenesulfonate) in polymer light-emitting diodes. Applied Physics Letters, 77(14): 2255–2257.

De, A., Sen, P., Poddar, A. and Das, A. (2009). Synthesis, characterization, electrical transport and magnetic properties of PEDOT–DBSA–Fe$_3$O$_4$ conducting nanocomposite. Synthetic Metals, 159(11): 1002–1007.

Dhawan, S. K. and Trivedi, D. C. (1993). Investigations on the effect of 5-sulfosalicylic acid on the properties of polyaniline. Synthetic Metals, 58: 309–324.

Dominis, A. J., Spinks, G. M. and Wallace, G. G. (2003). Comparison of polyaniline primers prepared with different dopants for corrosion protection of steel. Progress in Organic Coatings, 48: 43–49.

Ehinger, K., Summerfield, S., Bauhofer, W. and Roth, S. (1984). DC and microwave conductivity of iodine-doped polyacetylene. Journal of Physics C: Solid State Physics, 17(21): 3753.

Elschner, A., Kirchmeyer, S., Lovenich, W., Merker, U. and Reuter, K. (2010). PEDOT: Principles and Applications of an Intrinsically Conductive Polymer. CRC Press.

Frackowiak, E., Khomenko, V., Jurewicz, K. t., Lota, K. and Beguin, F. (2006). Supercapacitors based on conducting polymers/nanotubes composites. Journal of Power Sources, 153(2): 413–418.

Ganash, A. A., Al-Nowaiser, F. M., Thabaiti, S. A. Al and Hermas, A. A. (2011). Comparison study for passivation of stainless steel by coating with polyaniline from two different acids. Progress in Organic Coatings, 72: 480–485.

Genis, E. M. and Tsintavis, C. (1985). Redox mechanism and electrochemical behaviour or polyaniline deposits, Journal of Electroanalytical Chemistry and Interfacial Electrochemistry, 195(1): 109–128.

Genis, E. M., Syed, A. A. and Tsintavis, C. (1985). Electrochemical study of polyaniline in aqueous and organic medium. Redox and Kinetic Properties. Molecular Crystals and Liquid Crystals, 121: 181–186.

Genis, E. M., Boyle, A., Lapkossky, M. and Tsintavis, C. (1990). Polyaniline: A historical survey. Synthetic Metals, 36: 139–182.

Glarum, S. H., Marshall, J. H. and Electrochem, J. (1987). Soc. The Impedance of Poly(aniline) Electrode Films, 134, 142.

Grgur, B. N., Milica, G., Stankovic, V. B. M. and Popovic, Z. K. (2006). Corrosion of metals with composite polyaniline coatings. Progress in Organic Coatings, 56: 214–219.

Günes, S., Neugebauer, H. and Saiciftci, N. S. (2007). Conjugated polymer-based organic solar cells. Chemical Reviews. 107(4): 1324–1338.

Gustafsson, G., Cao, Y., Treacy, G., Klavetter, F., Colaneri, N. and Heeger, A. (1992). Flexible light-emitting diodes made from soluble conducting polymers. Nature, 357(6378): 477–479.

Haba, Y., Segal, E., Narkis, M., Titelman, G. and Siegmann, A. (2000). Polyaniline–DBSA/polymer blends prepared via aqueous dispersions. Synthetic Metals, 110(3): 189–193.

Han, C. and Elsenbaumer, R. (1989). Protonic acids: generally applicable dopants for conducting polymers. Synthetic Metals, 30(1): 123–131.

Heeger, A. J. (2001). Semiconducting and metallic polymers: the fourth generation of polymeric materials (Nobel lecture). Angewandte Chemie International Edition, 40(14): 2591–2611.

Hou, J., Zhu, G., Xu, J. K. and Huang, Y. (2012). Epoxy resin modified with PEDOT/PSS and corrosion protection of steel. In Advanced Materials Research, Trans Tech Publications, 560: 947–951.

Huang, J., Moore, J. A., Acquave, H. and Kaner, R. B. (2005). Mechanochemical route to the conducting polymer polyaniline. Macromolecules, 38: 317–321.

Iroh, J. O., Zhu, Y., Shah, K., Levine, K., Rajagopalan, R., Uyar, T., Donley, M., Mantz, R., Johnson, J., Voevodin, N. N., Balbyshev, V. N. and Khramov, A. N. (2003). Electrochemical synthesis: a novel technique for processing multi-functional coatings. Progress in Organic Coatings, 47: 365–375.

Janata, J. and Josowicz, M. (2003). Conducting polymers in electronic chemical sensors. Nature Materials, 2(1): 19–24.

Jonas, F. and Schrader, L. (1991). Conductive modifications of polymers with polypyrroles and polythiophenes. Synthetic Metals, 41(3): 831–836.

Jönsson, S., Birgerson, J., Crispin, X., Greczynski, G., Osikowicz, W., Van Der Gon, A. D., Salaneck, W. R. and Fahlman, M. (2003). The effects of solvents on the morphology and sheet resistance in poly (3, 4-ethylenedioxythiophene)–polystyrenesulfonic acid (PEDOT–PSS) films. Synthetic Metals, 139(1): 1–10.

Joo, J. and Lee, C. (2000). High frequency electromagnetic interference shielding response of mixtures and multilayer films based on conducting polymers. Journal of Applied Physics, 88(1): 513–518.

Jozefowicz, M. E., Laversanne, R., Javadi, H. H. S., Epstein, A. J., Pouget, J. P., Tang, X. and MacDiarmid, A. G. (1989). Multiple lattice phases and polaron-lattice—spinless-defect competition in polyaniline. Physical Review B 39, 12958(R).

Karim, M. R., Lee, C. J. and Lee, M. S. (2007). Synthesis of conducting polypyrrole by radiolysis polymerization method. Polymers for Advanced Technologies, 18: 916–920.

Karpfen, A. and Höller, R. (1981). Cis-trans isomerism in infinite polyacetylenes: an Ab initio study. Solid State Communications, 37(2): 179–182.

Kaufman, J., Colaneri, N., Scott, J., Kanazawa, K. and Street, G. (1985). Evolution of polaron states into bipolarons in polypyrrole. Molecular Crystals and Liquid Crystals, 118(1): 171–177.

Kawai, T., Kuwabara, T. and Yoshino, K. (1991). Electrochemical doping of conducting polymer in solution phase. Technology reports of the Osaka University, 41(203052): 93–97.

Kim, W., Mäkinen, A., Nikolov, N., Shashidhar, R., Kim, H. and Kafafi, Z. (2002). Molecular organic light-emitting diodes using highly conducting polymers as anodes. Applied Physics Letters, 80(20): 3844–3846.

Kim, J. H., Joo, C. W., Lee, J., Seo, Y. K., Han, J. W., Oh, J. Y., Kim, J. S., Yu, S., Lee, J. H. and Lee, J. I. (2016). Highly Conductive PEDOT: PSS Films with 1, 3-Dimethyl-2-Imidazolidinone as transparent electrodes for organic light-emitting diodes. Macromolecular Rapid Communications, 37(17): 1427–1433.

Kim, Y. H., Sachse, C., Machala, M. L., May, C., Müller-Meskamp, L. and Leo, K. (2011). Highly conductive PEDOT: PSS electrode with optimized solvent and thermal post-treatment for ITO-free organic solar cells. Advanced Functional Materials, 21(6): 1076–1081.

Kinlen, P., Liu, J., Ding, Y., Graham, C. and Remsen, E. 1998. Emulsion polymerization process for organically soluble and electrically conducting polyaniline. Macromolecules, 31(6): 1735–1744.

Kinlen, P. J., Menon, V. and Ding, Y. (1999). A mechanistic investigation of polyaniline corrosion protection using the scanning reference electrode technique. Journal of the Electrochemical Society, 146: 3690–3695.

Kirchmeyer, S. and Reuter, K. (2005). Scientific importance, properties and growing applications of poly (3, 4-ethylenedioxythiophene). Journal of Materials Chemistry, 15(21): 2077–2088.

Kivelson, S. and Heeger, A. (1988). Intrinsic conductivity of conducting polymers. Synthetic Metals, 22(4): 371–384.

Koul, S., Chandra, R. and Dhawan, S. (2000). Conducting polyaniline composite for ESD and EMI at 101 GHz. Polymer, 41(26): 9305–9310.

Kumar, S. A., Meenakshi, K. S., Sankaranarayanan, T. S. N. and Srikanth, S. (2008). Corrosion resistant behaviour of PANI–metal bilayer coatings. Progress in Organic Coatings, 62: 285–292.

Lakshmi, K., John, H., Mathew, K., Joseph, R. and George, K. (2009). Microwave absorption, reflection and EMI shielding of PU–PANI composite. Acta Materialia, 57(2): 371–375.

Lefebvre, M., Qi, Z., Rana, D. and Pickup, P. G. (1999). Chemical synthesis, characterization, and electrochemical studies of poly (3, 4-ethylenedioxythiophene)/poly (styrene-4-sulfonate) composites. Chemistry of Materials, 11(2): 262–268.

Le, D. P., Yoo, Y. H., Kim, J. G., Cho, S. M. and Son, Y. K. (2009). Corrosion characteristics of polyaniline-coated 316L stainless steel in sulphuric acid containing fluoride. Corrosion Science, 51: 330–338.

Lei, Y., Deng, P., Li, J., Lin, M., Zhu, F., Ng, T. -W., Lee, C. -S. and Ong, B. S. (2016). Solution-Processed Donor-acceptor polymer nanowire network semiconductors for high-performance field-effect transistors. Scientific reports, 6.

Louwet, F., Groenendaal, L., Dhaen, J., Manca, J., Van Luppen, J., Verdonck, E. and Leenders, L. (2003). PEDOT/PSS: synthesis, characterization, properties and applications. Synthetic Metals, 135: 115–117.

Lu, Y. (1988). Solitons and polarons in conducting polymers. 1988: World Scientific.

Lu, Y., Shi, G., Li, C. and Liang, Y. (1998). Thin polypyrrole films prepared by chemical oxidative polymerization. Journal of Applied Polymer Science, 70: 2169–2172.

Luo, K., Shi, N. and Sun, C. (2006). Thermal transition of electrochemically synthesized polyaniline. Polymer Degradation and Stability, 91: 2660.

MacDiarmid, A. G., Chiang, J. C., Halpern, M., Huang, W. S., Lin, Mu, S., Somasiri, N. L. D., Wu, W., Stuart, I. and Yaniger, S. I. (1985). "Polyaniline": interconversion of metallic and insulating forms. Molecular Crystals and Liquid Crystals, 121(1-4): 173–180.

MacDiarmid, A. G., Somasiri, N. L. D., Salaneck, W. R., Lundstrsrn, I., Liedberg, B., Hasan, M. A., Erlandsson, R. and Konrasson, P. (1985). Investigation of the electronic structure of conducting polymers by electron energy-loss spectroscopy. Springer Series in Solid State Sciences, 63: 173–178.

MacDiarmid, A. G., Yang, L., Huang, W. and Humphrey, B. (1987). Polyaniline: Electrochemistry and application to rechargeable batteries. Synthetic Metals, 18(1): 393–398.

MacDiarmid, A. G. (2001). Synthetic metals: a novel role for organic polymers. Synthetic Metals, 125(1): 11–22.

MacDiarmid, A. G. (2001). Synthetic metals: A novel role for organic polymers (Nobel lecture). Angewandte Chemie International Edition, 40(14): 2581–2590.

Machida, S., Miyata, S. and Techagumpuch, A. (1989). Chemical synthesis of highly conducting polypyrrole. Synthetic Metals, 31: 311–318.

Marianovic, G. C., Dondur, V., Miloievic, M., Moiovic, M., Mentus, S., Radulovic, A., Vukovic, Z. and Steiskal, J. (2009). Synthesis and characterization of conducting self-assembled polyaniline nanotubes/zeolite nanocomposite. Langmuir, 25: 3122–3131.

Martina, V., Riccardis, F. D., Carbone, D. and Mele, C. (2011). Electrodeposition of polyaniline–carbon nanotubes composite films and investigation on their role in corrosion protection of austenitic stainless steel by SNIFTIR analysis. Journal of Nanoparticle Research, 13(11): 6035–6047.

Martyak, N. M., MacAndrew, P., EMcCaskie, J. and Dijon, J. (2002). Electrochemical polymerization of aniline from an oxalic acid medium. Progress in Organic Coatings, 45: 23–32.

Martyak, N. M., MacAndrew, P., EMcCaskie, J. and Dijon, J. (2002). Corrosion of polyaniline-coated steel in high pH electrolytes. Science and Technology of Advanced Materials, 3: 345–352.

Mastragostino, M., Arbizzani, C. and Soavi, F. (2002). Conducting polymers as electrode materials in supercapacitors. Solid State Ionics, 148(3): 493–498.

McCullough, R. D. and Williams, S. P. (1993). Toward tuning electrical and optical properties in conjugated polymers using side-chains: highly conductive head-to-tail, heteroatom functionalized polythiophenes. Journal of the American Chemical Society, 115(24): 11608–11609.

Menke, S. M., Friend, R. H. and Credgington, D. (2016). polymer light emitting diodes, in the wspc reference on organic electronics: organic semiconductors: Fundamental aspects of materials and applications. World Scientific, 243–276.

Michalik, A. and Rohwerder, M. (2005). Conducting polymers for corrosion protection: a critical view. Zeitschrift für Physikalische Chemie, 219: 1547–1559.

Miasik, J. J., Hooper, A. and Tofield, B. C. (1986). Conducting polymer gas sensors. Journal of the Chemical Society, Faraday Transactions 1: Physical Chemistry in Condensed Phases, 82(4): 1117–1126.

Moraes, S. R., Vilca, D. H., Motheo, A. J. (2003). Corrosion protection of stainless steel by polyaniline electrosynthesized from phosphate buffer solutions. Progress in Organic Coatings, 48: 28–33.

Mortimer, R. J., Dyer, A. L. and Reynolds, J. R. (2006). Electrochromic organic and polymeric materials for display applications. Displays, 27(1): 2–18.

Nardes, A. M., Kemerink, M., De Kok, M., Vinken, E., Maturova, K. and Janssen, R. (2008). Conductivity, work function, and environmental stability of PEDOT: PSS thin films treated with sorbitol. Organic Electronics, 9(5): 727–734.

Narayanasamy, B. and Rajendran, S. (2010). Electropolymerized bilayer coatings of polyaniline and poly(N-methylaniline) on mild steel and their corrosion protection performance. Progress in Organic Coatings, 67: 246–254.

Novák, P., Müller, K., Santhanam, K. and Haas, O. (1997). Electrochemically active polymers for rechargeable batteries. Chemical Reviews, 97(1): 207–282.

Nowak, M., Rughooputh, S., Hotta, S. and Heeger, A. (1987). Polarons and bipolarons on a conducting polymer in solution. Macromolecules, 20(5): 965–968.

Nur Ain, A. R., Mohd. Sabri, M. G., Wan Rafizah, W. A., Nurul Azimah, M. A. and Wan Nik. (2018). Effect of ZnO-PEDOT:PSS incorporation in epoxy based coating on corrosion behaviour in immersed condition, ASM Sci. J. Special Issue (1) AiMS2018, 56–67.

Olad, A. and Naseri, B. (2010). Preparation, characterization and anticorrosive properties of a novel polyaniline/clinoptilolite nanocomposite. Progress in Organic Coatings, 67: 233–238.

Otero, T. F. and Sansieña, J. M. (1998). Soft and wet conducting polymers for artificial muscles. Advanced Materials, 10(6): 491–494.

Otero, T. and Sansinena, J. (1995). Artificial muscles based on conducting polymers. Bioelectrochemistry and Bioenergetics, 38(2): 411–414.

Ogurtsov, N. A. and Shapoval, G. S. (2006). Protective properties of electrochemical polyaniline coatings on low-carbon steel. Russian Journal of Applied Chemistry, 79: 605–609.

Ozyilmaz, A. T., Ozyilmaz, G. and Yigitoglu, O. (2010). Synthesis and characterization of poly(aniline) and poly(o-anisidine) films in sulphamic acid solution and their anti-corrosion properties. Progress in Organic Coatings, 67: 28–37.

Patil, R. C. and Radhakrishnan, S. (2006). Conducting polymer-based hybrid nano-composites for enhanced corrosion protective coatings. Progress in Organic Coatings, 57: 332–336.

Park, J. W., Na, W. and Jang, J. (2016). One-pot synthesis of multidimensional conducting polymer nanotubes for superior performance field-effect transistor-type carcinoembryonic antigen biosensors. RSC Advances, 6(17): 14335–14343.

Park, K. S., Schougaard, S. B. and Goodenough, J. B. (2007). Conducting-polymer/iron-redox-couple composite cathodes for lithium secondary batteries. Advanced Materials, 19(6): 848–851.

Prasannan, A., Somanathan, N., Hong, P. -D. and Chuang, W. -T. (2009). Studies on polyaniline–polypyrrole copolymer micro emulsions. Materials Chemistry and Physics, 116: 406–414.

Popovic, M. M. and Grgur, B. N. (2005). Corrosion studies on electrochemically deposited PANI and PANI/epoxy coatings on mild steel in acid sulfate solution. Progress in Organic Coatings, 52(4): 359–365.

Pud, A., Ogurtsov, N., Korzhenko, A. and Shapoval, G. (2003). Some aspects of preparation methods and properties of polyaniline blends and composites with organic polymers. Progress in Polymer Science, 28(12): 1701–1753.

Radhakrishnan, S., Siju, C. R., Mahanta, D., Patil, S. and Madras, G. (2009). Conducting polyaniline–nano-TiO$_2$ composites for smart corrosion resistant coatings. Electrochimica Acta, 54: 1249–1254.

Rajagopalan, R. and Iroh, J. O. (2002). Electrodeposition of adherent polyaniline-polypyrrole composite coatings on low carbon steel. The Journal of Adhesion, 78: 835–860.

Raotole, P. R., Gaikwad, A. B. and Patil, P. P. (2006). Electrochemical synthesis of corrosion protective polyaniline coatings on mild steel from aqueous salicylate medium. Science and Technology of Advanced Materials, 7(7): 732–744.

Ren, Y. J., Chen, J. and Zeng, C. L. (2010). Corrosion protection of type 304 stainless steel bipolar plates of proton -exchange membrane fuel cells by doped polyaniline coating. Journal of Power Sources, 195: 1914–1919.

Ruhi, G., Bhandari, H. and Dhawan, S. K. (2015). Corrosion resistant polypyrrole/flyash composite coatings designed for mild steel substrate. American Journal of Polymer Science, 5(1A): 18–27.

Sadki, S., Schottland, P., Brodie, N. and Sabourand, G. (2000). The mechanism of pyrrole electrochemical polymerization. Chemical Society Reviews, 29: 283–293.

Sathiyanarayanan, S., Muralidharan, S., Venkatachari, G. and Raghavan, M. (2004). Corrosion protection by electropolymerised and polymer pigmented coatings—a review. Corrosion Reviews, 22: 157.

Sathiyanarayanan, S., Muthukrishnan, S., Venkatachari, G. and Trivedi, D. C. (2005). Corrosion protection of steel by polyaniline (PANI) pigmented paint coating. Progress in Organic Coatings, 53(4): 297–301.

Saini, P., Sharma, R. and Akodia, S. (2016). Graphene oxide and reduced graphene oxide coated polyamide fabrics for antistatic and electrostatic charge dissipation. World Journal of Textile Engineering and Technology, 2: 1–5.

Salunkhe, P. B. and Rane, S. P. (2016). Treatise on conducting polymers for corrosion protection—Advanced approach, Paint India, 61.

Samui, A. B., Patankar, A. S., Rangarajan, J. and Deb, P. C. (2003). Study of polyaniline containing paint for corrosion prevention. Progress in Organic Coatings, 47: 1–7.

Samui, A. B. and Phadnis, S. M. (2005). Polyaniline–dioctyl phosphate salt for corrosion protection of iron. Progress in Organic Coatings, 54(3): 263–267.

Sazou, D. and Georgolios, C. (1997). Formation of conducting polyaniline coatings on iron surfaces by electropolymerization of aniline in aqueous solutions. Journal of Electroanalytical Chemistry, 429: 81–93.

Shinde, V., Chaudhari, S., Patil, P. and Sainkar, R. (2003). Electrochemical polymerization of 2,5-dimethylanilineon low carbon steel. Materials Chemistry and Physics, 20 December 2003, 82(3): 622–630.

Shirakawa, H., Louis, E. J., MacDiarmid, A. G., Chiang, C. K. and Heeger, A. J. (1977). Synthesis of electrically conducting organic polymers: halogen derivatives of polyacetylene, (CH) x. Journal of the Chemical Society, Chemical Communications, (16): 578–580.

Shirakawa, H., Ito, T. and Ikeda, S. (1978). Electrical properties of polyacetylene with various cis-trans compositions. Die Makromolekulare Chemie, 179(6): 1565–1573.

Shirakawa, H. (2001). The discovery of polyacetylene film: the dawning of an era of conducting polymers (Nobel lecture). Angewandte Chemie International Edition, 40(14): 2574–2580.

Shirakawa, H., McDiarmid, A. and Heeger, A. (2003). Twenty-five years of conducting polymers. Chemical Communications, 2003(1): 1–4.

Shukla, S. K., Quraishi, M. A. and Prakash, R. (2008). A self-doped conducting polymer "polyanthranilic acid": An efficient corrosion inhibitor for mild steel in acidic solution. Corrosion Science, 50: 2867–2872.

Skotheim, T. A. and Reynolds, J. (2007). Handbook of Conducting Polymers, 2 Volume Set. CRC press.

Snook, G. A., Kao, P. and Best, A. S. (2011). Conducting-polymer-based supercapacitor devices and electrodes. Journal of Power Sources, 196(1): 1–12.

Somani, P. R. and Radhakrishnan, S. (2003). Electrochromic materials and devices: present and future. Materials Chemistry and Physics, 77(1): 117–133.

Song, M. -K., Kim, Y. -T., Kim, B. -S., Kim, J., Char, K. and Rhee, H. -W. (2004). Synthesis and characterization of soluble polypyrrole doped with alkylbenzenesulfonic acids. Synthetic Metals, 141(3): 315–319.

Sotzing, G. A. and Lee, K. (2002). Poly (thieno [3, 4-b] thiophene): A p- and n-dopable polythiophene exhibiting high optical transparency in the semiconducting state. Macromolecules, 35(19): 7281–7286.

Tallman, D., Pae, Y. and Bierwagen, G. (1999). Conducting polymers and corrosion: polyaniline on steel. Corrosion, 55(8): 779–786.

Tamburri, E., Orlanducci, S., Toschi, M. L. Terranova and Passeri, D. (2009). Growth mechanisms, morphology, and electroactivity of PEDOT layers produced by electrochemical routes in aqueous medium. Synthetic Metals, 159(5): 406–414.

Tung, T. T., Kim, T. Y. and Suh, K. S. (2011). Nanocomposites of single-walled carbon nanotubes and poly (3, 4-ethylenedioxythiophene) for transparent and conductive film. Organic Electronics, 12(1): 22–28.

Vera, R., Schrebler, R. and Cury, P. (2007). Corrosion protection of carbon steel and copper by polyaniline and poly(ortho-methoxyaniline) films in sodium chloride medium. Electrochemical and morphological study. Journal of Applied Electrochemistry, 37: 519–525.

Wang, D. -W., Li, F., Zhao, J., Ren, W., Chen, Z.-G., Tan, J., Wu, Z. -S., Gentle, I., Lu, G. Q. and Cheng, H. -M. (2009). Fabrication of graphene/polyaniline composite paper via *in situ* anodic electropolymerization for high-performance flexible electrode. ACS Nano, 3(7): 1745–1752.

Wang, T., He, Sun, D., Guo, Y., Ma, Y., Hu, Y., Li, G., Xue, H., Jing, T. and Sun, X. (2011). Synthesis of mesoporous carbon-silica-polyaniline and nitrogen-containing carbon-silica films and their corrosion behavior in simulated proton exchange membrane fuel cells environment. Journal of Power Sources, 196: 9552–9560.

Wang J. G., Neoh, K. G. and Kang, E. T. (2004). Comparative study of chemically synthesized and plasma polymerized pyrrole and thiophene thin films. Thin Solid Films, 446: 205–217.

Wang, J., Yang, J., Xie, J. and Xu, N. (2002). A novel conductive polymer–sulfur composite cathode material for rechargeable lithium batteries. Advanced Materials, 14(13-14): 963–965.

Wang, X., Zhi, L. and Müllen, K. (2008). Transparent, conductive graphene electrodes for dye-sensitized solar cells. Nano letters, 8(1): 323–327.

Wang, Y. and Jing, X. (2005). Intrinsically conducting polymers for electromagnetic interference shielding. Polymers for Advanced Technologies, 16(4): 344–351.

Wrobleski, D., Benicewicz, B., Thompson, K. and Bryan, C. (1993). Corrosion resistant coatings from conducting polymers. Los Alamos National Lab., NM (United States).

Wakizaka, D., Fushimi, T., Ohkita, H. and Ito, S. (2004). Hole transport in conducting ultrathin films of PEDOT/PSS prepared by layer-by-layer deposition technique. Polymer, 45(25): 8561–8565.

Wudl, F., Angus, R. O., Lu, F. L., Allemand, P. M., Vachon, D., Nowak, M., Liu, Z. X., Schaffer, H. and Heeger, A. J. (1987). Poly-p-phenyleneamineimine: synthesis and comparison to polyaniline. Journal of the American Chemical Society, 10912: 3677–3684.

Yagan, A., Pekmez, N. O. and Yidiz, A. (2007). Inhibition of corrosion of mild steel by homopolymer and bilayer coatings of polyaniline and polypyrrole. Progress in Organic Coatings, 59: 297–303.

Yang, X., Li, B., Wang, H. and Hou, B. (2010). Anticorrosion performance of polyaniline nanostructures on mild Steel. Progress in Organic Coatings, 69: 267–271.

Ye, S., Besner, S., Dao, LeH and Vijh, A. K. (1995). Electrochemistry of poly(aniline-*co*-*N*-butylaniline) copolymer: Comparison with polyaniline and poly(*N*-butylaniline). Journal of Electroanalytical Chemistry, 381: 71–80.

Yeh, J. M., Liou, S. J., Lai, C. Y., Wu, P. C. and Tsai, T. Y. (2001). Enhancement of corrosion protection effect in polyaniline via the formation of polyaniline–clay nanocomposite materials. Chemistry of Materials, 13: 1131–1136.

Zhang, R., Barnes, A., Ford, K. L., Chambers, B. and Wright, P. V. (2003). A new microwave 'smart window' based on a poly (3, 4-ethylenedioxythiophene) composite. Journal of Materials Chemistry, 13(1): 16–20.

Zhang, W., Zhao, B., He, Z., Zhao, X., Wang, H., Yang, S., Wu, H. and Cao, Y. (2013). High-efficiency ITO-free polymer solar cells using highly conductive PEDOT:PSS/surfactant bilayer transparent anodes. Energy and Environmental Science, 6(6): 1956–1964.

Zhang, J., Vlachopoulos, N., Hao, Y., Holcombe, T. W., Boschloo, G., Johansson, E. M., Grätzel, M. and Hagfeldt, A. (2016). Efficient blue-colored solid-state dye-sensitized solar cells: enhanced charge collection by using an *in situ* photoelectrochemically generated conducting polymer hole conductor. ChemPhysChem, 17(10): 1441–1445.

Zhu, H., Hu, J., Zhong, L. and Gana, F. (2008). Cathodic polarization behavior of self-supporting polyaniline film Electrode. Russian Journal of Electrochemistry, 44(10): 1120–1126.

Zotti, G., Comisso, N., D'Aprano, G. and Leclerc, M. (1992). Electrochemical deposition and characterization of poly (2, 5-dimethoxyaniline): A new highly conducting polyaniline with enhanced solubility, stability and electrochromic properties. Advanced Materials, 4(11): 749–752.

3

Poly(Aniline-co-Pentafluoroaniline)/ SiO$_2$ Composite Based Anticorrosive Coating

3.1 Introduction

Conducting polymers such as polypyrrole, polyaniline and their copolymers have attracted consideration for their potential in corrosion protection of metals in aggressive environments; due to their low cost, ease of synthesis, nontoxic property, chemical and thermal stability (Bernard et al. 2001; Ates 2016; Ruhi et al. 2015; Sambyal et al. 2015; Kumar et al. 2013; Shazly et al. 2012; Chang et al. 2012).

The interest of researchers in polyaniline could possibly be linked to the numerous applications that exist for electronic conducting polymers and, on the other hand, to the fact that aniline is a cheap product and also a very stable material. Industrialists are also getting interested and the leading world chemical industries have research programmes based on PANI. The concern is related to its polymerization mechanism, the structure of the material in its oxidation states, the redox mechanisms, the electronic and ionic conduction mechanisms, the role of the doping ions, protons, etc. Still, researchers have not completely understood all the observed phenomena and the existing models need to be elaborated to explain the experimental results. Polyaniline (PANI) ranks highest among electrically conducting polymers. Its high conductivity and chemical variability make it suitable for a number of applications like sensors, ESD, EMI shielding and corrosion inhibition. In the course of polymerization, PANI has the ability to create thin conducting films with very good adhesion on various base materials. In principle, polyaniline and its analogs can be correlated to a class of polymers, which can be described by the formula presented in Figure 3.1.

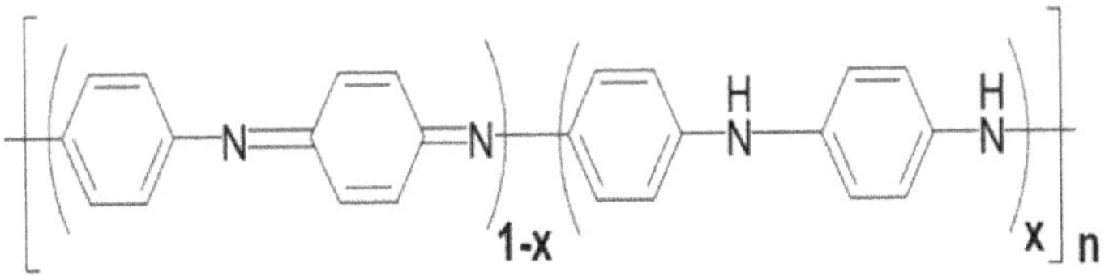

Figure 3.1: General formula for emeraldine base

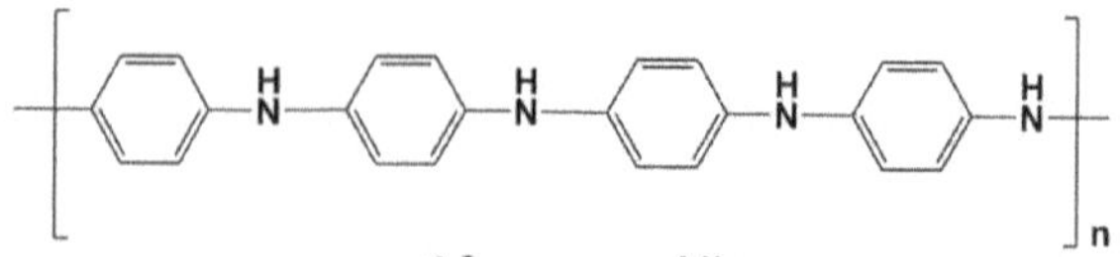

a) Leucoemeraldine

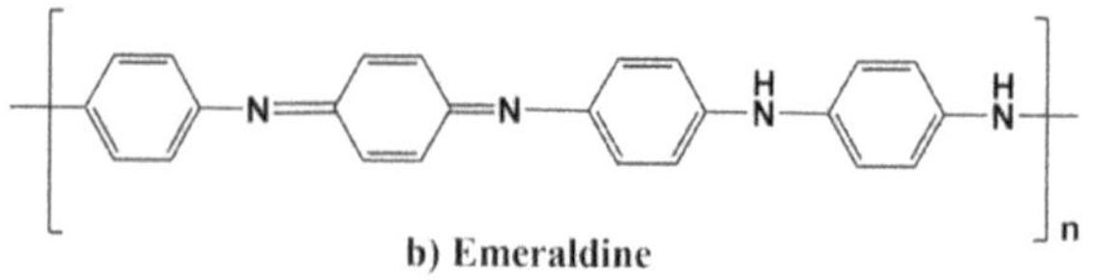

b) Emeraldine

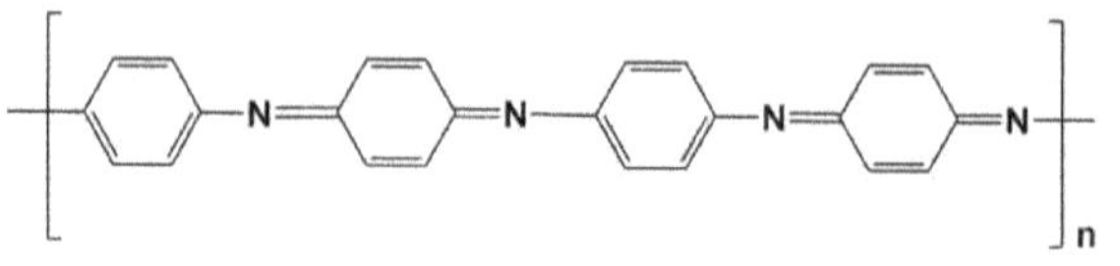

c) Pernigraniline

Figure 3.2: Three main forms of polyaniline

Polyaniline exists in three principal forms as shown in Figure 3.2. All these forms exhibit interesting electrochemical and typical spectroscopic behavior. Fully reduced form is leucoemeraldine (yellowish-white) (a) whereas emeraldine (blue powder) (b) is a partially oxidized form which consists of partially reduced and partially oxidized moieties. Pernigraniline (c) (red-purple, partially crystalline powder) composed of oxidized units, easily undergoes hydrolytic type degradation via chain scission. Emeraldine base (dark-blue powder with metallic gloss)—a semi-oxidized form of PANI—is stable in air and can be stored for a long time without chemical changes. Emeraldine is the most extensively studied form of polyaniline. In comparison to other conducting polymers, in the case of polyaniline, the conjugation is not only due to π-electrons of aromatic rings interaction but also due to the lone pair of electrons present in nitrogen atoms. Moreover, in emeraldine base, comparatively strong interactions exist between amine and imine groups in neighbor chains via hydrogen bonds, which makes it a better conjugated system having unique properties.

Partial dissolution of PANI base is possible only when solvent-polymer hydrogen bond interactions replace interchain interactions. Emeraldine base is soluble in solvents like NMP (N-methyl pyrrolidinone), DMSO (di-methyl sulphoxide), DMF (dimethyl formamide) and DMA (N,N-dimethylacetamide) to a certain extent. Gelation process is sometimes observed when a higher concentration of emeraldine base is taken in an organic solvent like NMP. To inhibit this process, a mixture of solvents can be applied (combination of electron donors and acceptors that interact with amine and imine groups respectively). Conjugation as well as the presence of hydrogen bonds cause not only insolubility in a majority of common solvents but are also the reason why PANI is infusible. At high temperatures (above 400°C) a polymer gradually decomposes without melting. Polyaniline, in its base form, exhibits the electrical conductivity lower than 10^{-7} S/cm. The poor conductivity is limited by the band gap between HOMO and LUMO levels, i.e., 3.8 eV. However, emeraldine base can be converted to a doped conducting polyaniline form by a simple protonation process which increases its conductivity to 1–100 S/cm.

3.2 Mechanism of oxidative polymerization of aniline

Mechanism of oxidative polymerization is investigated by many authors independently on the proposed mechanism. It is believed that the polymerization is initiated by the oxidation of the monomer to a radical cation as given in Figure 3.3 (which is stabilized by resonance).

The electrophilic substitution of the formed radical cation to a neutral molecule of aniline leads to chain propagation leading to dimerization of two radical cations (Figure 3.4).

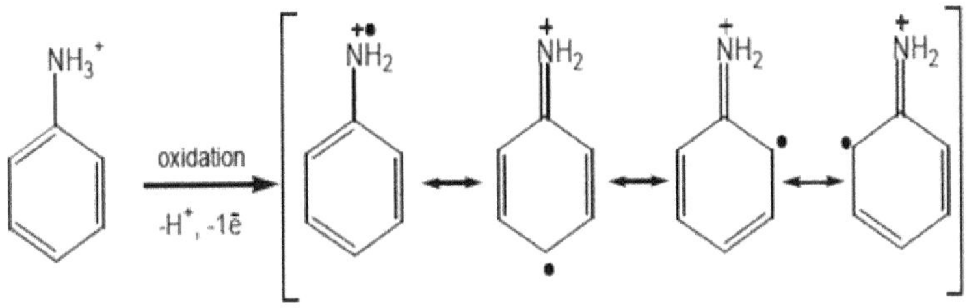

Figure 3.3: Formation of anilinium radical cation

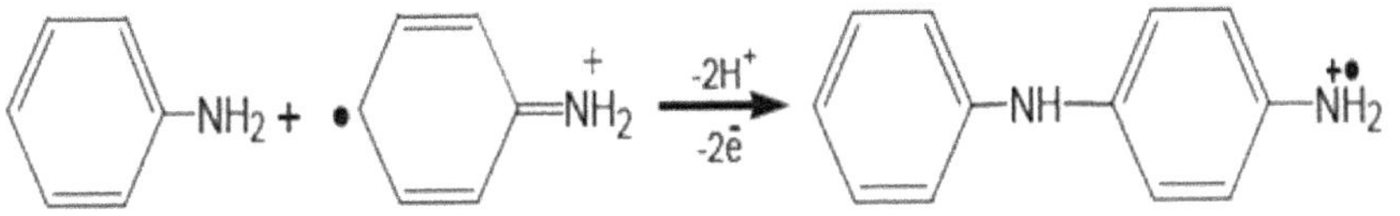

Figure 3.4: The coupling of two radical cations

In the propagation step, the dimer oxidizes to a radical cation and then it can couple with either the radical cation formed by the oxidation of the monomer (Figure 3.5) or with another dimer-type radical cation.

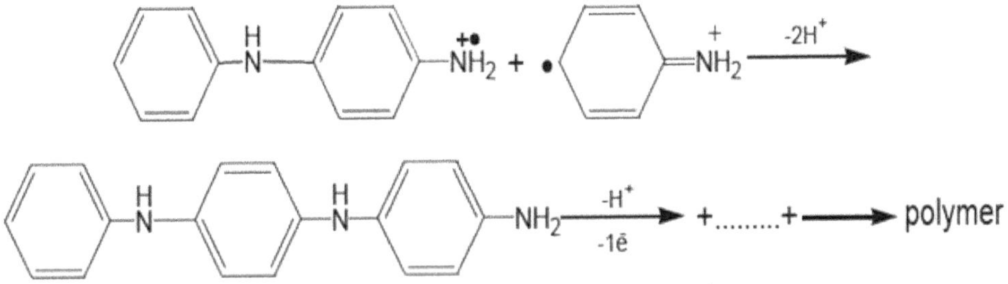

Figure 3.5: Propagation of the polyaniline chain

Emeraldine base form of the polyaniline can be converted into emeraldine salt form by simply treating it with protonic acid as shown in Figure 3.6.

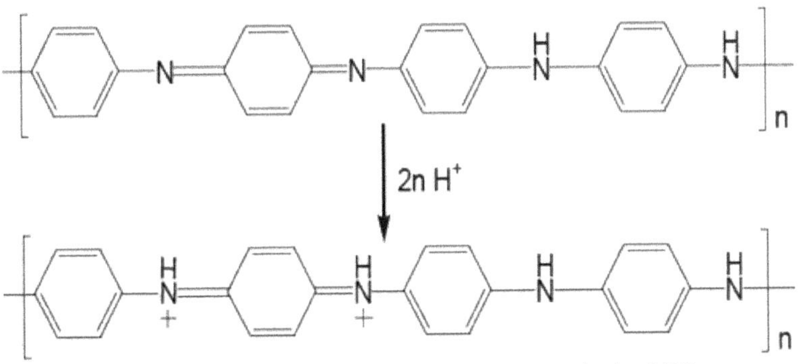

Figure 3.6: Protonation of the emeraldine base to conducting PANI

Polyaniline is highly conducting and can be readily converted between its various oxidation states. These oxidation states differ from each other by the number of quinoid rings, which range from zero to two in the elementary unit of four rings, with the other rings being benzenoid. The interconversion between the emeraldine base, with three benzenoid rings, and the leucoemeraldine

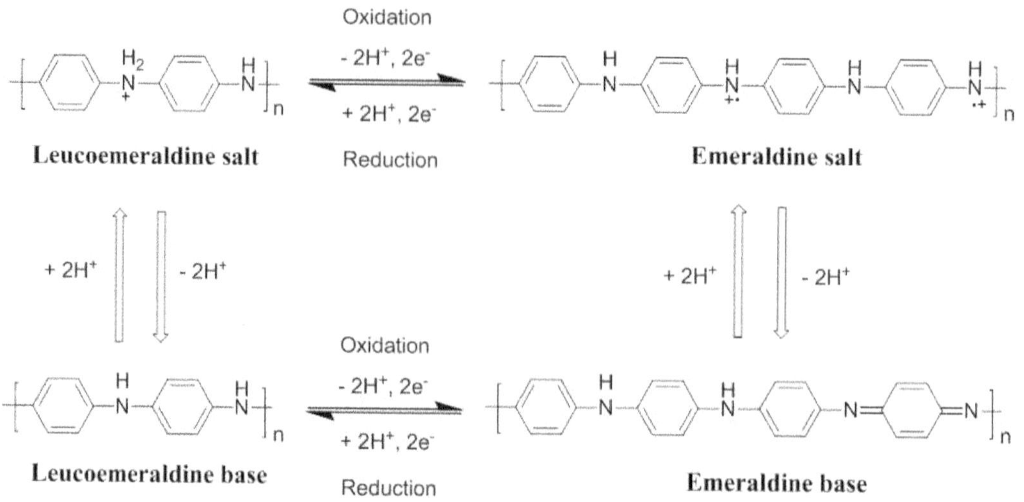

Figure 3.7: Scheme showing the interconversion of polyaniline

base (LB), with four benzenoid rings; and the interconversion between the base and conducting salt can be seen in Figure 3.7. In all cases, electroneutrality of the polymer is maintained by the presence of counter anions.

There have been several reports of corrosion studies of polyaniline on iron or mild steel. In the past decade, the use of polyaniline as anti-corrosive coatings had been extremely explored as the potential candidate to replace materials containing toxic elements such as chromium (Iroh 2002; Shah and Iroh 2004; Saidman 2002). Conducting polymers as either film forming corrosion inhibitors or as protective coating have attracted good attention due to their brilliant anti-corrosion ability and their environment friendly properties (Kamaraj et al. 2012). Polyaniline based coatings are found to be highly protective for mild steel surface against corrosion (Iribarren et al. 2005; Baldissera et al. 2012; Sivaraman et al. 2006). It has already been shown that conducting polymers provide better protection to metals as compared to insulating polymers such as polystyrene and epoxy in a corrosive environment (Wei et al. 1995). A lot of studies have been reported in literature on the application of homopolymer polyaniline for the corrosion prevention of mild steels (Mobin and Tanveer 2012; Santos et al. 1998). Even though homopolymers of aniline as corrosion protection coatings have claimed success, there are still a lot of problems associated with these materials which prohibit them as a replacement for conventional coating systems. The major drawbacks related to the conducting homopolymers have been a concern in processing these materials and the limited number of available coatings monomer. In addition, insolubility and infusibility of these materials make the deposition of the coating on active metals difficult. The direct electrochemical deposition of conducting polymers can be carried out in order to develop the coating on the metal surface. However, this method is not feasible with active metals that oxidize at the deposition potential. Further, the numbers of conjugated π-bond coatings monomer that are important for electrical conductivity are also limited and found to be a major drawback. Synthesis of copolymers containing different monomer molecules has long been employed to modify the physical and chemical properties of polymer coatings. Conducting copolymers (Bhandari et al. 2011; Bhandari et al. 2010; Mahulikar et al. 2011; Fang et al. 2007) have extensively been used for protection of metals in aggressive environment, and their coatings on the metal surface are found to be effective to control the rate of corrosion. The incorporation of co-monomers with hydrophobic groups could lower the rate of water uptake or another group may improve the stability and adherence properties (Rawat et al. 2015). These facts motivated the studies subjecting to the development of new design of polymeric materials using copolymers with desired properties. The coatings of these materials have shown better corrosion performance than individual homopolymers coatings. Tanveer and Mobin (2012) and Yalcinkaya et al. (2010) reported that anti-corrosive coating, based on copolymer of aniline and substituted aniline was found to be better than the coating based on homopolymer of aniline in different corrosive media such as 0.1 M HCl, 5.0% NaCl solution, seawater, distilled water and open atmosphere. With respect to polyaniline based copolymers, on the basis of various literature, it is found that the copolymerization of aniline with substituted aniline provides a suitable synthetic method to prepare new conducting materials with desired properties (Tanveer and Mobin 2014; Sato et al. 1994; Yang et al. 2010). Yao et al. (2009) had synthesized aniline-p-phenyldiamine copolymer, and reported its corrosion protection performance using electrochemical measurements.

It was found that the polyaniline based copolymer coating protect the metals from corrosion by three mechanisms functioning simultaneously, i.e., via improvement of barrier properties, redox property of polyaniline and generation of p-n junctions that prevent easy charge transport when the coating is damaged by scratch. Although, by copolymerization some desired properties can be achieved in coating, however, there are still few drawbacks of the copolymer such as poor mechanical properties, poor adherence on the metal surface and porosity (Wessling 1994; Song and Choi 2013; Cardoso et al. 2007). Subsequently, protective properties of the copolymer coating may be lost on prolonged exposure to the corrosive medium. In addition, porosity and anion exchange behavior of copolymers could be unfavorable, particularly when pitting corrosion occurs due to small corrosive anions (e.g., chlorides). In order to increase the efficiency of the conductive polymer as an effective and efficient protective coating on metals, several strategies have been used such as incorporation of inorganic fillers in conducting polymer matrix. Composites based on conducting polymer-inorganic fillers have also attracted more attention. Different metals and metal oxide particles have so far been incorporated into the conducting polymer matrix to produce a host of composite materials. These composite materials have exposed better mechanical, physical and chemical properties due to combining the properties of conducting polymers and inorganic particles (Chen et al. 2013; Merisalu et al. 2015; Mert 2016; Gangopadhyay and De 2000; Sengouo and Deshmukh 2015; Niedbala 2011; Wang et al. 2012; Chen et al. 2012). The purpose of development of polymer-inorganic composite based materials was to achieve the best properties of each component in polymer composites, and eliminate their drawbacks; getting in an ideal way a synergic effect that results in the development of new materials with new properties. Recently, many researchers have tried to improve corrosion protection properties of polyaniline by developing its composites with inorganic fillers such as glass flake (Sathiyanarayanan et al. 2008), metal oxide (Chen et al. 2010), zeolite minerals (Oladand Naseri 2010) and Fe$_2$O$_3$ (Nooshabadi et al. 2015), etc. Recently, Jadhav and Gelling (2015) examined the anti-corrosive performance of conducting polymer/TiO$_2$ composites for protection of cold-rolled steel. Lenz et al. (2002) reported that nanocomposites based on polypyrrole/TiO$_2$ film displayed developed behavior as compared to the original polypyrrole films. Such performance was due to the minute porosity of the polymer through filling by TiO$_2$ particles. Shi et al. (2008) have reported the anti-corrosion performance of mild steel using organic coating containing TiO$_2$ and SiO$_2$. Radhakrishnan et al. (2009) investigated that the coating developed from polyaniline-TiO$_2$ nanocomposites presented outstanding corrosion resistance and these nanocomposites were found to be much superior to polyaniline in the corrosive environment. Kirubaharan et al. (2012) reported the anti-corrosive performance of the coating based on TiO$_2$ and SiO$_2$ nanoparticles, and their results revealed that silica nanoparticles based coating were found to be superior as compared to TiO$_2$ nanoparticles based coating. Shi et al. (2009) demonstrated the effect of nanoparticles such as SiO$_2$, Zn, Fe$_2$O$_3$ and halloysite clay on corrosion and mechanical properties of the epoxy coating. Results revealed that the SiO$_2$ nanoparticles were found to be superior in improving both the anti-corrosive and young modulus of the epoxy coating. Gonzalez et al. (2011) reported the use of silica tubes as nanocontainers for corrosion inhibitor storage. All the corrosion studies were carried out in a saline environment and the results revealed that the coating silica nanotubes were excellent corrosion inhibitors as compared to the coating without silica particles. Among various inorganic fillers, silica (SiO_2) is found to be excellent reinforcing inorganic filler for conducting polymers and have remarkable anti-corrosive properties. As the filler in polymer coatings, it improves the resistance of the organic coating toward diffusion of chloride ions and considerably reduces the corrosion process. It was found from the literature that the coatings prepared from conducting polymer-SiO_2 composites synthesized by *in situ* chemical polymerization have shown an excellent corrosion resistance, and were found to be much superior to conducting polymers in aggressive environments (Herrasti and Oc´on 2001). The novelty of these coatings lies in the generation of corrosion protection by enhancing the barrier properties and preventing easy charge transport of corrosive ions when coatings are damaged by scratches or scribes.

Corrosion protective coatings on the mild steel surface by electrochemical deposition of conducting polymers composites have been widely studied (Kilmartin et al. 2002; Meneguzzi et al. 1999; Camalet et al. 1998). Le et al. (2009) used electrochemical polymerization method to obtain the different polyaniline coatings onto 316 L stainless steel using cyclic voltammetry. On the other hand, Moraes et al. (2002, 2003) synthesized polyaniline by chemical as well as electrochemical methods. The chemically synthesized polyaniline was solubilized in N-methyl-pyrrolidone (NMP) and was applied on stainless steel. By electrochemical via a film was deposited on the steel surface by cyclic voltammetry. The efficiency of polyaniline films as anti-corrosive applications was then studied. The authors concluded that chemically prepared polyaniline was able to protect the stainless steel more efficiently than electrochemically deposited film. The drawback of electrodeposited conducting polymers coating is lack of durability when they are exposed to corrosive medium for a longer period. Also, chemical deposition methods are not found to be environmental-friendly as deposition is carried out in presence of hazardous organic solvents.

McAndrew et al. (1998) proposed that the conducting polymer film acts as a barrier coating. Regardless of the barrier properties of conducting polymer films, any barrier coating can only protect the metal if it is firmly intact with the metal substrate. Once the substrate is exposed to the environment, corrosion may take place within the damaged area because the corrosive ions can penetrate into the substrate (Beentjes et al. 2004). The detachment of the coating may be avoided if it shows sufficient adhesion to the substrate. Epoxy-based coatings are well-known and are superior in several respects to conventional paints due to their better scratch hardness and better adhesion on the metal surface. However, epoxy coatings are found to often fail on exposure to corrosive environments for a longer period. In order to improve the efficiency and adhesion property of conducting polymer-based coating on the metal surface, the use of powder coating technique has been performed. The significant benefits of using powder coating technique are to attain durable, environmentally friendly and inexpensive coating that gives excellent finish in a single coat.

Development of conducting copolymer composites to be used as corrosion protection with hydrophobic properties, greater durability and high wear resistance is expected to stir a major revolution in the world of corrosion. Hydrophobic substrates have the ability to remove the water droplets, dust particles and other contaminants from the surface. Recently, Liu et al. (2016) investigated the super-hydrophobic/icephobic coatings based on silica nanoparticles modified by self-assembled monolayers. A super-hydrophobic surface is obtained from nanocomposite materials based on silica nanoparticles and self-assembled monolayers of perfluoro-octyltriethoxysilane. The purpose of incorporating perfluoro-octyltriethoxysilane and silica nanoparticles in the composite matrix were to introduce hydrophobic properties and durability in the coating. Similarly, Dhawan et al. (2015) reported highly hydrophobic anti-corrosive coating based on polyaniline-SiO_2 composites using perfluoro-octanoic acid as a dopant. The purpose of using perfluoro-octanoic acid as a dopant was to introduce the hydrophobic character in the coating. Arturi et al. (2016) investigated hydrophobic properties and durability of fluoropolymer-TiO_2 coatings. The authors reported that fluoropolymer was responsible for introducing hydrophobicity in the coating matrix while its durability was increased due to the presence of inorganic filler such as TiO_2 in the coating matrix.

Therefore, the present chapter describes the development of highly resilient, inexpensive and environment-friendly hydrophobic epoxy coating based on conducting copolymer composites. SiO_2 particles were incorporated in conducting copolymer based on aniline and pentafluoro aniline. The synthesized poly(aniline-co-pentafluoro aniline)/SiO_2 composites were formulated with epoxy resin. Different physico-mechanical properties, as well as anti-corrosion performance of epoxy formulated copolymer-silica composites coatings, was demonstrated by using mild steel substrate in 3.5 wt.% NaCl medium.

3.2.1 Preparation of Poly(AN-co-PFA)/SiO₂ composites

Preparation of Poly(AN-co-PFA)/SiO_2 composite is carried out in two steps. The first step involves the synthesis of silica particles and the second step involves the incorporation of silica in the copolymer matrix by *in situ* chemical oxidative polymerization. There are various methods reported by many researchers to synthesize SiO_2 particles such as reverse microemulsion (Tan et al. 2011), flame synthesis (Vansant et al. 1995) and sol-gel method (Hench and West 1990; Stober et al. 1968). The major drawbacks of the reverse microemulsion approach are quite expensive and cause difficulties in removal of surfactants in the final products. Silica can also be synthesized by high temperature flame decomposition of metal-organic precursors. The major disadvantage of this method is the difficulty in controlling the particle size, morphology and phase composition. Out of all these methods, sol-gel method has been the most widely used method for synthesis of silica particles. Synthesis of silica using sol-gel method is carried out in the presence of ammonia as catalyst and ethanol as the solvent by hydrolysis of tetra-ethylorthosilicates (TEOS). The formation of SiO_2 particles is confirmed by FTIR spectroscopy and SEM analysis.

Synthesized SiO_2 particles have been incorporated into conducting copolymer matrix which is carried out by chemical *in situ* oxidative copolymerization. The procedure of *in situ* polymerization is carried out by dispersing the inorganic fillers directly in the monomer solution prior to a polymerization process. The copolymerization of freshly distilled aniline (AN) and 2,3,4,5,6 pentafluoro aniline (PFA) can be performed by chemical oxidative polymerization using ammonium peroxydisulfate (APS) as an oxidant and orthophosphoric acid as a dopant. For the preparation of poly(AN-co-PFA)/SiO_2 composite, the aqueous mixture of aniline (0.1 M), PFA (0.01 M), ortho phosphoric acid (0.2 M) and SiO_2 (20 g) is homogenized using a high speed blender for about 30–40 minutes to form an emulsion. The emulsion solution is then transferred to double-walled glass reactor under constant stirring.

The polymerization is initiated by the drop-wise addition of aqueous solution of APS (0.1 M). The polymerization is carried out at a temperature of 0–5°C for a period of 5–6 hours. Synthesized copolymer composite is isolated from the reaction mixture by filtration and washed with distilled water to remove oxidant and oligomers, and followed by drying in the vacuum oven at about 60°C. The flow chart of complete preparation of poly(AN-co-PFA)/SiO_2 composite is shown in Scheme 3.1.

3.2.2 Preparation of Poly(aniline-co-phenetidine)/SiO₂ composites

In our earlier work, poly(aniline-co-phenetidine)/SiO_2 composite was synthesized by using the chemical oxidative polymerization techniques. Both the monomers were taken in equimolar concentrations and adsorbed on SiO_2 nanoparticles. The reaction was carried out in a triple wall reactor at –2°C. The resultant slurry was mixed in an aqueous solution of orthophosphoric acid (0.2 M). 0.1 M ammonium persulfate solution was added in drop-wise manner to initiate the polymerization reaction. The obtained suspension was stirred for 4–5 hours to complete the polymerization. The synthesized copolymer composites were filtered and washed with distilled water. The final product was dried at 60°C under vacuum conditions to obtain powder copolymer composite. Schematic of the synthesis of aniline with o-phenetidine encapsulated with SiO_2 nanoparticles is shown in Scheme 3.2.

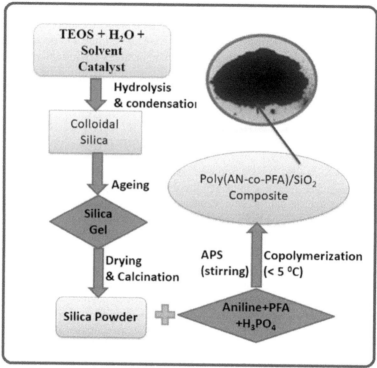

Scheme 3.1: Flow chart of preparation of poly(AN-co-PFA)/SiO$_2$ composite

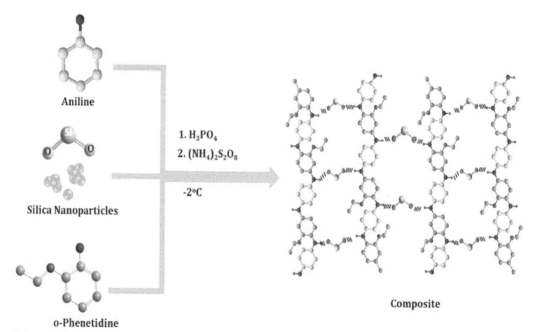

Scheme 3.2: Schematic of the synthesis of poly(aniline-co-phenetidine)/SiO$_2$ composite by chemical oxidative copolymerization process (Surface and Coatings Technology, doi.org/10.1016/j.surfcoat.2015.12.038)

3.2.3 Preparation of Poly(aniline-co-o-toluidine)/Flyash composites

In another work, copolymer composite of aniline and o-toluidine with different filler like flyash was synthesized by chemical oxidative polymerization as shown in Scheme 3.3. Monomers were freshly distilled in a vacuum and encapsulated on flyash particles with constant stirring at 60°C. A pre-cooled aqueous solution of phosphoric acid (0.2 M) at 0–5°C was prepared. Monomers encapsulated on flyash particles were added to the above aqueous solution. To initiate polymerization reaction, the aqueous solution of APS (0.1 M) was added drop-wise in the reaction mixture and stirred continuously for 4–5 hours. The generation of dark green-blue precipitates in the reaction thus confirms the synthesis of copolymer composite of aniline and o-toluidine.

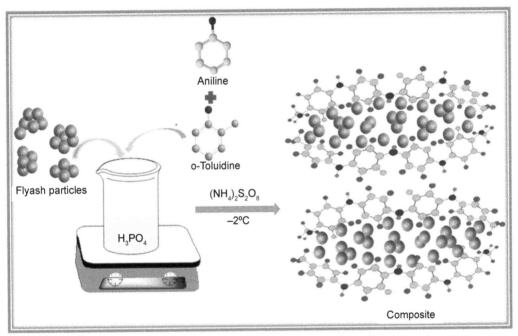

Scheme 3.3: Synthesis of poly(aniline-co-o-toluidine)/flyash composite by chemical oxidative polymerization process (Surface and Coatings Technology, doi.org/10.1016/j.surfcoat. 2015.04.013)

3.3 Development of epoxy formulated copolymer composites coating on mild steel

All the corrosion studies are performed on mild steel panels coated with polymer, copolymer and their composites with silica. Various methods have been reported in literature for development of coating (conducting polymers-inorganic fillers based composites) on the metal substrate which are mentioned in Table 3.1.

Among all the methods for development of coating on the metal substrate, powder coating is widely used by many researchers. The coating on mild steel panels was carried out by using different wt. ratio of the poly(AN-co-PFA)/SiO$_2$ in epoxy resin. Before developing the coating on mild steel panels, surface treatments followed by cleaning of panels were carried out. Cleaning was carried out by 1/0, 2/0, 3/0 and 4/0 grade emery papers, and the panels were then thoroughly cleaned using

Table 3.1: Examples of methods of corrosion protection coating and their results

Type of polymer coating and substrate	Method of coating	Results	References
Epoxy-Polyaniline	Powder coating	Coatings withstood hot saline medium for about 1,000 hours of immersion even after intentionally damaged by scratching/indentation.	Radhakrishnan et al. (2009)
Poly (aniline-co-o-toluidine) Mild steel	Solution evaporation method	96.58% corrosion protection efficiency was achieved.	Tanveer et al. (2014)
PANI+ Sulfonated chitosan (S-CTS) on steel	Chemical synthesis	Corrosion inhibitive efficiency was found to be 92.3%.	Yi et al. (2013)
Anodic Alumina with PANI+TiO_2 on Al alloy AA2024T3	Potentiostatic deposition	The coating containing TiO_2 nanoparticles protected the substrate against the corrosion better than pure PANI.	Zubillaga et al. (2009)
PANI+TiO_2 on steel	Chemical oxidative polymerization	The corrosion current of PANI+TiO_2 coatings on steel was better than the single component system (pure PANI or nano-TiO_2). The corrosion protection property of PANI+TiO_2 coatings depended on TiO_2 content and it was better when TiO_2/aniline weight ratio is 0.05.	Rathod et al. (2013)
PANI+SiO_2 on Al	Dipping	The corrosion protection ability of PANI+SiO_2 coating decreased when the TEOS (precursor of SiO_2) increased.	Yu et al. (2012)
Epoxy formulated PANI-SiO_2 doped with perfluorooctanoic acid Mild steel	Powder coating	Coating was found to be highly durable and protective in saline medium.	Dhawan et al. (2015)
Epoxy formulated poly(aniline-co-phenetidine)/SiO_2	Powder coating	Better corrosion protection and durability were shown by copolymer composites in saline medium.	Sambyal et al. (2015)
PANI-Silica Stainless steel	Cyclic voltammetry	Charge transfer resistance of nanocomposite coated steel was higher than the pure polyaniline coated and bare steel.	Amirdehi et al. (2014)
Super-hydrophobic PANI-Silica (TMS) 316SS surface	Spin coating	The long-term EIS results revealed that the high corrosion resistance of the coating was stable even after 240 h of immersion in 3.5% NaCl, and no drastic degradation was observed as compared to bare metal.	Syed et al. (2017)

trichloroethylene and acetone to remove any contamination on the metal surface. The powder polymer was mixed with the epoxy formulation in different proportions ranging from 1.0% to 4.0 wt.%. The epoxy powder coating formulation of composition: resin {epoxy (bisphenol A + polyester) (70 wt.%), flow agent (D-88) (2.3 wt.%), degassing agent (benzoin) (0.7 wt.%) and fillers (TiO_2 and $BaSO_4$) (27 wt.%) was used for the coating purpose. The different weight percentage (1.0, 2.0, 3.0 and 4.0) of the synthesized copolymer composite was blended with epoxy resin using a laboratory ball mill. The blended powder coating formulation was applied on mild steel panels using an electrostatic spray gun held at 67.4 kV potential with respect to the substrate (grounded). The powder coated mild steel panels were cured at 130–140°C for 25 minutes. Mild steel panels coated with epoxy resin were designated as EC, epoxy coatings with different wt.% loadings of poly(AN-co-PFA) were designated as PF1 (1.0 wt.%), PF4 (4.0 wt.%) and epoxy coatings with different wt.% loadings of poly(AN-co-PFA)/ SiO_2 composites were designated as PFS1 (1.0 wt.%), PFS2 (2.0 wt.%), PFS3 (3.0 wt.%) and PFS4 (4.0 wt.%). Similarly, for the poly(aniline-co-phenetidine)/SiO_2 composite and poly(aniline-co-o-toluidine)/flyash composite coatings were designated as PPS1 (1.0 wt.%), PPS2 (2.0 wt.%), PPS3 (3.0 wt.%) , PPS4 (4.0 wt.%) and OPF1 (1.0 wt.%), OPF2 (2.0 wt.%), OPF3 (3.0 wt.%) and OPF4 (4.0 wt.%) respectively.

3.4 Characterization of epoxy formulated copolymer composite coated substrate

3.4.1 FTIR spectroscopy

Interaction of two different co-monomers and incorporation of silica particles in the copolymer chain is investigated by using FTIR spectroscopy. Figures 3.8 and 3.9 show the FTIR spectra of SiO_2, flyash, poly(AN-co-PFA), poly(AN-co-PFA)/SiO_2, poly(aniline-co-phenetidine)/SiO_2 composite and poly(aniline-co-o-toluidine)/flyash composite. The FTIR spectra of SiO_2 particles shows the characteristic peaks at 1,099 cm^{-1} and 802 cm^{-1} are assigned to the stretching and bending vibration

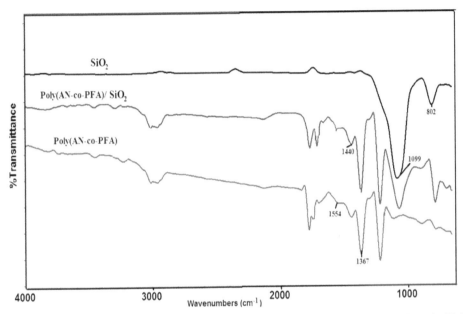

Figure 3.8: FTIR spectra of SiO_2, poly(AN-co-PFA) and poly(AN-co-PFA)/SiO_2 composites (American Journal of Polymer Science, doi:10.5923/j.ajps.20160603.03)

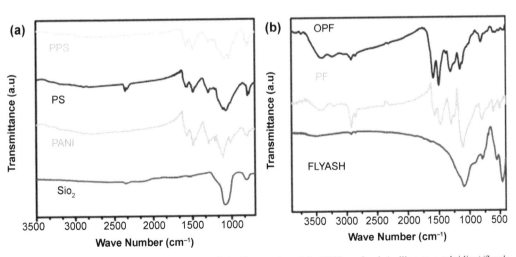

Figure 3.9: FTIR spectra of SiO_2, flyash, PANI, poly(aniline-co-phenetidine)/SiO_2 and poly(aniline-co-o-toluidine)/flyash composite (Surface and Coatings Technology, doi.org/10.1016/j.surfcoat.2015.12.038)

of Si–O–Si respectively (Chang et al. 2014). These peaks can also be observed in poly(AN-co-PFA)/SiO$_2$, this indicating the interaction of SiO$_2$ particles in copolymer chain. Mahulikar et al. (2011) reported the FTIR spectra of polyaniline which shows the different characteristic peaks such as 829 cm^{-1} due to presence of out-of-plane C-H bond, 1,153 cm^{-1} due to aromatic C-N-C bond; 1,463 cm^{-1} due to aromatic C = C double bond, and 1,605 cm^{-1} which is attributed to the nitrogen bond between benzenoid and quinonoid rings (Mahulikar et al. 2011). In case of poly(AN-co-PFA) and poly(AN-co-PFA)/SiO$_2$, main characteristics bands at 1,554 and 1,438–1,460 cm^{-1} (stretching mode of C = N and C = C), the bands at 1,155 cm^{-1} (C–N stretching mode of benzenoid ring) indicates the polymerization of aniline moiety in copolymer chain. In addition, poly(AN-co-PFA) and poly(AN-co-PFA)/SiO$_2$ shows the characteristic strong peaks at 1,128 cm^{-1} and 1,367 cm^{-1} due to C-F stretching mode (Hami et al. 2005; Silverstein and Webster 2002) indicating the interaction of co-monomers (PFA and aniline) in the copolymer chain.

Poly(aniline-co-phenetidine)/SiO$_2$ composite shows all the characteristics peak of polyaniline and presence of two peaks at 1,031 cm^{-1} and 1,294 cm^{-1} which confirms the existence of aromatic ether in the copolymer. Poly(aniline-co-o-toluidine)/flyash composite exhibits peaks at 1,100 cm^{-1} and 660 cm^{-1} that arises due to asymmetric stretching of Si–O–Si bonds and Al–O bonds of flyash. It has been observed that peaks in copolymer composites have shifted to higher wave number. This shifting is arising due to the presence of bulkier moieties (methyl, ethoxy) in the copolymer chain.

3.4.2 Thermogravimetric analysis (TGA)

Thermal stability of copolymer composites and fillers materials is investigated by thermogravimetric analysis. In this technique, the change in physio-chemical properties at elevated temperature is measured as a function of increasing temperature. Figure 3.10 shows the thermogravimetric curves of SiO$_2$, poly(AN-co-PFA) and poly(AN-co-PFA)/SiO$_2$ composites. These samples are heated from 25 to 700°C under a constant heating rate of 10°C/minutes and inert atmosphere of nitrogen gas.

Bhandari et al. (2008) reported the thermogravimetric analysis of polyaniline and its different copolymers with o-substituted aniline. Their studies revealed that the thermal stability of polyaniline and their copolymer depend on the introduction of bulky groups in comonomer moieties and size of counter ion in the copolymer chain. Introduction of bulky groups and large counter ion bring about a change in the thermal stability of the copolymer which can again be correlated to the steric hindrance brought by the introduction of the substituted group and bulky dopant in the aniline system. SiO$_2$

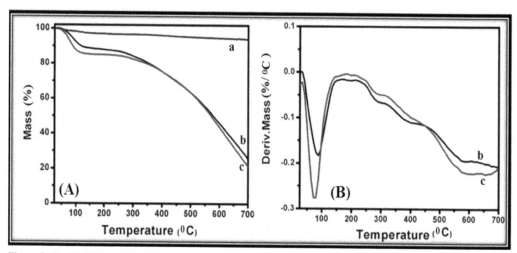

Figure 3.10: (A) TGA and (B) DTG curves of (a) SiO$_2$, (b) poly(AN-co-PFA) and (c) poly(AN-co-PFA)/SiO$_2$ composite (American Journal of Polymer Science, doi:10.5923/j.ajps.20160603.03)

particles showed excellent thermal stability up to 700°C and only 0.1% mass loss was observed in it as shown in Figure 3.10A (curve a).

TGA curve of poly(AN-co-PFA) and poly(AN-co-PFA)/SiO$_2$ composite indicates the first mass loss percentage at 100–110°C is owing to the loss of water and other volatile species. The second weight loss from about 190°C to 280°C is attributed to the loss of dopant ions from the copolymer matrix. This can be seen clearly from the DTG curves of the polymer (Figure 3.10B). Further mass loss is due to the onset of degradation of copolymer and copolymer composites. From the comparison of the TGA traces of poly(AN-co-PFA) and poly(AN-co-PFA)/SiO$_2$ composites, it is observed that the first stage of weight loss is similar whereas variation is observed in the second stage of weight loss. It is observed that the poly(AN-co-PFA) is thermally stable up to 190°C, whereas thermal stability of poly(AN-co-PFA)/SiO$_2$ composites is higher than the copolymer. It indicates that the thermal stability of the composite is enhanced due to the incorporation of SiO$_2$ particles in the copolymer matrix.

Jin et al. (2011) studied TG analysis of epoxy/SiO$_2$ nanocomposites and it was observed that the addition of SiO$_2$ nanoparticle improved the thermal stability of epoxy resin by 30%. Sambyal et al. (2015) studied the thermograms of powder coating formulations of epoxy and epoxy with different wt.% loadings of poly(aniline-co-phenetidine)/SiO$_2$ composites as shown in Figure 3.11a. Like poly(AN-co-PFA)/SiO$_2$ composites, the two steps thermal degradation process was also observed for poly(aniline-co-phenetidine)/SiO$_2$ composite due to the removal of moisture and residual solvent from the composite and thermal breakdown of copolymer chain. However, thermograms of the powder coating formulations showed a totally different pattern of thermal decomposition as compared to the copolymer composite. The neat epoxy and epoxy with different wt.% loadings of poly(aniline-co-phenetidine)/SiO$_2$ composite revealed one-step thermal decomposition process which indicated that the incorporation of copolymer composite did not change the degradation mechanism in the epoxy resin. The thermal degradation temperature of the coating was found to be enhanced on increasing different wt.% loadings of poly(aniline-co-phenetidine)/SiO$_2$ composite. Similar results were obtained for poly(aniline-co-o-toluidine)/flyash composite coatings as shown in Figure 3.11b. All the composites demonstrated one-step thermal degradation. The loading of copolymer composite in epoxy matrix enhanced the thermal stability of the coatings. The degradation temperature of the coating shifted almost 60°C higher in comparison to epoxy. Furthermore, Sambyal et al. (2015) explained that the enhanced thermal degradation was due to the presence of the filler material (SiO$_2$) present in the composite which hindered the thermal mobility of the copolymer chain by acting as physical interlocking points in the cured epoxy matrix. The improved thermal stability in conducting

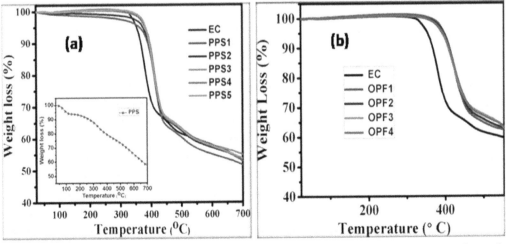

Figure 3.11: (a) TGA curves neat epoxy (EC) and epoxy with of poly(aniline-co-o-Phenetidine)/SiO$_2$ composite powder coating formulations. (b) poly(aniline-co-o-toluidine)/flyash composite coating formulations (OPF) (Surface and Coatings Technology, doi.org/10.1016/j.surfcoat.2015.12.038)

polymer composites systems relative to the neat conducting polymer had also been reported by Zhang et al. (2011) and Mavinakuli et al. (2010). In agreement with the data provided in these works of literature, the observed higher thermal of conducting polymer-inorganic fillers composites suggested the existence of an interfacial interaction between inorganic fillers and the polymer shell.

3.4.3 Micro-structural analysis

Scanning Electron Microscopy (SEM) is employed to study the surface morphology of copolymers and its composites. Figure 3.12 shows the SEM micrograph of SiO_2, poly(AN-co-PFA) and poly(AN-co-PFA)/SiO_2 respectively. SEM image of SiO_2 particles shows the spherical morphology having a smooth surface as shown in Figure 3.12a.

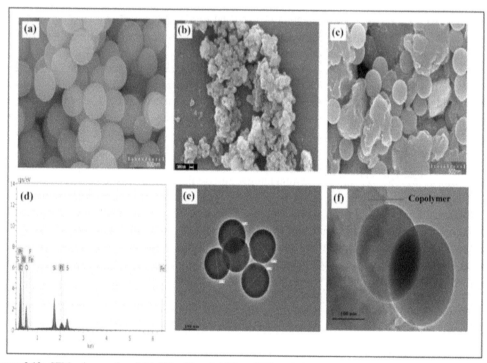

Figure 3.12: SEM micrographs of (a) SiO_2, (b) poly(AN-co-PFA) and (c) poly(AN-co-PFA)/SiO_2 composite, (d) EDS spectra of poly(AN-co-PFA)/SiO_2, (e) TEM images of (e) SiO_2 particles and (f) poly(AN-co-PFA)/SiO_2 composite (American Journal of Polymer Science, doi:10.5923/j.ajps.20160603.03)

Similar morphology of SiO_2 particles had also been reported by Sambyal et al. (2015). The authors also reported the morphology of composites based on poly(aniline-co-Phenetidine) and SiO_2 particles (Figure 3.13). It was observed that the copolymer composites showed a regular clustered morphology having embedded SiO_2 nanoparticles as shown in Figure 3.12a and Figure 3.13a.

Babazadeh et al. (2015) had carried out the fabrication of polyaniline nanocomposites based on silica nanoparticles via *in situ* chemical oxidative polymerization method. The results revealed that the silica nanoparticles showed a remarkable effect on the morphology of nanocomposites. Similar morphology of silica particles had also been reported by Gu et al. (2013).

Moreover, they reported that the polyaniline-silica nanocomposites showed rough surface as compared to the silica particles due to the polymerization that occurred on the surface of silica nanoparticles to form the core-shell structure as shown in Figure 3.14. Sambyal et al. (2015) had carried out the microstructural characterization of flyash, and poly(aniline-co-o-toluidine)/flyash composite.

Figure 3.13: SEM images of (a) SiO₂ (b) poly(aniline-co-phenitidine)/SiO₂ composite (Surface and Coatings Technology, doi.org/10.1016/j.surfcoat.2015.12.038)

Figure 3.14: SEM images of (a) silica particles and (b) polyaniline/silica doped with H₃PO₄ (Industrial and Engineering Chemistry Research, doi.org/10.1021/ie400275n)

The flyash shows spherical morphology and has diameter upto few micrometers (ranges from 50 nm to 5 μm). The micrograph of poly(aniline-co-o-toluidine)/flyash composite illustrates the uniform distribution of flyash particles in the copolymer matrix as shown in Figure 3.15, while the inset image shows the flyash particles covered with the copolymer. The electron diffraction spectrum of composite also confirms the presence of elements like carbon, oxygen, silicon, aluminum, sulfur and nitrogen in the copolymer composite. The SEM image of poly(AN-co-PFA) shows cauliflower like morphology (Figure 3.12b). Meanwhile, the SEM image of poly(AN-co-PFA)/SiO₂ composite shows that the silica particles are found to be well-dispersed in the copolymer matrix which indicates good compatibility between SiO₂ and copolymer matrix as shown in SEM (Figure 3.12c). According to literature, it is observed that morphology of conducting copolymer/metal oxide composites depends

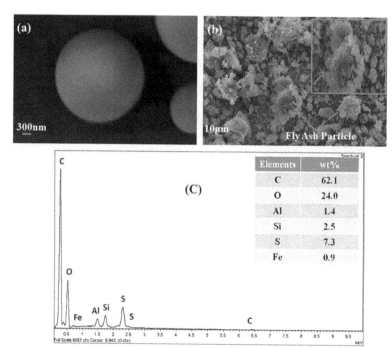

Figure 3.15: SEM micrographs of (a) flyash and (b) poly(aniline-co-o-toluidine)/flyash composite. Figure C shows the EDS spectrum of the copolymer composite (Surface and Coating Technology, doi.org/10.1016/j.surfcoat.2015.04.013)

on various factors such as nature of comonomers used and its concentration in copolymer chain, dopants, reaction conditions and nature of fillers, etc.

The TEM images of poly(aniline-co-phenetidine)/SiO_2 composite shows a uniform distribution of silica nanoparticles in the copolymer matrix as depicted by the red arrows (silica nanoparticles) and yellow arrows (copolymer matrix). The TEM images also confirm the average size of silica nanoparticles range from 50 nm to 150 nm as shown in Figure 3.16.

Transmission Electron Microscopy (TEM) image shows that the dimension of SiO_2 particles is observed in the range of 150–200 nm as shown in Figure 3.12e. The TEM image of poly(AN-co-PFA)/SiO_2 composite (Figure 3.12f) also indicates better and uniform dispersion of silica particles in the copolymer matrix. Figure 3.12d shows the Energy Dispersive X-Ray Spectroscopy (EDS) analysis

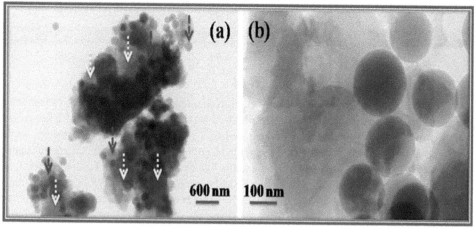

Figure 3.16: TEM images of poly(aniline-co-phenetidine)/SiO_2 composite (PPS) at low and high magnification (Surface and Coatings Technology, doi.org/10.1016/j.surfcoat. 2015.12.038)

of poly(AN-co-PFA)/SiO$_2$, which shows the different peaks of carbon, oxygen, nitrogen, fluorine, phosphorous and silicon that indicates the presence of PFA and SiO$_2$ molecules in the copolymer matrix.

3.4.4 Surface wettability test

The surface wettability test is carried out by contact angle measurement. It is a useful technique for characterizing the wettability of coated surfaces. A low contact angle value ($< 90°$) is indicative of the wetting ability of the coating. Water has been used as a test liquid to establish whether a surface is hydrophilic (angle $< 90°$) or hydrophobic (angle $> 90°$) in nature.

It was noticed that surface contact angle of epoxy coating formulated with poly(AN-co-PFA) and poly(AN-co-PFA)/SiO$_2$ composite was observed to be 95°, which revealed the formation of hydrophobic coating on the metal surface as shown in Figure 3.17, while the contact angle of the epoxy coating with water droplets was 73°. The contact angle of the coating based on epoxy formulated polyaniline and polyaniline-SiO$_2$ composite was found to be less than 90° that showed hydrophilic behavior of the coating. For the coating to be hydrophobic, the contact angle must be equal to or greater than 90°. Recently Liu et al. (2016), reported that the contact angle value in silica nanoparticles coated substrate was 13°. When the silica nanoparticles were modified with perfluorooctyltriethoxysilane, contact angle value increased to 163° which showed super-hydrophobic nature of the coating. In addition, recently Shi et al. (2017) also reported high hydrophobicity performance of the coating based on hybrid polyaniline/SiO$_2$ composites modified by the dodecyltrimethoxysilane. The contact angle value was observed to be 95° and such coating had a good self-repairing function. Xu and Langmuir (2012) developed the transparent sol-gel coatings modified with hollow, silica nanoparticle sols filled with 3-aminopropyltriethoxysilane. After the heat treatment, the coatings had shown high transparency and super hydrophobic performance. Using the same route, Shang et al. (2013) demonstrated the hydrophobic nature of silica particles functionalized with vinyl groups.

Recently Syed et al. (2017) studied the surface wettability and anti-corrosive behavior of polyaniline-silica and polyaniline-trimethylsiloxane-silica composites based coating. It was observed thatpolyaniline-silica based coating was found to be hydrophilic in nature, while trimethylsiloxane functionalized silica-polyaniline composites showed super-hydrophobic behavior and were found to be better corrosion resistant materials as compared to polyaniline-silica composites based coating. This was due to the synergistic effect of super-hydrophobicity and redox catalytic behavior of polyaniline.

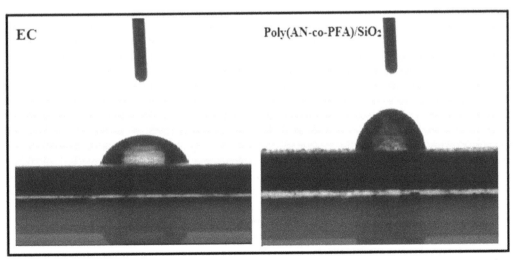

Figure 3.17: Contact angle images of (a) Epoxy coating and (b) epoxy with Poly(AN-co-PFA)/SiO$_2$ coating (American Journal of Polymer Science, doi:10.5923/j.ajps.20160603.03)

In case of the coatings based on epoxy modified with poly(AN-co-PFA)/SiO$_2$ composites, the incorporation of pentafluoro aniline as comonomer introduces the hydrophobic character in conducting copolymer chain, which is considered to be an important parameter required for effective anti-corrosive properties of coatings (Dhawan et al. 2002). Wankhede et al. (2013) and Brassard et al. (2011) studied the super-hydrophobic coating based on fluoro containing additives. Brassard et al. (2011) demonstrated the fabrication of fluorinated silica particles having an enhanced contact angle when added to the coating formulations. On the basis of literature, it was clearly revealed that water repellent properties in a coating system can be introduced by either using functionalized fillers or incorporation of fluoropolymer.

3.4.5 *Physico-mechanical testing of coating*

Shi et al. (2009) reported mechanical properties of the epoxy coating containing different nanoparticles such as SiO$_2$, Zn, Fe$_2$O$_3$ and halloysite clay. The mechanical properties were found to be improved with the incorporation of nanoparticles in the epoxy resin. Similarly, Becker et al. (2002) suggested that the nanoparticles tend to occupy the pinholes and voids in the thin-film coating, and acted as the bridges in the interconnected matrix causing a reduction of the total free volume and enhancement of the cross-linking density of the cured epoxy. Shi et al. also reported that among the various nanoparticles, SiO$_2$ nanoparticles were found to be the most effective fillers for enhancing mechanical properties of epoxy resin. Zhang et al. (2006) also confirmed that the toughness and thermal properties of epoxy resin were improved by the incorporation of silica.

The adhesion of the coating to mild steel surface can be estimated by cross-cut adhesion test. This test is carried out at a room temperature as per ASTM D3359-09 (Test Method B). In order to carry out this test, a lattice pattern with either six or eleven cuts in each direction is made on the substrate. A pressure-sensitive tape is applied over the lattice and subsequently removed, and the adhesion of coating on the metal surface is then evaluated. Baldissera and Ferreira (2012) demonstrated the adherence test of the epoxy coating based on the different forms of polyaniline. The results showed that epoxy coating based on doped as well as the undoped form of polyaniline displayed a very good adhesion on the metallic substrate while coating based on polyaniline fiber showed a low adhesion on the metal surface.

Figure 3.18A shows the results of cross-cut tape test carried out on mild steel panels coated with epoxy and epoxy with different loading of poly(AN-co-PFA) and poly(AN-co-PFA)/SiO$_2$ composites. The result revealed that the small flakes of coating were found to be detached along edges and at an intersection of cuts in epoxy and epoxy formulated poly(AN-co-PFA) coated mild steel. The area affected in the epoxy coated panel was found to be about 5% (Grade 4B), while epoxy coating with different loading of poly(AN-co-PFA)/SiO$_2$ showed completely smooth edges of cuts and none of the square of the lattice was detached (Grade 5B). Results indicated that mechanical integrity of the epoxy-based conducting copolymer was enhanced by incorporation of SiO$_2$ particles in the copolymer matrix.

Performance of the epoxy-based copolymer composite coating on the metal surface is evaluated using the Mandrel Bend Test as per ASTM D522M/D522-93a. This test is carried out by placing the coated test samples over a mandrel (size 6.25 mm) with the uncoated side in contact, and with at least 50 mm overhang on either side. Using a steady pressure of the fingers bend the panel approximately 180 degrees around the mandrel at a uniform velocity in a time 1.0 seconds. Remove and examine the panel for cracking that is visible to the unaided eye. Figure 3.18B shows the results obtained by performing the Mandrel Bend Test for epoxy coating and epoxy having copolymer composite coating. It is observed from the Figure 3.18B (a) that the slight cracks on the metal surface is noticed on the panels coated with epoxy resin, this indicates the detachment of coating from the metal surface. On the other hand, no cracks are observed in panel coated with epoxy formulated poly(AN-co-PFA) and poly(AN-co-PFA)/SiO$_2$ composites as shown in Figure 3.18B (b) and 3.18B (c). This indicates that

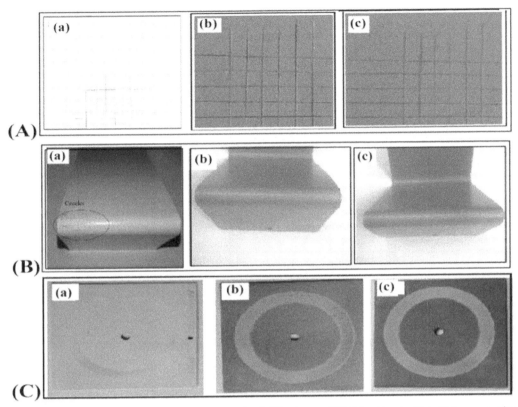

Figure 3.18: Images of (A) cross-cut adhesion test, (B) Mandrel Bend test and (C) Taber Abrasion Test of (a) epoxy coating (b) epoxy with poly(AN-co-PFA) coating and (c) epoxy with poly(AN-co-PFA)/SiO₂ coating (American Journal of Polymer Science, doi:10.5923/j.ajps.20160603.03)

the copolymers and copolymer composites in epoxy resin are able to improve the binding properties of the coating on the metal surface.

Abrasion resistance test of epoxy-copolymer coated panels can be carried out by Taber abrasion apparatus as per ASTM D 4060-95. This test is performed on the rigid panels (dimension 10 cm²) having a uniform coating. The specimens area disk of 100 mm in diameter or a plate (100 mm) square with rounded corners and with a 6.3-mm hole centrally located on each panel. The test is performed at 25°C and 50% relative humidity for 24 hours. The organic coating is applied at a uniform thickness to a plane. After curing, the surface is abraded by rotating the panel under 1.0 kg weighted abrasive CS10 wheels. A test of 1,000 cycles is done to observe the weight loss in coatings. Abrasion resistance is calculated as a loss in weight at a specified number of abrasion cycles, as a loss in weight per cycle or a number of cycles required to remove a unit amount of coating thickness.

Coating loss in epoxy coated panels was found to be 64 mg/1,000 cycle, the epoxy coating containing copolymer showed 61.9 mg/1,000 cycle, while the remarkable reduction in coating loss was observed in copolymer-silica nanocomposite based epoxy coating which was found to be 48 mg/1,000 cycle as shown in Table 3.2. Results revealed that the abrasion resistance of the coating was improved by the addition of copolymer-silica nanocomposites in epoxy resin. The coating industries have developed scratch-resistant coatings by incorporating silica nanoparticles into the organic matrix (Lawrence et al. 2002).

In addition, the scratch resistance test can also be performed with coated samples. Gläsel et al. (2000) have developed scratch and abrasion-resistant coatings based on siloxane encapsulated SiO₂ nanoparticles. Various coating industries have developed scratch-resistant coatings by incorporating silica nanoparticles into an organic matrix. By using this technique, the scratch

Table 3.2: Comparative physico-mechanical test results of epoxy and epoxy formulated coatings

Sample name	Cross-cut adhesion test	Mandrel bend test	Taber Abrasion test coating loss (in mg/1,000 cycle)
EC	Grade 4B	Slight Cracks Observed	64.0
PF1	Grade 4B	Slight Cracks Observed	63.1
PF4	Grade 4B	Slight Cracks Observed	61.9
PFS1	Grade 5B	No cracks Observed	54.6
PFS2	Grade 5B	No cracks Observed	53.2
PFS3	Grade 5B	No cracks Observed	49.8
PFS4	Grade 5B	No cracks Observed	48.0

resistance of the coating can be enhanced due to an enrichment of the nanoparticles in the coating matrix. Gu et al. (2013) reported the mechanical strength of epoxy coating based on polyaniline-silica nanocomposites. Their studies revealed that strong interaction between amine groups of polyaniline chain with functionalized silica was responsible for the enhanced mechanical strength of the coating. On the basis of literature, it was observed that the inorganic fillers play an important role in enhancing the mechanical integrity of the coating as evident from the various physico-mechanical testings of the copolymer-silica composite based coatings.

3.4.6 Corrosion studies of the coated mild steel substrate

3.4.6.1 Salt spray test

Performance of coating in NaCl medium is investigated by the salt spray test. This test is carried out on epoxy coated, epoxy formulated poly(AN-co-PFA) and poly(AN-co-PFA)/SiO$_2$ composites coated mild steel panels in 5.0 wt.% NaCl (pH of 6.5–7.2) for 120 days as per ASTM B117 method. Coated mild steel panels are prepared by coating on fully finished mild steel specimen of dimensions 15.0 cm × 10.0 cm × 0.12 cm. All the coated mild steel panels are made with a scribe mark across the panel and placed in the salt spray chamber for 120 days. Figure 3.19 demonstrates the photographs of epoxy coated (EC), epoxy formulated PF4 and copolymer composite (PFS1, PFS2, PFS3 and PFS4) coated mild steel panels after exposure to the salt spray fog of 5.0 wt.% NaCl solution. Results show that the spread of corrosion along the scribe mark for epoxy coated steel panel are observed only for 60 days of exposure to salt spray chamber.

Loss of adherence of the epoxy coating during prolonged exposure to salt fog was indicated by the appearance of rust as well as severe blistering and pinholes on the coating surface. The incorporation of copolymers (PF1 and PF4) and copolymer composites (PFS1, PFS2, PFS3 and PFS4) in epoxy coatings improved the corrosion resistance performance of the epoxy resin which could be clearly observed from the photographs (Figure 3.19) obtained after the salt spray tests.

The mild steel panels coated with epoxy resin containing 1.0 and 4.0 wt.% loading of copolymers (PF1 and PF4) showed less expansion of corrosion along the scribe. Similar behavior was observed in samples coated with PFS1 and PFS2. Further, almost no expansion of corrosion along the scribe mark was detected for the samples coated with PFS3 and PFS4 (Figure 3.21), hence these samples revealed better corrosion resistance performance with very less expansion of corrosion along the scribe mark as shown in Figure 3.19. The salt spray test results clearly indicated that the poly(AN-co-PFA)/SiO$_2$ composite efficiently constrained the spread of corrosion near scribe mark that exposed the steel surface. It was assumed that the copolymer, poly(AN-co-PFA) and its composite with SiO$_2$ particles might have supported the adhesion of the epoxy coating to the metal substrate and improved the corrosion resistance under accelerated test conditions. Additionally, the results revealed

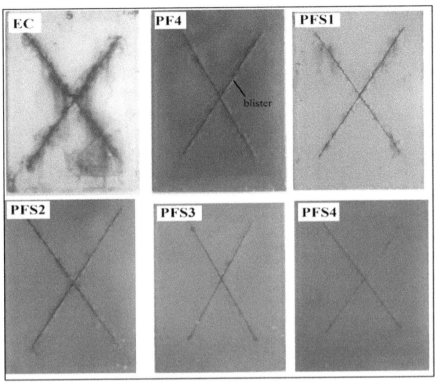

Figure 3.19: Photographs of (a) epoxy coated (EC) and epoxy with PF4 and different loading of poly(AN-co-PFA)/SiO₂ composite coated mild steel specimens exposed to salt spray fog after 120 days (American Journal of Polymer Science, doi:10.5923/j.ajps.20160603.03)

that the inorganic filler such SiO_2 particles also played an important role in improving the corrosion protection performance of the coating. As compared to epoxy and epoxy with poly(AN-co-PFA) coating, poly(AN-co-PFA)/SiO_2 based epoxy coating provided an efficient protective function against delamination around scribed areas and ensured the powerful galvanic function of the primer. On the basis of these results, it was presumed that the conducting copolymer-silica composites content was increased in epoxy resin and hydrophobic surface with increased crosslink density was formed, which did not allow corrosive ions to enter the coating/metal interface due to which the corrosion resistance performance of the epoxy modified copolymer coating increased as compared to the epoxy coating.

Ghanbari and Attar (2015) studied the anti-corrosion performance of epoxy/silica nanocomposites. Salt spray test results showed that the incorporation of 4–6 wt.% SiO_2 nanoparticles revealed better corrosion performance as compared to neat epoxy resin. Similarly, Sambyal et al. (2015) also reported the salt spray test for mild steel samples coated with epoxy resin containing the different loading (i.e., 1.0 wt.% to 4.0 wt.%) of poly(aniline-co-phenetidine)/SiO_2 composites for 60 days, it was observed that the poly(aniline-co-phenetidine)/SiO_2 composites based epoxy coating exhibited very minimal extension of corrosion around the scribe marks as compared to epoxy coating. PANI/o-toluidine/composite coatings were also subjected to the salt spray analysis for 120 days as shown in Figure 3.20. All the copolymer coatings showed excellent anti-corrosive properties in the highly corrosive condition. There was no extension of corrosion along the cut mark. Especially 2.0 wt.% loading coating exhibited superior anti-corrosive properties in highly aggressive conditions.

Becker et al. (2005) suggested that composites materials tend to occupy the pinholes in the epoxy coating and work as bridges in the interconnected matrix, causing a reduction of the total free volume and an enhancement of the cross-linking density of the epoxy resin (Babazadeh et al. 2015).

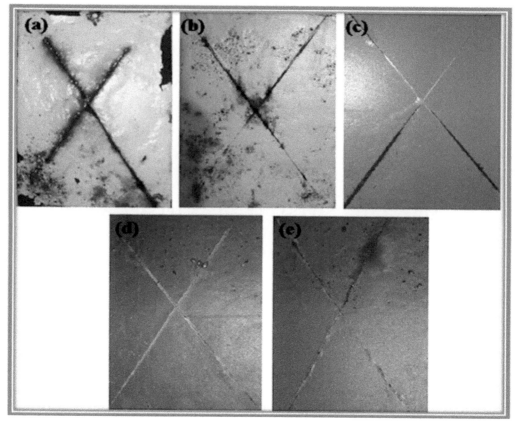

Figure 3.20: Photographs of (a) epoxy coated (EC) and epoxy with (b) 1.0% (OPF1), (c) 2.0% (OPF2), (d) 3.0% (OPF3) and (e) 4.0% (OPF4) loading of poly(aniline-co-o-toluidine)/flyash composite coated steel specimens exposed to salt spray fog after 120 days (Surface and Coatings Technology, doi.org/10.1016/j.surfcoat. 2015.04.013)

Radhakrishnan et al. (2009) investigated the anti-corrosive performance of epoxy powder coating containing polyaniline. They performed the salt spray test of epoxy-polyaniline coating for 700 hours. The results indicated that the epoxy coatings without polyaniline showed fast degradation with rust formation while those with polyaniline showed no rust formation in the scratched region. Wicks et al. (2007) suggested the corrosion resistance of an organic coating without defects mainly depends on its barrier properties. Amongst the parameter that contributes to barrier properties of the polymer, the crosslink density is very important. Higher the crosslink density, lower is the diffusion of ions and better is its barrier properties (Forsgren and Schweitzer 2006).

Amirdehi et al. (2014) explained that the barrier property of polyaniline could be improved by the incorporation of silica nanoparticles in it. They had achieved up to 79 to 87% corrosion protection efficiency using polyaniline-silica nanocomposites in the acidic environment. High corrosion protection performance shown by polyaniline-silica nanocomposites was related to the increase in the barrier to diffusion, inhibition of charge transport by the silica nanoparticles, redox properties of polyaniline, as well as large surface area available for the liberation of dopant due to nanosize filler.

3.4.6.2 Electrochemical studies of the coating

3.4.6.2.1 Open Circuit Potential (OCP) versus time

The corrosion protection behavior of bare and coated mild steel electrodes coated with epoxy, epoxy formulated copolymers and a copolymer composite was evaluated by OCP versus time measurements.

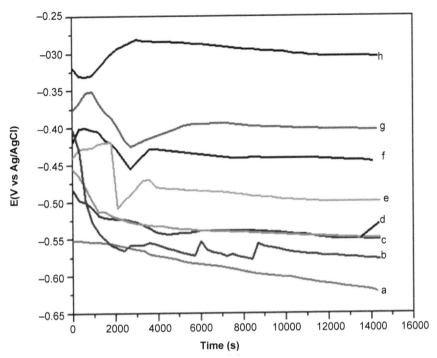

Figure 3.21: OCP versus time curves of (a) bare mild steel, (b) epoxy coating, (c) PF1 coating, (d) PFS1, (e) PF4, (f) PFS2, (g) PFS3, (h) PFS4 coated mild steel exposed to 3.5 wt.% NaCl solution at room temperature 25 ± 2°C (American J of Polymer Science, doi: 10/5923/j.ajps.20160603.03)

The coated specimens were dipped in 3.5 wt.% NaCl solution and the OCP was noted versus time at room temperature (25 ± 2°C). The protection time was characterized by the time during which the OCP of the covered electrode remained in the passive state of mild steel before it dropped to the corrosion potential of unprotected mild steel.

Radhakrishnan et al. (2009) demonstrated the OCP studies of epoxy coating containing polyaniline. OCP values were determined from the Tafel plots after different times of exposure. This indicated that there was practically no change in the OCP value which remained on the anodic side with respect to steel.

Sathiyanarayanan et al. (2006) studied the corrosion protection behavior of steel coated with epoxy blended polyaniline coating in different corrosive conditions, i.e., 3.0% NaCl, 0.1 N H_3PO_4 and 0.1 N HCl. It was observed that the open circuit potential values of the bare steel in 3.0% NaCl, 0.1 N H_3PO_4 and 0.1 N HCl are −0.574, −0.551 and −0.549 mV versus SCE respectively. The variation of OCP of coated steel with a period of immersion in 3.0% NaCl, 0.1 N H_3PO_4 and 0.1 N HCl are shown. The OCP values of coated steel in 3% NaCl and in 0.1 N H_3PO_4 were decreased initially and shifted to noble direction after 28 days immersion, which indicated that the polyaniline based coating was able to passivate the iron in 3% NaCl and in 0.1 N H_3PO_4 media after prolonged immersion. Meanwhile, the OCP values had shifted to more negative values with immersion time in 0.1 N HCl medium, which indicated that there no passivation effect occurred. It was observed from the various literature (Fahlman et al. 1998; Santros et al. 1998; Talo et al. 1997) that the OCP of PANI coated steel is higher than bare metal in NaCl medium due to passivation of metal.

The OCP versus time curves of uncoated and poly(AN-co-PFA)-silica composite coated mild steel in NaCl solution is shown in Figure 3.21. The trend of OCP variation for uncoated mild steel displays a continuous shifting of OCP towards negative potential. A steady-state OCP is not noticed for the metal. This is due to the existence of common corrosion on mild steel surface in the presence of saline medium as shown in Figure 3.21a.

The OCP curves of epoxy and epoxy formulated copolymers (i.e., PF1 and PF4) coated mild steel are initially high at around –0.40 V, –0.48 V and –0.44 V respectively, which specifies that these coated metal substrates retains its passive state at the initial stage of immersion. The tendency of OCP distinction for epoxy coating displays a sharp decrease of OCP toward negative potential within a few minutes of immersion this is owing to the fast diffusion of chloride ions through the coating. However, the potential reduces sharply within a few minutes and becomes nearly steady until the completion of the test. This is due to the penetration of the large concentration of chloride ions in the epoxy-metal interface and thus epoxy cannot protect mild steel any longer.

The OCP curve of the coating having epoxy formulated copolymers (PF1 and PF4) also displayed a similar fashion of the variation of potential. The potential reduced sharply and then shifted towards the slightly positive direction followed by the attainment of steady state value until the end of dipping period (Figure 3.21c and 3.21e). This behavior specified that the copolymer based coatings were more capable to protect the mild steel from corrosion as compared to the epoxy coating. On the other hand, a coating based on epoxy formulated copolymer composites, the OCP decreased sharply within few minutes of immersion. After that, a significant shifting of OCP to positive potential was detected with the passage of time as shown in Figure 3.21f, 3.21g and 3.21h. The OCP of the coated mild steel samples such as PFS1, PFS2, PFS3 and PFS4 shifted toward negative potential followed by a sharp positive shift until the end of the immersion period. The positive shifting of OCP mainly demonstrated the passive state of underlying metal due to the good corrosion protection ability of the surface film (Atles 2016). On the other hand, the epoxy with copolymer composites coated mild steel the OCP shifted progressively to more anodic side as compared to epoxy coated mild steel, as well as epoxy formulated copolymers (PF1 and PF4) coated mild steel.

The occurrence of high positive OCP value is attributed to the effective barrier property shown by copolymer composites for diffusion of corrosive ions into the metal surface. The corrosion protection behavior exhibited by epoxy formulated copolymer composite is due to the synergistic influence of the barrier property of the coating and the formation of a passive oxide layer owing to redox reaction at the mild steel-coating interface. The barrier property works as long as the coating remains undamaged and constrains the diffusion of chloride ions into the metal surface. With the passage of time, the electrolyte reaches the metal surface through the pores present in the coating which leads to the dissolution of metal. During this period the barrier property of the coating weakens. Therefore, it is presumed that the presence of aniline and pentafluoroaniline moieties in the copolymer chain offer instantaneous repassivation owing to the occurrence of redox reactions on the metal-coating interface. This phenomenon is evident from the significant shifting of potential toward a positive direction. The incorporation of silica particles as a reinforcing material in conducting copolymer enhanced the mechanical integrity of the epoxy coating in a corrosive environment, which leads to the efficient corrosion protection shown by metal for a prolonged period.

3.4.6.2.2 Tafel extrapolation measurement

Corrosion protection performance of different organic coatings is examined by electrochemical studies using Tafel extrapolation method. This method is an important tool to understand the protective properties of the coatings and is employed to compare the corrosion resistant properties of various surface coatings.

The electrochemical parameters associated with Tafel plots such as corrosion potential (E_{corr}) and corrosion current density (i_{corr}) are shown in Table 3.3 and Table 3.4. The results indicate that the coated mild steel electrode exhibited remarkably lower i_{corr} values and more positive corrosion potential (E_{corr}) with respect to uncoated mild steel (Chan et al. 2010). This behavior proposes that the coatings favor the generation of more stable passive layer and protects the underlying metal from corrosion.

Table 3.3: Comparative corrosion test results obtained from Tafel extrapolation method and EIS studies for the coated samples exposed to 3.5% NaCl solution for 24 hrs.

Test samples	Tafel test results			EIS measurement results	
	i_{corr} (μA/cm^2)	E_{corr} (mV)	P.E (%)	R_{pore} (Ωcm^2)	C_c (F/cm^2)
EC	84	−687	--	7.4 X 10^3 ± 3.4%	6.6 X 10^{-6} ± 4.0%
PF1	28	−671	60	9.7 X 10^4 ± 4.5%	1.4 X 10^{-7} ± 3.9%
PF4	21	−570	75	2.9 X 10^6 ± 3.0%	4.7 X 10^{-8} ± 3.7%
PFS1	29	−665	65	1.1X 10^5 ± 3.5%	5.2 X 10^{-7} ± 3.3%
PFS2	16	−583	81	1.5 X 10^6 ± 3.1%	2.6 X 10^{-9} ± 4.9%
PFS3	9	−575	89	9.8 X 10^6 ± 4.9%	3.3 X 10^{-10} ± 4.1%
PFS4	2	−467	98	5.8 X 10^7 ± 3.2%	6.4 X 10^{-10} ± 4.0%

Table 3.4: Comparative corrosion test results obtained from Tafel extrapolation method and EIS studies for the coated samples exposed to 3.5% NaCl solution for 30 days

Test Samples	Tafel test results			EIS measurement results	
	i_{corr} (μA/cm^2)	E_{corr} (mV)	P.E. (%)	R_{pore} (Ωcm^2)	C_c (F/cm^2)
EC	124	−710	--	9.9 X 10^2 ± 4.2%	9.6 X 10^{-5} ± 5.0%
PF1	65	−708	45	1.1 X 10^3 ± 3.5%	8.2 X 10^{-6} ± 5.2%
PF4	40	−645	67	8.9 X 10^3 ± 4.1%	4.0 X 10^{-7} ± 5.1%
PFS1	60	−708	52	8.1 X 10^3 ± 3.5%	9.0 X 10^{-6} ± 4.9%
PFS2	35	−685	71	4.2 X 10^4 ± 3.2%	8.1 X 10^{-8} ± 5.2%
PFS3	27	−677	78	8.4 X 10^5 ± 4.5%	2.6 X 10^{-9} ± 4.2%
PFS4	5	−667	96	7.8 X 10^6 ± 4.1%	8.2 X 10^{-9} ± 5.0%

Dhawan et al. (2015) studied the anti-corrosive behavior of mild steel coated with epoxy resin formulated with polyaniline as well as polyaniline-silica composites using the Tafel Extrapolation Method in 3.5% NaCl solution. It was observed that the corrosion current density values (i_{corr}) were found to decrease with increasing the concentration of polyaniline-silica composites in epoxy resin. Furthermore, polyaniline-silica based coating showed better corrosion protection efficiency as compared to polyaniline alone in the saline medium at the same concentration. The results obtained from Tafel extrapolation study is shown in Figure 3.22.

Recently, Cheng et al. (2018) investigated the corrosion protection performance of coating based on epoxy/silica, epoxy/Poly(o-ethoxyaniline) (Epoxy/POEA) and epoxy/Poly(o-ethoxyaniline)/silica (Epoxy/POEA/Silica) composites. The corrosion protection ability of the coating was found to be in following order: epoxy < epoxy-silica < epoxy-Poly(o-ethoxyaniline) < epoxy-Poly(o-ethoxyaniline)/ silica coating. In case of uncoated steel, when it was exposed in corrosive medium corrosive ions easily penetrated into the metal. Meanwhile in coated samples, due to the interaction between Poly(o-ethoxyaniline) and silica nanoparticles, when the corrosive medium initiated to contact with the coating film the micro/nanostructures on the surface of the coating had a certain barrier effect to the corrosive ions. This increased the barrier effect of the composite coating to the corrosive medium and made the corrosion protection of the epoxy-poly(o-ethoxyaniline)/silica containing coating efficient. This was mainly attributed to the addition of SiO$_2$ nanoparticles which improved the barrier properties by enhancing the uniformity and density of the coating surface.

Sambyal et al. (2015) demonstrated the corrosion studies of mild steel coated with an epoxy coating containing different loading of poly(aniline-co-Phenetidine)/SiO$_2$ composites by using Tafel polarization method in 3.5 wt.% NaCl medium for 24 h and 20 days of immersion, Tafel plots are

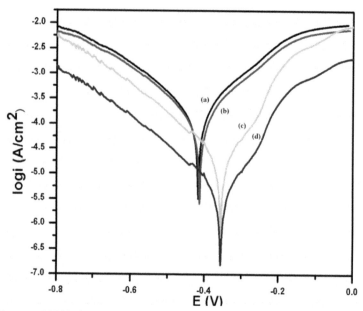

Figure 3.22: Tafel curves of (a) blank mild steel (b) epoxy coated mild steel (c) and (d) are PANI and PANI-Silica coated mild steel electrode with 6.0% loading level in epoxy resin in 3.5% NaCl solution (American Journal of Polymer Science, doi:10.5923/s.ajps.201501.02)

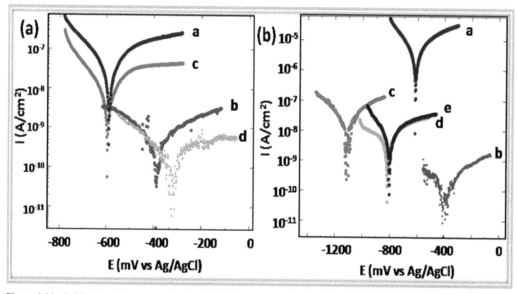

Figure 3.23: Tafel plots of (a) epoxy with (a) 1.0% (PPS1), (b) 2.0% (PPS2), (c) 3.0% (PPS3), (d) 4.0% (PPS4) and (e) 5.0% (PPS5) loading of poly(aniline-co-phenetidine)/SiO_2 composite coated steel substrate immersed in 3.5% NaCl solution for 24 hours at room temperature $25 \pm 2°C$. Figure b shows the respective plots after 20 days of immersion in 3.5% NaCl solution. (Sambyal et al. Surface and Coatings Technology, doi.org/10.1016/j.surfcoat.2015.12.038)

shown in Figure 3.23. The results demonstrated a better corrosion resistance exhibited by coatings containing copolymer composites. Moreover, by increasing the loading percentage of copolymer composites in epoxy resin, the corrosion current density was found to decrease which indicated that the copolymer-silica composites based coatings can provide an effective protection to the metal surface from the corrosive electrolyte.

The copolymer composites contain o-ethoxy aniline moiety in the polymer chain. According to the literature by Sathiyanarayanan et al. (1992), the ethoxy groups change the polymer structure and make it denser, this features of copolymers are very valuable for corrosion protection purpose. The initial corrosion resistance of the composite coatings was significantly high in 3.5% NaCl solution. Sambyal et al. (2015) also reported the Tafel plots of coated mild steel samples after 20 days of immersion as shown in Figure 3.23b.

The E_{corr} of all test specimens is observed to be shifted towards more negative potential as shown in Figure 3.23 as compared to the E_{corr} of specimens exposed for 24 hours. The shifting of the E_{corr} values is due to the diffusive chloride ions that weakens the protective property of the coatings. According to the i_{corr} values of the test, specimens increases with the exposure time which indicates that with the passage of time the diffusive chloride ions make pathways to the metal surface through the pores of the coatings. The i_{corr} values of mild steel coated with an epoxy coating containing 2.0 wt.% loading of poly(Aniline-co-o-ethoxy aniline)-silica composites (0.9×10^{-10} A/cm²) is found to be almost four orders of magnitude less compared to the mild steel coated with neat epoxy resin.

Poly(aniline-co-o-toluidine)/flyash composite coated steel substrate is also subjected to Tafel polarization analysis. Figure 3.24(a) and 3.24(b) shows Tafel plot of 4 hours and 24 hours after the immersion in 3.5% NaCl solution. All the coatings exhibit low i_{corr} values and positive E_{corr} values in comparison to the bare steel. The polarization resistance for 2 wt.% and 3 wt.% loading is found to be 3,060 KΩ/cm², 3,580 KΩ/cm². Such high values of polarization resistance indicate the development of stable barrier passive coating and shield the metal substrate. After the 24 hours immersion in 3.5%, NaCl solution epoxy coating shows a negative shift for E_{corr} values. This signifies the diffusion of aggressive diffusive ion in the coating through the pores present in the coating and leads to the degradation of the coating due to the prolonged exposure to the corrosive conditions. However, copolymer coatings do not show any shift for i_{corr} and E_{corr} values, especially OPF2 and OPF3 coatings which have maintained the low i_{corr} and high polarization resistance values after its immersion in the NaCl solution. This superior anti-corrosive property arises due to the presence of

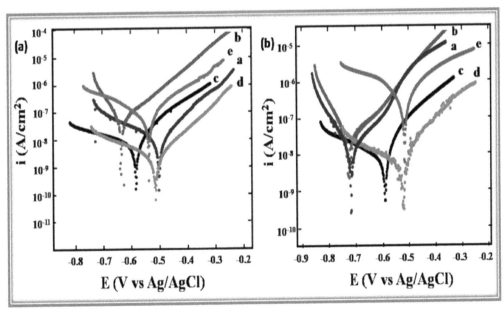

Figure 3.24: Tafel plots of (a) epoxy coated steel and epoxy with (b) 1.0% (OPF1), (c) 2.0% (OPF2), (d) 3.0% (OPF3) and (e) 4.0% (OPF4) loading of poly(aniline-co-o-toluidine)/flyash composite coated steel substrate exposed to 3.5% NaCl solution for 4 hours at room temperature 25 ± 2°C. Figure b shows the respective plots after 24 hours of exposure to 3.5% NaCl solution (Sambyal et al. (2015). Surface and Coatings Technology, doi.org/10.1016/j.surfcoat. 2015.04.013)

plenty of π-bond and quaternary nitrogen in copolymer composites that leads to the protection of the underlying metal substrate.

Tafel polarization studies on epoxy coated (EC) and epoxy formulated showed different wt.% loadings of copolymer composite coated mild steel sample (PFS1, PFS2, PFS3 and PFS4) which was carried out after dipping the mild steel samples in 3.5 wt.% NaCl solution for a different interval such as 24 h to 30 days at room temperature (25 ± 2°C) as shown in Figure 3.25a. The i_{corr} value of epoxy coated mild steel electrode was 84.0 μA/cm² that decreased to 16.0 μA/cm² in PFS2, 9.0 μA/cm² in PFS3 and 2.0 μA/cm² in PFS4 coated samples. The i_{corr} value was found to decrease when the loading of poly(AN-co-PFA)/SiO₂ composite in epoxy resin was increased. As a result, the corrosion protection efficiency (% P.E.) as calculated from equation 1.0 was 89% and 98.0% for the mild steel samples coated with PFS3 and PFS4 respectively as shown in Table 3.4.

Tafel plots of mild steel electrodes coated with epoxy formulated copolymers (PF1 and PF4) demonstrated that the i_{corr} value of these copolymer coatings was high compared to the mild steel electrode coated with epoxy formulated poly(AN-co-PFA)/SiO₂ composite at the similar loading level and time interval (see Table 3.3). This indicated that the incorporation of SiO₂ in poly(AN-co-PFA) improves the corrosion protection performance of metal. Furthermore, it was also noticed that the i_{corr} value of epoxy formulated poly(AN-co-PFA) and poly(AN-co-PFA)/SiO₂ coated electrodes was less than the epoxy coated specimen as shown in Table 3.3. This indicated that the barrier property of polymer coating was superior to that of epoxy coated mild steel. Moreover, lower i_{corr} values also

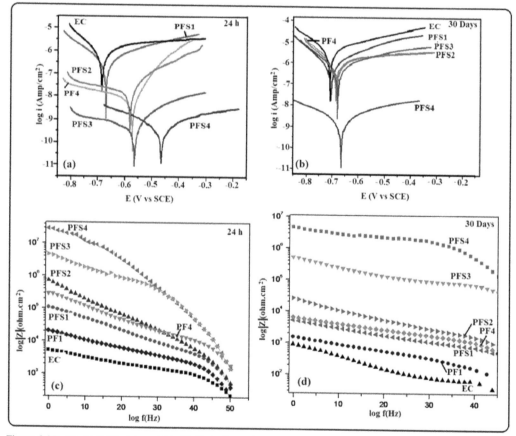

Figure 3.25: (a), (b) Tafel plots and (c), (d) Bode Plots of epoxy coated and epoxy with different loading of Poly(AN-co-PFA) and Poly(AN-co-PFA)/SiO₂ composites coated mild steel electrode exposed to 3.5 wt.% NaCl solution for 24 h and 30 days at room temperature 25 ± 2°C. (Bisht et al. (2016). American Journal of Polymer Science, doi:10.5923/j.ajps.20160603.03)

indicated the formation of a protective film on the metal surface. The i$_{corr}$ values of mild steel electrode were found to reduce when the loading level of poly(AN-co-PFA)/SiO$_2$ in epoxy resin was increased.

Some remarkable results are noticed from the Tafel plots of mild steel electrode coated with epoxy and epoxy formulated copolymers (i.e., PFS1, PFS2, PFS3 and PFS4) at 30 days of exposure to 3.5% NaCl solution as shown in Table 3.4 and Figure 3.25b. A negative shift of potential is due to the occurrence of active corrosion process at the coating-metal interface. This phenomenon results in the degradation of the coating due to the prolonged immersion of electrodes in corrosive solution. However, the results demonstrate the weakening of the barrier property of the epoxy coating with the exposure time.

In the presence of corrosive environment, the interfacial bonding between epoxy-metal is damaged due to the entrance of electrolyte into the interface which breaks the chemical bonds and results in the failure of the epoxy coating. Corrosion protection efficiency of epoxy formulated copolymers coated samples on 30 days of exposure is 45% and 67% respectively. On the other hand, the i$_{corr}$ values and protection efficiency of PFS4 coated sample does not reveal any substantial change after a prolonged exposure, i.e., 30 days to the saline solution as shown in Table 3.4. The corrosion protection efficiency of PFS4 coated sample is found to be 96% after 30 days of exposure. This shows that the corrosion resistance property of the epoxy coatings with copolymer composite is not much affected with time. The copolymer/silica composite present in the epoxy resin acts as a proficient physical barrier and protects the underlying metal surface against the entrance of corrosive ions. Moreover, the presence of silica particles in copolymer provides additional strengthening to the coating in the corrosive medium as evident from the physico-mechanical studies of copolymer composites coating.

3.4.6.2.3 Electrochemical Impedance Spectroscopy (EIS)

Electrochemical Impedance Spectroscopy (EIS) is the method in which the impedance of an electrochemical system is studied as a function of the frequency of an applied A.C. wave. EIS is a nondestructive electrochemical technique for the evaluation of quality and corrosion protection performance of coatings. Impedance measurements are carried out by dipping the test specimens in 3.5 wt.% NaCl solution at open circuit potential settings for 24 h to 30 days at room temperature. Baldissera et al. (2012) reported the electrochemical studies of the epoxy coating containing polyaniline in 3.5% NaCl medium. The results indicated that polyaniline based epoxy coating showed better corrosion protection than blank epoxy coating. Shazly and Turaif (2012) studied the anti-corrosion performance of polyaniline-based coating of buried steel. The results revealed that the buried steel coated with polyaniline showed better corrosion resistance against NaCl, H$_2$SO$_4$ and water by a factor up to 1.88, 1.89 and 1.54 respectively as compared to bare buried steel.

Pud et al. (1999) had performed the corrosion protection performance of mild steel coated by polyaniline primers with epoxy top coat and epoxy alone. A hole was drilled through coatings to the metal surface, and the results revealed that the corrosion rate for mild steel coated by polyaniline/epoxy was significantly lower compared to the epoxy coating alone. Similar results were also observed for scribed samples with polyaniline based primers (Wessling 1994; Kinlen et al. 1997).

Tanveer et al. (2011) studied the anti-corrosion performance of polyaniline and polypyrrole based copolymers with o-toluidine such as poly(aniline-co-o-toluidine) and poly(pyrrole-co-o-toluidine). The corrosion tests were performed in 0.1 M HCl and 5.0 wt.% NaCl solution, and the results showed that polyaniline based copolymers displayed better anti-corrosion performance as compared to pure polyaniline as well as polypyrrole-based copolymers. Another work demonstrated by Tanveer et al. (2014) indicated that the copolymer based on poly(aniline-co-o-toluidine) and poly(aniline-co-2,3 xylidine) showed better anti-corrosion performance in the various corrosive medium such as 0.1 M HCl, 5.0 wt.% NaCl, artificial seawater and distilled water as compared to their respective homopolymers such as polyaniline, poly(o-toluidine) and poly(2,3 xylidine).

Similar work was demonstrated by Srikanth et al. (2008) by using the copolymer based on aniline and N-methyl aniline on carbon steel. The corrosion studies were performed in 0.1 M HCl solution using potentiodynamic polarization measurements. Copolymer based on poly (aniline-co-N-methylaniline) coating showed high protection efficiency as compared to other coatings based on their respective homopolymers. Another work was performed by Hur et al. (2006) with polyaniline-based copolymer such as poly(aniline-co-2-chloroaniline) films. The copolymer film was synthesized by electrochemical deposition on 304 L SS from an acetonitrile solution. Different electrochemical studies, i.e., potentiodynamic polarization and EIS measurements showed that copolymer films have a significant protective performance against corrosion of stainless steel in 0.5 M HCl solution. The results also indicated that polyaniline and its copolymer with 2-chloroaniline films were able to provide effective anodic protection in addition to barrier properties for cathodic reaction, while poly(2-chloroaniline) film had provided only barrier properties.

Similarly, Bereket et al. (2005) carried out electrodeposition of polyaniline, poly(2-iodoaniline) and poly(aniline-co-2-iodoaniline) on steel surfaces. The higher value of the polarization resistance obtained in this work was assigned to the protective properties of poly(aniline-co-2-iodoaniline). Apart from copolymers based on aniline and ortho-substituted aniline, Srikanth et al. (2008) investigated the protection efficiency of a thin film electrochemically synthesized as poly(m-toluidine), poly(N-methyl aniline), and its copolymer such as poly(aniline-co-N-methyl aniline) on carbon steel in 0.1 M HCl. Their studies revealed that the copolymers based on poly(aniline-co-N-methyl aniline) showed higher corrosion protection performance than its homologous homopolymers. This was attributed to the compact continuous dense morphology of poly(aniline-co-N-methyl aniline) which provided better protection than other coatings.

The impedance graphs obtained for epoxy coated mild steel electrode (EC) and epoxy formulated with different wt.% loadings of copolymer composite (PFS1, PFS2, PFS3 and PFS4) coated electrode are presented in Bode plots. In general, Bode plots reveal the different steps of degradation of coatings on the metal surface. The unexposed coating systems act as a pure dielectric, separating the mild steel substrate from the corrosive medium such as NaCl solution. This nature of the coating system results in pure capacitive behavior (Sambyal et al. 2015). Figure 3.26a and Figure 3.26b show the Bode plots of epoxy coated and epoxy having different loading of copolymer composites coated mild steel electrode at 24 h of immersion in 3.5 wt.% NaCl solution. According to the literature, the impedance modulus at low frequency ($|Z|_{0.01 \text{ Hz}}$) is a suitable parameter for characterization of the corrosion protection performance of coatings (Park et al. 2002; Chen et al. 2007).

The electrical resistance of a coating system is measured in terms of pore resistance (R_p), which indicates the performance of the surface coating. Whereas, coating capacitance (C_c), which is related to water uptake tendency of the coating, is an important parameter to measure the mechanical integrity of the coating in a corrosive medium. The high magnitude of impedance in the lower frequency region of Bode curves denotes the higher barrier properties of the surface films (Scully and Hensley 1994). Sathiyanarayanan et al. (2006) have investigated the pore resistance value of steel coated with epoxy-polyaniline coating, which is greater than 10^7 Ωcm^2 and indicates better protection of iron in a saline medium.

Zu et al. (2015) carried out the electrochemical studies polyaniline-silica based nanocomposites. They observed that the PANI-silica had a larger charge transfer resistance than that of bulk PANI. Bisht et al. (2016) observed that in epoxy coated mild steel sample, the pore resistance value was found to be around 7.4×10^3 Ωcm^2 which increased to 5.8×10^7 Ωcm^2 on incorporation poly(AN-co-PFA)/SiO$_2$ composites in epoxy resin. The pore resistance value for the epoxy and epoxy formulated with different loading of poly(AN-co-PFA) and poly(AN-co-PFA)/SiO$_2$ composite coated samples was measured using EIS data fitting and listed in Table 3.2. The results indicated that the R_p value increased when the loading of poly(AN-co-PFA)/SiO$_2$ composite in epoxy resin increased. Moreover, R_p value of the epoxy containing poly(AN-co-PFA)/SiO$_2$ composite coating was found to be higher than epoxy containing poly(AN-co-PFA) coating. The high pore resistance value was characteristic of good protective coatings on mild steel surface (Wang et al. 2015). Thus, the epoxy coating containing

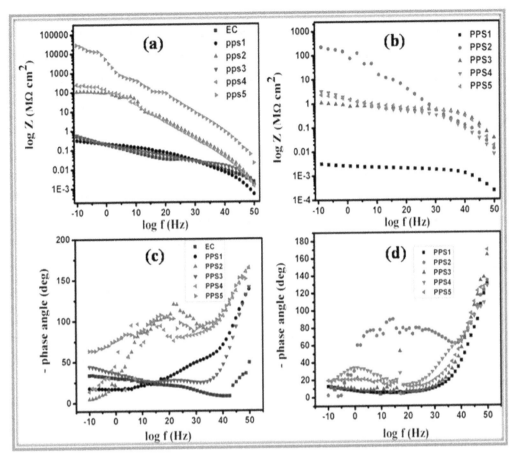

Figure 3.26: Bode curves of (a) epoxy with 1.0% (PPS1), 2.0% (PPS2), 3.0% (PPS3), 4.0% (PPS4) and 5.0% (PPS5) loading of poly(aniline-co-phenetidine)/SiO$_2$ composite coated steel substrate immersion in 3.5% NaCl solution for 24 hours at room temperature 25 + 2°C. Figure (b) shows the respective plots after 20 days of immersion. Figure (c) and (d) show the phase angle plots for 24 hours and 20 days, respectively (Sambyal et al. Surface and Coatings Technology, doi.org/10.1016/j.surfcoat.2015.12.038)

copolymer-silica composite exhibited the lowest ion diffusion and best barrier property in corrosive environment (Kumar et al. 2006).

It was noticed that these coatings had a good resistance against the diffusion of corrosive ions at the beginning of dipping in corrosive medium, and had very low possible paths for water molecules and other ions to diffuse from coating-water interface to the metal-coating interface. Literature had also revealed that when incorporating the conducting polymers like polyaniline (Bhadra et al. 2009) and their derivatives (Chou et al. 2001) in epoxy coating, the coating resistance values had increased remarkably which indicated better protection performance as shown by conducting polymers containing coatings than pure epoxy coatings (Tavandashti et al. 2009). Therefore, the results indicated that the introduction of a comonomer such as PFA and silica in epoxy matrix further enhanced the anti-corrosive performance of epoxy-coatings.

However, the evaluation of corrosion resistance of coatings for a comparatively longer period of immersion is equally essential. Therefore, impedance parameters of the coatings are also measured for 30 days of immersion in 3.5% NaCl solution and its results are shown in Table 3.4. Among the test samples, the epoxy coated mild steel electrode show low pore resistance and high coating capacitance

(C_c) due to its the high water uptake and poor barrier property on 30 days of immersion as shown in Table 3.4 (Floyd et al. 2009). The R_p value increases with the increase in the loading level of copolymer-silica composites from 1.0 wt.% to 4.0 wt.% in epoxy resin as compared to epoxy coated mild steel sample. The mild steel sample coated with epoxy formulated with 4.0 wt.% loading also show the highest value of pore resistance (7.8×10^6 Ωcm^2) and the lowest value of Cc (8.2×10^{-9} F/cm^2), thus displaying its superior corrosion protection performance even on longer immersion time. The coating capacitance (C_c) occurs in the decreasing order as EC, PFS1, PFS2, PFS3 and PFS4 as shown in Table 3.4. The low coating capacitance is observed for the epoxy coating containing 1.0 to 4.0 wt.% loading of poly(AN-co-PFA)/SiO$_2$ composite, which is due to the low penetration of electrolyte into the surface coating. Moreover, it is also observed that the R_p value of PF1 and PF4 coated mild steel is lower than PFS1 and PFS4, which indicates that the poor barrier property is shown by epoxy coating containing poly(AN-co-PFA) as compared to poly(AN-co-PFA)/SiO$_2$ composite.

Similar behavior of Bode plots (Figure 3.26) was also observed for the mild steel samples coated with the epoxy coating containing poly(aniline-co-Phenetidine)/SiO$_2$ composite in 3.5 wt.% NaCl medium (Sambyal et al. 2015). It was observed that mild steel samples coated with epoxy formulated poly(aniline-co-Phenetidine)/SiO$_2$ composite showed are markably high magnitude of impedance as compared to neat epoxy coated mild steel.

Bode plots of mild steel samples coated with epoxy formulated poly(aniline-co-Phenetidine)/SiO$_2$ composite were also studied for 20 days of immersion in a saline medium. It was observed that the values of R$_{pore}$ remained very low and the Cc was maintained very high for the neat epoxy coating (EC) as shown in Figure 3.26. Meanwhile poly(aniline-co-Phenetidine)/SiO$_2$ composite based coating showed high R$_{pore}$ value as compared to neat epoxy coating, even after 20 days of exposure to the saline medium. The Bode curve of mild steel samples coated with the epoxy having 2.0 wt.% loading of poly(aniline-co-Phenetidine)/SiO$_2$ composite (PPS2) also showed a high magnitude of impedance in the low-frequency region, which indicated its efficient barrier behavior against the diffusion of the electrolyte. The diffusion element was only observed after 20 days of immersion. The start of diffusion was basically due to the appearance of porosities in the coating resulting in the weakening of the barrier property of the coating.

Epoxy coatings are well-known for high adhesive strength due to the presence of polar groups in the structure, which is chemically bonded to the metal surface (Sato et al. 1994; Yang et al. 2010). In the presence of corrosive medium, the interfacial bonding between epoxy/metal is destroyed because of the entry of electrolyte into the interface, which leads to the failure of organic coatings. Therefore, in copolymer composite system SiO$_2$ particles act as reinforcing materials, which improves the mechanical integrity of the coating. SiO$_2$ particles in coating provide additional barrier protection to the substrate and prevent the penetration of electrolyte into the interface.

Figure 3.27 (Sambyal et al. 2015) shows the Nyquist and bode for the poly(aniline-co-o-toluidine)/flyash composite after the immersion in the 3.5% NaCl solution. The R$_{pore}$ value of the composite (OPF2 and OPF3) coating is higher in comparison to the neat epoxy coating. This is attributed to the presence of the copolymer in the epoxy matrix. OPF2 is almost double the value of R$_{pore}$ (23,576 Ω) to that of epoxy coating, and shows superior barrier properties of the copolymer composite coatings. Bode analysis is carried out to analyze the impedance of coatings in the low frequency. Bode plots for OPF2 and OPF3 show high impedance values in low-frequency region. The barrier properties of copolymer coatings are demonstrated by the high impedance modulus in the low-frequency region. The service life of the coating depends upon their interaction with aggressive corrosive ions on diffusion after the prolonged immersion. With the lapse of time, the protective coating system weakens, and values of R$_{pore}$ decreases with the increment in the Cc value. However, an interesting result is observed for copolymer composite coatings. Especially OPF2 and OPF3 maintain a high value of R$_{pore}$ and Cc value is maintained at significantly low value throughout the immersion time. This means the water intake of the coating system is very minimal and coating acts as a barrier system.

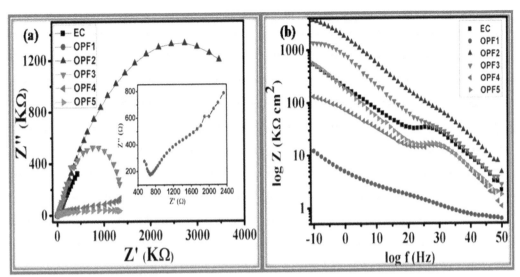

Figure 3.27: (a) Nyquist and (b) Bode plots of epoxy coated (EC), epoxy with 1.0% (OPF1), 2.0% (OPF2), 3.0% (OPF3), 4.0% (OPF4) and 5.0% (OPF5) loading of poly(aniline-co-o-toluidine)/flyash composite coated specimens immersed in 3.5% NaCl solution for 4 hours at room temperature (25 ± 2°C) (Sambyal et al. Surface and Coatings Technology, doi. org/10.1016/j.surfcoat. 2015.04.013)

3.5 Mechanism of corrosion protection of mild steel coated with polyaniline based copolymer composites

Conducting polymers shows a dual protection mechanism. Firstly, they act as a barrier against aggressive penetrative ion, and secondly they provide anodic protection by shifting the corrosion potential of the substrate to the passive region (Skotheim 1997; Kinlen et al. 1997; Deshpande and Sazou 2016). Barrier protection is commonly used to explain the corrosion protection of organic coatings. The protection is achieved by isolating the metal surface from the corrosive environment. The protection is effective as long as the barrier remains undamaged. Scratches or bare metal surface allows contact with corrosive ions, which leads to corrosion process. Application of coating on the metal surface produces a kind of barrier effect.

Beck (1998) studied the corrosion protection behavior of iron by electrochemically deposited polyaniline. It was observed that metal coated with an organic coating (thickness > 1.0 μm) showed reduced corrosion which was due to barrier effect of polyaniline. Likewise, Weslling noted that a decrease in corrosion current density was attributed to thicker polyaniline coatings. Pud et al. investigated corrosion protection of mild steel coated with polyaniline/epoxy primers and epoxy alone, the coating was damaged by drilling a hole on the metal surface. The results showed that the corrosion rate for polyaniline/epoxy coated mild steel was remarkably lower as compared to epoxy coating alone (Pud et al. 1999).

DeBerry (1995) investigated that the electrochemically grown polyaniline (PANI-ES) provided a form of anodic protection to stainless steels in sulphuric acid media. When metal or alloy substrate was exposed to corrosive conditions at anode, oxidation of iron took place with the liberation of free electrons as shown in Figure 3.28. Reduction of conducting polymers took place by consuming these free electrons and converting into a fully reduced leucoemeraldine form (PANI-LS). So, this process delayed the corresponding cathodic reaction but with the lapse of time leucoemeraldine form reoxidized to emeraldine form by the action of atmospheric or dissolved oxygen from the electrolyte. Further, it led to the formation of a passive layer to the metal substrate which acted as a barrier.

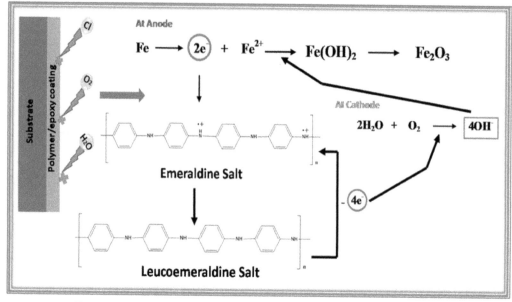

Figure 3.28: Schematic representation of corrosion inhibition mechanism of polyaniline

Color version at the end of the book

Cook et al. (2004) explained the corrosion protection mechanism of polyaniline. Their study indicated that protection of metal takes place via an inhibitory action of polyaniline. The stable form of polyaniline (PANI-ES) in electrical contact with the metal was polyaniline leuco-base (PANI-LB). During the reduction of the PANI-ES, dopant anions such as sulfonates and phosphonates would release, which are also capable of preventing the corrosion of mild steel in corrosive condition. The independence of E_{corr} with respect to the coating type suggested mixed inhibition. Because dopant anions were released at the defect edges, protection possibly initiated at cathodic sites. Inhibition of the cathodic reaction at the defect edges would then slow the rate at which hydroxyl ions were produced and would thus decrease the rate of coating delamination. This was a key factor in preventing the spread of corrosion at coating defects. Anodic effects required transport of the released dopant ions to the net anodic site in the canter of the defect.

Kilmartin et al. (2002) reported that both electrochemically developed polyaniline and poly-ortho-methoxyaniline had the same effect on the behavior of stainless steels in a similar environment. The literature demonstrated that the copolymer based coatings were expected to behave similarly to their parent polymers, i.e., polyaniline. Yao et al. (2009) demonstrated anti-corrosion properties of epoxy coating formulated with copolymer based on poly(aniline-co-*p*-phenylenediamine). The results revealed that the anti-corrosion behavior of the epoxy coating is attributed to electrochemical protection as well as barrier effects. The electrochemical protection was caused by increase in the corrosion potential and the formation of a passive layer on the surface of the metal, while paint film prevented the entrance of oxygen and water into the metal surface.

Bhandari et al. (2008) studied the corrosion inhibition performance of poly(o-isopropyl aniline) and its copolymer with aniline in acidic medium. It was observed that the corrosion inhibition efficiency was found to increase in the case of the homopolymer of o-isopropyl aniline and their copolymers with aniline. Presence of delocalized π-electrons formed coordination type of bonds with the iron surface. Iron in acidic medium was negatively charged, the existence of quarternary ammonium nitrogen acted as an anchoring unit which made the polymer chain to attach to the surface. The bulkiness of substitution made the cluster network on iron surface anchor more strongly, resulting in a good coverage of surface and enhanced corrosion inhibition efficiency.

The mechanism of corrosion protection by conducting polymers present in insulating coatings is mainly dependent on three factors such as the formation of the passive layer, barrier effect and their adhesion on the metal surface. According to the literature (Kinlen et al. 1997), it was considered that the conducting moieties of these polymers acted as an active coating in the reaction taking place across the polymer coated and metal-electrolyte interface. The formation of the iron oxide layer at the coating metal interface improved the corrosion protection by the dispersion of copolymer in coatings.

When copolymer-coated substrate is immersed in corrosive media, oxidation of Fe to Fe^{2+} and Fe^{3+} takes place through the formation of the passive oxide layer. In this case, the oxidation is accompanied by the reduction of copolymer-ES (emeraldine salt) to copolymer LS (leuco-emeraldine salt), formation of a passive layer takes place in which the copolymer based coating is able to provide better corrosion resistance. Here, in the case of epoxy coatings containing poly(AN-co-PFA) shows two-step mechanism. Furthermore, the presence of silica particles in epoxy-copolymer formulation plays two important roles for enhancing the corrosion protection efficiency of mild steel. First, silica particles improve the quality of epoxy coating by reducing its porosity in corrosive medium, hence improving the barrier performance of coating. Secondly, silica particles improve the adherence as well as mechanical integrity of the epoxy-copolymer coating to the underlying substrate, as evident from physico-mechanical tests of silica-based coating. The literature demonstrates (Shi et al. 2009; Huang et al. 2011) that SiO_2 particles in epoxy resin tends to occupy holidays in the thin film coating and serves to bridge more molecules in the interconnected matrix; leading to an increased cross-linking density of the cured epoxy as well as improved corrosion protection for the steel substrate.

3.6 Conclusion

Epoxy resin has been preferred due to the fact that paints formulated using these resin are generally applied in marine and industrial maintenance. Many of the formulation containing conventional coating based metal is presently subject to environmental constraints and their use may not be acceptable in future. The main objective of this study is to demonstrate that the addition of conducting copolymer-silica based composite to an epoxy coating, which leads to an improved protection of mild steel.

The corrosion test studies of the epoxy coating containing copolymer/nanofillers composite showed a greater corrosion protection property as compared to epoxy and epoxy formulated copolymer coating in 3.5% NaCl medium. The Tafel parameters indicated low corrosion current for coatings with 3.0 and 4.0 wt.% loading of the copolymer-silica composite in 3.5 wt.% NaCl solution. These coatings had revealed superior anti-corrosion performance, even for prolonged exposure to 3.5 wt.% NaCl solution. The results of mechanical testing such as abrasion resistance bend test and cross-cut adhesion test for epoxy formulated copolymer/nanofillers composite coating were found to be superior to epoxy coating and epoxy formulated copolymer composite coating. Apart from the prevention of corrosion, it was also observed that the epoxy formulated with copolymer/nanofillers composite coating showed better mechanical properties as compared to epoxy and epoxy with copolymer composite coating. It was observed that the mechanical properties of the epoxy-based copolymer/nanofillers composite coatings were found to be significantly improved with the loading of copolymer/nanofillers composite composites in coating due to the presence of nanofillers (SiO_2, Flyash) particles in the polymer matrix, which reinforced the polymer by providing additional mechanical strength. The synergistic interaction between copolymers and SiO_2 in the copolymer composite resulted in the enhanced thermal, mechanical, hydrophobic and anti-corrosive properties of the coating. Hence, such a system can be used as an effective and efficient anti-corrosive coating on mild steel where the probability of corrosion is greater.

References

Amirdehi, M. F. and Afzali, D. (2014). Deposition of polyaniline/silica nanocomposite coating on stainless steel; study of its corrosion properties. Advanced Materials Research, 829: 605–609.

Arturi, K. R., Jepsen, H., Callsen, J. N., Søgaard, E. G. and Simonsen, M. E. (2016). Superhydrophilicity and durability of fluoropolymer-TiO_2 coatings. Progress in Organic Coatings, 90: 132–138.

Babazadeh, M., Zalloi, F. and Olad, A. (2015). Fabrication of conductive polyaniline nanocomposites based on silica nanoparticles via *in-situ* chemical oxidative polymerization technique. Synthesis and Reactivity in Inorganic, Metal-Organic, and NanoMetal Chemistry, 45(1): 86–91.

Baldissera, A. F. and Ferreira, C.A. (2012). Coatings based on electronic conducting polymers for corrosion protection of metals. Progress in Organic Coatings, 75(3): 241–247.

Başak Doğru Mert. (2016). Corrosion protection of aluminum by electrochemically synthesized composite organic coating. Corrosion Science, 103: 88–94.

Beck, F. (1988). Electrodeposition of polymer coatings. Electrochimica Acta, 33(7): 839–850.

Becker, O., Varley, R. and Simon, G. (2002). Morphology, thermal relaxations and mechanical properties of layered silicate nanocomposites based upon high-functionality epoxy resins. Polymer, 43(16): 4365–4373.

Beentjes, P. C. J. (2004). Durability of Polymer Coated Steel in Diluted Acetic Acid Environment. Ph.D. thesis. Delft University of Technology.

Bernard, M. C. and Hugot-LeGoff, A. (2001). Polyaniline films for protection against corrosion. Synthetic Metals, 119(1-3): 283–284.

Bereket, G., Hur, E. and Sahin, Y. (2005). Electrodeposition of polyaniline, poly(2-iodoaniline), and poly(aniline-co-2-iodoaniline) on steel surfaces and corrosion protection of steel. Applied Surface Science, 252(5): 1233–1244.

Bhadra, S., Khastgir, D., Singha, N. K. and Lee, J. H. (2009). Progress in preparation, processing and applications of polyaniline. Progress in Polymer Science, 34(8): 783–810.

Bhandari, H., Sathiyanarayana, S., Choudhary, V. and Dhawan, S. K. (2008). Synthesis and characterizations of proccessible polyaniline derivatives for corrosion inhibition. Journal of Applied Polymer Science, 111(5): 2328–2339.

Bhandari, H., Choudhary, V. and Dhawan, S. K. (2010). Enhancement of corrosion protection efficiency of iron by poly (aniline-co-amino-naphthol-sulphonic acid) nanowires coating in highly acidic medium. Thin Solid Films, 519(3): 1031–1039.

Bhandari, H., Choudhary, V. and Dhawan, S. K. (2011). Influence of self-doped poly(aniline-co-4-amino-3-hydroxy-naphthalene-1-sulfonic acid) on corrosion inhibition behaviour of iron in acidic medium. Synthetic Metals, 161(9-10): 753–762.

Bisht, B. M. S., Bhandari, H., Sambyal, P., Gairola, S. and Dhawan, S. K. (2016). Highly durable and novel anticorrosive coating based on epoxy reinforced with poly(Aniline-co-Pentafluoroaniline)/SiO_2 composite. American Journal of Polymer Science, 6(3): 75–85.

Brassard, J. D., Sarkar, D. K. and Perron, J. (2011). Synthesis of monodisperse fluorinated silica nanoparticles and their superhydrophobic thin films. ACS Applied Materials & Interfaces, 3(9): 3583–3588.

Camalet, J. L., Lacroix, J. C., Aeiyach, S., Chane-Ching, K. and Lacaze, P. C. (1998). Electrosynthesis of adherent polyaniline films on iron and mild steel in aqueous oxalic acid medium. Synthetic Metals, 93(2): 133–142.

Cao, G. Z. (2001). Organic–inorganic hybrid coatings for corrosion protection. Journal of Non-Crystalline Solids, 290(2-3): 153–162.

Cardoso, M. J. R., Lima, M. F. S. and Lenz, D. M. (2007). Polyaniline synthesized with functionalized sulfonic acids for blends manufacture. Materials Research, 10(4): 425–429.

Chang, C. -H., Huang, T. -C., Peng, C. -W., Yeh, T. -C., Lu, H. -I., Hung, C. -J., Weng, C. -J., Yang, T. -I. and Yeh, J. -M. (2012). Novel anticorrosion coatings prepared from polyaniline/graphene composites. Carbon, 50(14): 5044–5051.

Chang, K. C., Hsu, C. H., Peng, C. W., Huang, Y. Y., Yeh, J. M., Wan, H. P. and Hung, W. C. (2014). Preparation and comparative properties of membranes based on PANI and three inorganic fillers. Express Polymer Letters, 8(3): 207–218.

Chen, X., Shen, K. and Zhang, J. (2010). Preparation and anticorrosion properties of polyaniline-SiO_2-containing coating on Mg-Li alloy. Pigment & Resin Technology, 39(6): 322–326.

Chen, Y., Wang, X. H., Li, J., Lu, J. L. and Wang, F. S. (2007). Long-term anticorrosion behaviour of polyaniline on mild steel. Corrosion Science, 49(7): 3052–3063.

Chen, Y., Kang, G., Xu, H. and Kang, L. (2016). Two composites based on $CoMoO_4$ nanorods and PPy nanoparticles: fabrication, structure and electrochemical properties. Synthetic Metals, 215: 50–55.

Chou, T. P., Chandrasekaran, C., Limmer, S. J., Seraji, S., Wu, Y., Forbess, M. J., Nguyen, C., Chen, H., Zhang, X., Zhang, P. and Zhang, Z. (2012). Facile approach in fabricating superhydrophobic SiO₂/polymer nano-composite coating. Applied Surface Science, 261: 628–632.

Cheng, H., Hu, C., Wang, X. and Ziqiang, H. (2018). Synthesis and characterization of Poly(o-ethoxyaniline)/nano silica composite and study of its anticorrosion performance. International Journal of Electrochemical Science, 13: 196–208.

Cook, A., Gabriel, A. and Laycock, N. (2004). On the mechanism of corrosion protection of mild steel with polyaniline. Journal of the Electrochemical Society, 151(9): B529–B535.

DeBerry, D. W. (1985). Modification of the electrochemical and corrosion behavior of stainless steels with an electroactive coating. Journal of the Electrochemical Society, 132(5): 1022–1026.

Deshpande, P. P. and Sazou, D. (2016). Corrosion Protection of Metals by Intrinsically Conducting Polymers.: CRC Press.

Dewan, A. K., Valenzuela, D. P., Dubey, S. T. and Dewan. (2002). Industrial and Engineering Chemistry Research, 41: 914–921.

Dhawan, S. K., Kumar, S. A., Bhandari, H., Bisht, B. M. S. and Khatoon, F. (2015). American Journal of Polymer Science, 5(1A): 7–17.

El-Shazly, A. H. and Al-Turaif, H. A. (2012). Improving the corrosion resistance of buried steel by using polyaniline coating. International Journal of Electrochemical Science, 7: 211–221.

Fang, J., Xu, K., Zhu, L., Zhou, Z. and Tang, H. (2007). Corrosion Science, 49: 4232–4242.

Fahlman, M., Guan, H., Smallfield, J. A. O. and Epstein, A. T. (1998). In Technical Papers of the Annual Technical Conference-Society of Plastics Engineers Incorporated, 2: 1238–1241.

Forsgren, A. and Schweitzer, P. A. (2006). Corrosion Control through Organic Coatings, CRC Press, Boca Raton.

Gangopadhyay, R. and De, A. (2000). Conducting polymer nanocomposites: a brief overview. Chemistry of Materials, 12(3): 608–622.

Ghanbari, A. and Attar, M. M. (2015). A study on the anticorrosion performance of epoxy nanocomposite coatings containing epoxy-silane treated nano-silica on mild steel substrate. Journal of Industrial and Engineering Chemistry, 23: 145–153.

Gläsel, H. J., Bauer, F., Ernst, H., Findeisen, M., Hartmann, E., Langguth, H., Mehnert, R. and Schubert, R. (2000). Preparation of scratch and abrasion resistant polymeric nanocomposites by monomer grafting onto nanoparticles, 2 Characterization of radiation-cured polymeric nanocomposites. Macromolecular Chemistry and Physics, 201(18): 2765–2770.

Gonzalez, C. A., Silva, R. C., Menchaca, C., Guzman, S. S. and Uruchurtu, J. (2011). Use of silica tubes as nanocontainers for corrosion inhibitor storage. Journal of Nanotechnology, DOI: 10.1155/2011/461313.

Gu, H., Guo, J., He, Q., Tadakamalla, S., Zhang, X., Yan, X., Huang, Y., Colorado, H. A., Wei, S. and Guo, Z. (2013). Flame-retardant epoxy resin nanocomposites reinforced with polyaniline-stabilized silica nanoparticles. Industrial & Engineering Chemistry Research, 52(23): 7718–7728.

Hench, L. L. and West, J. K. (1990). The Sol-Gel process, Chemical Reviews, 90(1): 33–72.

Herrasti, P. and P. Oc′on. (2001). Polypyrrole layers for steel protection. Applied Surface Science, 172(3-4): 276–284.

Huang, T. C., Su, Y. A., Yeh, T. C., Huang, H. Y., Wu, C. P., Huang, K. Y., Chou, Y. C., Yeh, J. M. and Wei, Y. (2011). Advanced anticorrosive coatings prepared from electroactive epoxy–SiO₂ hybrid nanocomposite materials. Electrochimica Acta, 56(17): 6142–6149.

Hur, E., Bereket, G. and Sahin, Y. (2006). Corrosion inhibition of stainless steel by polyaniline, poly (2-chloroaniline), and poly (aniline-co-2-chloroaniline) in HCl. Progress in Organic Coatings, 57(2): 149.

Iribarren, J. I., Cadena, F. and Liesa, F. (2005). Corrosion protection of carbon steel with thermoplastic coatings and alkyd resins containing polyaniline as conductive polymer. Progress in Organic Coatings, 52: 151–160.

Iroh, J. O. (2002). Conducting polymer coating: a viable alternative to chromate conversion coating. Surface Engineering, 17(4): 265–267.

Jadhav, N. and Gelling, V. (2015). Titanium dioxide/conducting polymers composite pigments for corrosion protection of cold rolled steel. Journal of Coatings Technology and Research, 12(1): 137–152.

Jin, H. Y., Yang, Y. Q., Xu, L. and Hou, S. E. (2011). Effects of spherical silica on the properties of an epoxy resin system. Journal of Applied Polymer Science, 121(2): 648–653.

Kirubaharan, K., Selvaraj, A. M., Maruthan, M. and Jeyakumar, K. D. (2012). Synthesis and characterization of nanosized titanium dioxide and silicon dioxide for corrosion resistance applications. Journal of Coatings Technology and Research, 9(2): 163–170.

Kamaraj, K., Karpakam, V., Azim, S.S. and Sathiyanarayanan, S. (2012). Electropolymerised polyaniline films as effective replacement of carcinogenic chromate treatments for corrosion protection of aluminium alloys. Synthetic Metals, 162(5-6): 536–542.

Kilmartin, P. A., Trier, L. and Wright, G. A. (2002). Corrosion inhibition of polyaniline and poly (o-methoxyaniline) on stainless steels. Synthetic Metals, 131(1-3): 99–109.

Kinlen, P. J., Silverman, D. C. and Jeffreys, C. R. (1997). Corrosion protection using polyanujne coating formulations. Synthetic Metals, 85(1-3): 1327–1332.

Kumar, A., Stephenson, L. D. and Murray, J. N. (2006). Self-healing coatings for steel. Progress in Organic Coatings, 55(3): 244–253.

Kumar, S. A., Bhandari, H., Sharma, C., Khatoonb, F. and Dhawan, S. K. (2013). A new smart coating of polyaniline–SiO$_2$ composite for protection of mild steel against corrosion in strong acidic medium. Polymer International, 62(8): 1192–1201.

Lawrence, G. A., Barkac, A. K., Chasser, M. A., Desaw, A. S., Hartman, E. M., Mavis, E., Hayes, E. D., Hockswender, R. T., Kuster, L. K., Montague, A. R., Nakajima, M., Olson, G. K., Richardson, S. J., Sadvari, J. R., Simpson, A. D., Tyebjee, S. and Wilt, F. T. (2002). US Patent 6387519.

Le, D. P., Yoo, Y. H., Kim, J. G., Cho, S. M. and Son, Y. K. (2009). Corrosion characteristics of polyaniline-coated 316L stainless steel in sulphuric acid containing fluoride. Corrosion Science, 51(2): 330–338.

Lenz, D. M., Ferreira, C. A. and Delamar, M. (2002). Distribution analysis of TiO$_2$ and commercial zinc phosphate in polypyrrole matrix by XPS. Synthetic Metals, 126(2-3): 179–182.

Liu, J., Janjua, Z. A., Roe, M., Xu, F., Turnbull, B., Choi, K. -So. and Hou, X. (2016). Super-hydrophobic/icephobic coatings based on silica nanoparticles modified by self-assembled monolayers. Nanomaterials, 6: 232.

Louis Floyd, F., Avudiappan, S., Gibson, J., Mehta, B., Smith, P., Provder, T. and Escarsega, J. (2009). Using electrochemical impedance spectroscopy to predict the corrosion resistance of unexposed coated metal panels. Progress in Organic Coatings, 66(1): 8–34.

Mahulikar, P. P., Jadhav, R. S. and Hundiwale, D. G. (2011). Performance of polyaniline/TiO$_2$ nanocomposites in epoxy for corrosion resistant coatings. Iranian Polymer Journal, 20(5): 367–376.

McAndrew, T. P. and Miller, S. A. (1998). Polyaniline in corrosion-resistant coatings. pp. 396–408. *In*: Bierwagen, G. P. (ed.). Organic Coatings for Corrosion Control. volume 689 of ACS Symposium Series, Book Chapter 32. American Chemical Society, Washington, DC, U.S.A., ISBN 0-84123549-X.

Meneguzzi, A., Ferreira, C. A., Pham, M. C., Delamar P. C. and Lacaze, M. (1999). Electrochemical synthesis and characterization of poly (5-amino-1-naphthol) on mild steel electrodes for corrosion protection. Electrochimica Acta, 44(12): 2149–2156.

Mavinakuli, P., Wei, S., Wang, Q., Karki, A. B., Dhage, S., Wang, Z., Young, D. P. and Guo, Z. (2010). Polypyrrole/silicon carbide nanocomposites with tunable electrical conductivity. The Journal of Physical Chemistry C, 114(9): 3874–3882.

Merisalu, M., Kahro, T., Kozlova, J., Niilisk, A., Nikolajev, A., Marandi, M., Floren, A., Alles, H. and Sammelselg, V. (2015). Graphene–polypyrrole thin hybrid corrosion resistant coatings for copper. Synthetic Metals, 200: 16–23.

Mobin, M. and Tanveer, N. (2012). Corrosion performance of chemically synthesized poly (aniline-co-o-toluidine) copolymer coating on mild steel. Journal of Coatings Technology and Research, 9(1): 27.

Moraes, S.R., Huerta-Vilca, D. and Motheo, A. J. (2002). Molecular Crystals and Liquid Crystals Science and Technology, Section A, 374(1): 391–396.

Moraes, S. R., Huerta-Vilca, D. and Motheo, A. J. (2003). Corrosion protection of stainless steel by polyaniline electrosynthesized from phosphate buffer solutions. Progress in Organic Coatings, 48(1): 28–33.

Niedbala, J. (2011). Surface morphology and corrosion resistance of electrodeposited composite coatings containing polyethylene or polythiophene in Ni-Mo base. Bulletin of Material Science, 34(4): 993.

Nooshabadi, M. S., Ghoreishi, S. M., Bidgoli, H. E. and Jafari, Y. J. (2015). Direct electrosynthesis of Polyaniline–Fe$_2$O$_3$ nanocomposite coating on aluminum alloy 5052 and its corrosion protection performance. Journal of Nanostructures, 5(4): 423–435.

Olad, A. and Naseri, B. (2010). Preparation, characterization and anticorrosive properties of a novel polyaniline/clinoptilolite nanocomposite. Progress in Organic Coatings, 67(3): 233–238.

Park, J. H., Lee, G. D., Nishikata, A. and Tsuru, T. (2002). Anticorrosive behavior of hydroxyapatite as an environmentally friendly pigment. Corrosion Science, 44(5): 1087–1095.

Pud, A., Shapoval, G. S., Kamarchik, P., Ogustov, N., Gromovaya, V. F., Myronyk, I. E. and Konstur, Y. V. (1999). Electrochemical behavior of mild steel coated by polyaniline doped with organic sulfonic acids. Synthetic Metals, 107(2): 111–115.

Radhakrishnan, S., Siju, C. R., Mahanta, D., Patil, S. and Madras, G. (2009). Conducting polyaniline–nano-TiO$_2$ composites for smart corrosion resistant coatings. Electrochimica Acta, 54(4): 1249–1254.

Radhakrishnan, S., Sonawane, N. and Siju, C. R. (2009). Epoxy powder coatings containing polyaniline for enhanced corrosion protection. Progress in Organic Coatings, 64(4): 383–386.

Rathod, R. C., Umare, S. S., Didolkar, V. K., Shambharkar, B. H. and Patil, A. P. (2013). Production and characterization of PANI/TiO$_2$ nanocomposites: anticorrosive application on 316LN SS. Transactions of the Indian Institute of Metals, 66: 97–104.

Rawat, N. K., Sinha, A. K. and Ahmad, S. (2015). Conducting poly (o-anisidine-co-o-phenyldiammine) nanorod dispersed epoxy composite coatings: synthesis, characterization and corrosion protective performance. RSC Advances, 5(115): 94933–94948.

Rao, K. S., Hami, K. E., Kodaki, T., Matsushige, K. and Makino, K. (2005). A novel method for synthesis of silica nanoparticles. Journal of Colloid and Interface Science, 289(1): 125–131.

Ruhi, G., Bhandari, H. and Dhawan, S. K. (2015). Corrosion resistant polypyrrole/flyash composite coatings designed for mild steel substrate. American Journal of Polymer Science, 5(1A): 18–27.

Saidman, S. B. (2002). The effect of pH on the electrochemical polymerisation of pyrrole on aluminium. Journal of Electroanalytical Chemistry, 534(1): 39–45.

Sambyal, P., Ruhi, G., Bhandari, H. and Dhawan, S. K. (2015). Advanced anti corrosive properties of poly (aniline-co-o-toluidine)/flyash composite coatings. Surface and Coatings Technology, 272: 129–140.

Sambyal, P., Ruhi, G., Bhandari, H. and Dhawan, S. K. (2015). Designing of smart coatings of conducting polymer poly (aniline-co-phenetidine)/SiO$_2$ composites for corrosion protection in marine environment. Surface and Coatings Technology, 303: 362–371.

Santos, J. R., Mattoso, L. H. C. and Motheo, A. J. (1998). Investigation of corrosion protection of steel by polyaniline films. Electrochimica Acta, 43(3-4): 309–313.

Sathiyanarayanan, S., Dhawan, S. K., Trivedi, D. C. and Balakrishnan, K. (1992). Soluble conducting poly ethoxy aniline as an inhibitor for iron in HCl. Corrosion Science, 33(12): 1831–1841.

Sathiyanarayanan, S., Muthkrishnan, S. and Venkatachar, G. (2006). Corrosion protection of steel by polyaniline blended coating. Electrochimica Acta, 51: 6313–6319.

Sathiyanarayanan, S., Azim, S. and Venkatachari, G. (2008). Corrosion protection coating containing polyaniline glass flake composite for steel. Electrochimica Acta, 53(5): 2087–2094.

Sato, M., Yamanaka, S., Nakaya, J. and Hyodo, K. (1994). Electrochemical copolymerization of aniline with o-aminobenzonitrile. Electrochimica Acta, 39(14): 2159–2167.

Scully, J. R. and Hensley, S. T. (1994). Lifetime prediction for organic coatings on steel and a magnesium alloy using electrochemical impedance methods. Corrosion, 50(9): 705–716.

Sengodu, P. and Deshmukh, A. D. (2015). Conducting polymers and their inorganic composites for advanced Li-ion batteries: a review. RSC Advances, 5(52): 42109–42130.

Shah, K. and Iroh, J. O. (2004). Adhesion of electrochemically formed conducting polymer coatings on Al-2024. Surface Engineering, 20(1): 53–58.

Shang, Q., Wang, M., Liu, H., Gao, L. and Xiao, G. (2013). Facile fabrication of water repellent coatings from vinyl functionalized SiO$_2$ spheres. Journal of Coatings Technology and Research, 10(4): 465–473.

Shi, H., Liu, F., Yang, L. and Han, E. (2008). Characterization of protective performance of epoxy reinforced with nanometer-sized TiO$_2$ and SiO$_2$. Progress in Organic Coatings, 62(4): 359–368.

Shi, S., Zhang, Z. and Yu, L. (2017). Hydrophobic polyaniline/modified SiO$_2$ coatings for anticorrosion protection. Synthetic Metals, 233: 94–100.

Shi, X., Nguyen, T. A., Suo, Z., Liu, Y. and Avci, R. (2009). Effect of nanoparticles on the anticorrosion and mechanical properties of epoxy coating. Surface and Coatings Technology, 204(3): 237–245.

Silverstein, R. M. and Webster, F. X. (2002). Spectrometric Identification of Organic Compounds, VI Edition, John Wiley and Son, Wiley India, pp. 165.

Sivaraman, P., Rath, S. K., Hande, V. R., Thakur, A. P., Patri, M. and Samui, A. B. (2006). Synthetic Metals, 156(16-17): 1057–1064.

Skotheim, T. A. (1997). Handbook of Conducting Polymers. CRC press.

Song, E. and Choi, J. -W. (2013). Conducting polyaniline nanowire and its applications in chemiresistive sensing. Nanomaterials, 3(3): 498–523.

Srikanth, A. P., Raman, V., Tamilselvi, S., Nanjundan, S. and Rajendran, N. (2008). Electropolymerization and corrosion protection of polyaniline and its copolymer on carbon steel. Anti-Corrosion Methods and Materials, 55(1): 3–9.

Stober, W., Fink, A. and Bohn, E. (1968). Controlled growth of monodisperse silica spheres in the micron size range. Journal of Colloid and Interface Science, 26(1): 62–69.

Syed, J. A., Tang, S. and Meng, X. (2017). Super-hydrophobic multilayer coatings with layer number tuned swapping in surface wettability and redox catalytic anti-corrosion application. Scientific Reports, 7: 4403. DOI: 10.1038/s41598-017-04651-3.

Talo, A., Passiniemi, P., Forsen, O. and Ylasaari, S. (1997). Corrosion protective polyaniline epoxy blend coatings on mild steel. Synthetic Metals, 102(1-3): 1394–1395.

Tan, T. T. Y., Liu, S., Zhang, Y., Han, M. -Y. and Selvan, S. T. (2011). Microemulsion preparative methods (overview). Comprehensive Nanoscience and Technology, 5: 399–441.

Tanveer, N. and Mobin, M. (2011). Corrosion protection of carbon steel by poly (aniline-co-o-toluidine) and poly (pyrrole-co-o-toluidine) copolymer coatings. Journal of Minerals and Materials Characterization and Engineering, 10(8): 735–753.

Tanveer, N. and Mobin, M. (2012). Anti-corrosive properties of poly (2-pyridylamine-co-aniline-co-2, 3-xylidine) terpolymer coating on mild steel in different corrosive environments. Progress in Organic Coatings, 75(3): 231–240.

Tanveer, N. and Mobin, M. (2014). Corrosion performance evaluation of chemically synthesized polyaniline and its co- and ter-polymer coatings on mild steel in different media. Chemical Science Review and Letters, 3(11s): 14–32.

Tavandashti, N. P., Sanjabi, S. and Shahrabi, T. (2009). Corrosion protection evaluation of silica/epoxy hybrid nanocomposite coatings to AA2024. Progress in Organic Coatings, 65(2): 182–186.

Vansant, E. F., Voort, P. V. D. and Vrancken, K. C. (1995). Characterization and Chemical Modification of the Silica Surface, Elsevier Science, New York, NY, USA.

Wang, M. H., Ruan, W. H., Huang, Y. F., Ye, L., Rong, M. Z. and Zhang, M. Q. (2012). A strategy for significant improvement of strength of semi-crystalline polymers with the aid of nanoparticles. Journal of Materials Chemistry, 22(11): 4592–4598.

Wang, M., Liu, M. and Fu, J. (2015). An intelligent anticorrosion coating based on pH-responsive smart nanocontainers fabricated via a facile method for protection of carbon steel. Journal of Materials Chemistry A, 3(12): 6423–6431.

Wankhede, R. G., Morey, S., Khanna, A. S. and Birbilis, N. (2013). Development of water-repellent organic–inorganic hybrid sol–gel coatings on aluminum using short chain perfluoro polymer emulsion. Applied Surface Science, 283: 1051–1056.

Wei, Y., Wang, J., Jia, X., Yeh, J. M. and Spellane, P. (1995). Polyaniline as corrosion protection coatings on cold rolled steel. Polymer, 36(23): 4535–4537.

Wessling, B. (1994). Passivation of metals by coating with polyaniline: Corrosion potential shift and morphological changes. Advanced Materials, 6(3): 226–228.

Wicks, Z. W., Jone, F. N. and Pappas, S. P. (2007). Organic Coatings: Science and Technology, Wiley, New Jersey.

Xu, L. and He, J. (2012). Fabrication of highly transparent superhydrophobic coatings from hollow silica nanoparticles. Langmuir, 28(19): 7512–7518.

Yang, O., Zhang, Y., Li, H., Zhang, Y., Liu, M., Luo, J., Tan, L., Tang, H. and Yao, S. (2010). Electrochemical copolymerization study of o-toluidine and o-aminophenol by the simultaneous EQCM and *in situ* FTIR spectroelectrochemisty. Talanta, 81(12): 664–672.

Yao, B., Wang, G., Li, X. and Zhang, Z. (2009). Anticorrosive properties of epoxy resin coatings cured by aniline/p-phenylenediamine copolymer. Journal of Applied Polymer Science, 112(4): 1988–1993.

Yalcinkaya, S., Tuken, T., Yazici, B. and Erbil, M. (2010). Electrochemical synthesis and corrosion behaviour of poly (pyrrole-*co*-o-anisidine-*co*-o-toluidine). Current Applied Physics, 10(3): 783–789.

Yi, Y., Liu, G., Jin, Z. and Feng, D. (2013). The use of conducting polyaniline as corrosion inhibitor for mild steel in hydrochloric acid. International Journal of Electrochemical Science, 8: 3540–3550.

Yu, Q., Xu, J., Liu, J., Li, B., Liu, Y. and Han, Y. (2012). Synthesis and properties of PANI/SiO$_2$ organic-inorganic hybrid films. Applied Surface Science, 263: 532–535.

Zhang, B., Xu, Y., Zheng, Y., Dai, L., Zhang, M., Yang, J., Chen, Y., Chen, X. and Zhou, J. (2011). A facile synthesis of polypyrrole/carbon nanotube composites with ultrathin, uniform and thickness-tunable polypyrrole shells. Nanoscale Research Letters, 6(1): 431.

Zhang, X., Xu, W., Xia, X., Zhang, Z. and Yu, R. (2006). Toughening of cycloaliphatic epoxy resin by nanosize silicon dioxide. Materials Letters, 60(28): 3319–3323.

Zubillaga, O., Cano, F. J., Azkarate, I. S., Molchan, I. S., Thompson, G. E. and Skeldon, P. (2009). Anodic films containing polyaniline and nanoparticles for corrosion protection of AA2024T3 aluminum alloy. Surface and Coatings Technology, 203(10-11): 1494–1501.

Zu, L., Cui, X., Jiang,Y., Hu, Z., Lian, H., Liu, Y., Jin, Y., Li, Y. and Wang, X. (2015). Preparation and electrochemical characterization of mesoporous polyaniline-silica nanocomposites as an electrode material for pseudocapacitors. Materials, 8(4): 1369–1383.

Poly(Aniline-co-Pentafluoroaniline)/ ZrO_2 Nanocomposite Based Anticorrosive Coating

4.1 Introduction

In order to increase the efficiency of the conductive polymer as a protective coating on metals, several strategies have been used. For example, incorporation of nanofillers in conducting polymer matrix is found to be one of the most powerful techniques to improve the barrier performance and mechanical integrity of the coating. In general, the unique combination of the nanomaterial and its characteristics include size, mechanical properties and low concentrations that is necessary to affect change in a polymer matrix. It is observed that the incorporation of nanomaterials like metal oxide nanoparticles in the epoxy coating is found to be an effective approach to enhance the barrier and mechanical performance of the coating, which also supports the probability of developing environmentally friendly anti-corrosion coatings that can last much longer compared to the traditional coatings. Significant researches have been done on nanocomposites coating for various applications, including on corrosion prevention. Incorporation of nanoparticles in the paint for mechanical property improvement, wear resistance, UV-protection, water repellency, gas barrier properties, etc., are in widespread commercial practice. For example, the scratch resistance of the paint can be improved by incorporating nanoparticles of alumina and silica. The barrier property of the epoxy coating is found to boost after incorporation of nanophase by effectively decreasing porosity and diffusion path, i.e., the incorporation of clay in polymer matrix nanocomposites increases water and gas barrier properties. The intermingling of nanomaterials is said to effectively fill tiny flaws providing more effective passive protection. Low wettability effectively prevents water on the substrate surface and exhibits excellent corrosion resistance in a wet environment. As a result of above particulars, nanoparticles incorporated coatings are expected to have a dramatic increase in resistance to corrosion of the substrate due to their improved hydrophobic, anti-wear and self-cleaning properties. It should be remembered that in applications, such as industrial coating, better performance is correlated factor, thanks to the hydrophobicity and wear/abrasion resistance of the coating. It is known that nanomaterial enhances these properties greatly.

Different inorganic nanofillers such as TiO_2 (Lenz et al. 2002; DoğruMert 2016), ZrO_2 (Bhattacharya et al. 1996), CeO_2 (Deflorian et al. 2011; Brusciotti et al. 2010), Al_2O_3 (Hawthorne et al. 2004) and SiO_2 (Kumar et al. 2013) nanoparticles have been incorporated with protective coatings to enhance the mechanical strength and corrosion protection performance of the coating.

Pareja et al. (2006) reported that the zirconia nanoparticles based barrier coating could be effectively used for protection of galvanized steel. Anwer et al. (2013) have developed polyaniline-zirconia nanocomposites by *in situ* oxidative polymerization method. The nanocomposites have showed better conductivity, and poorer isothermal and cyclic stability than pure polyaniline. However, zirconia nanoparticles have unique properties such as greater hardness, high density and better mechanical strength which make the polymer suitable for its wide range of applications.

As a filler material in polymer coatings, it can efficiently improve the resistance of the coatings toward chloride ions penetration by significantly reducing the rate of corrosion. ZrO_2 nanoparticles are specifically attractive due to their extraordinary mechanical strength, temperature resistance and chemical stability. Zirconium dioxide (ZrO_2), is a white crystalline oxide of zirconium and its most naturally occurring form, with a monoclinic crystalline structure (Septawendar et al. 2011), has resulted in a range of industrial and engineering applications that include high durability coating, medical prosthetics, catalytic agents, synthetic jewels, etc. (Piconi and Maccauro 1991; Zhang et al. 2000). Recently, Arun Jeeva Nijanthan et al. (2017) reported nanocoating of cobalt and nickel base alloy by zirconium using sol-gel process and validated the same with the electrophoretic deposition processes. The results revealed that there was an 84% increase in corrosion resistant and 30% improvement in wear resistance properties.

Gnedenkov et al. (2015) demonstrated the mechanical and anti-corrosive performance of coating based on silica and zirconium dioxide nanoparticles on magnesium alloy. The results revealed that the best corrosion inhibition and mechanical properties were observed in the coating based on zirconium dioxide nanoparticles. Mansour et al. (2017) and his coworkers investigated the physico-mechanical and morphological features of zirconia substituted hydroxyapatite nanocrystals. It was observed that the superior physico-mechanical and anti-corrosive performance of hydroxyapatite nanocrystal was due to the incorporation of zirconia nanoparticles. Brzezińska-Miecznik et al. (2014) have studied the addition of zirconia on hydroxyapatite (HAP) hot pressed materials, and the results presented extraordinary enhancement of hardness and mechanical strength compared to pure HAP.

A review by Milŏsev et al. (2018) gave details of conversion coatings based on zirconium and/or titanium. Authors reported that the conversion coating based on titanium or zirconium offered a great potential for various applications, especially for protection of metals from corrosion. Figueroa-Lara et al. (2017) studied the effect of zirconia nanoparticles in epoxy-silica hybrid adhesives to join aluminum substrates and reported that the addition of different amounts of zirconia nanoparticles increased the shear strength of the adhesively bonded aluminum joint. Medina et al. (2008) investigated the tensile properties and toughness of an epoxy resin containing zirconia nanoparticles. Mirabedini et al. (2012) and Behzadnasab et al. (2013) also investigated the effect of various combinations of zirconia and organo-clay nanoparticles on the mechanical and thermal properties of an epoxy nanocomposite coating. Authors investigated that the mechanical and thermal properties of epoxy were improved by incorporation of zirconia nanoparticles.

Montemor and Ferreira (2007) and Zheludkevich et al. (2005) investigated anti-corrosive performance of silane-based hybrid films modified with ceria-zirconia (CeO_2-ZrO_2) nanoparticles. Results demonstrated that CeO_2-ZrO_2 nanoparticles were very effective fillers, and showed improved barrier and corrosion protection properties of the silane coatings. Incorporation of zirconia nanoparticles led to improvement of the barrier properties of the coatings. Dorigato et al. (2010) demonstrated that the mechanical performances of an epoxy-based adhesive, which was improved by the addition of zirconia nanoparticles. They also concluded that the addition of zirconia nanoparticles could effectively improve epoxy adhesives, both by increasing their mechanical properties and by enhancing the interfacial wettability with an aluminum substrate.

Medina et al. (2008) incorporated zirconia nanoparticles to a diglycidyl ether of bisphenol-A based epoxy resin and reported the morphological and mechanical properties of the resulting composites. Tensile modulus, as well as fracture toughness of the epoxy resin, was found to increase on increasing the zirconia nanoparticles in the matrix. Masim et al. (2017) investigated that the polyaniline-zirconia nanocomposite not only showed an excellent antibacterial and phosphate adsorbent material but also exhibited remarkable anti-corrosive properties. Hu et al. (2015) demonstrated the anti-corrosive properties of epoxy coating formulated with conducting copolymer zirconia nanocomposites such as poly(o-toluidine)/ZrO$_2$. Results showed that epoxy coating containing poly(o-toluidine)/ZrO$_2$ nanocomposite coated steel revealed higher corrosion resistance than poly(o-toluidine)/epoxy composite coated steel and polyaniline/epoxy composite coated steel. Authors explained that the poly(o-toluidine) powders dispersed in the epoxy and polyamide system, improving the barrier and electrochemical protection properties of the epoxy coating. Furthermore, ZrO$_2$ nanoparticles increased the tortuosity of the diffusion pathway of the corrosion substances.

However, the main drawback of this material is lack of self-healing ability if the coating gets damaged. Self-healing polymers and inorganic nanofiller impregnated polymer have the ability to heal in response to damage wherever and whenever it takes place in the material. The introduction of self-healing properties into the epoxy matrix is an important and challenging topic. Various micro or nanocontainers loaded self-healing agents have been developed and incorporated into the epoxy matrix to impart self-healing ability.

Self-repair ability was studied in numerous reports. Zhong et al. (2013) studied the vanadate containing Zr-based coatings on AA6063, which supported self-healing properties that were confirmed by Electrochemical Impedance Spectroscopy in 3.5% NaCl medium. Govindaraj et al. (2015) demonstrated the corrosion resistance and self-healing behavior of zirconium-cerium conversion coating developed on AA2024 alloy. The authors observed that the zirconium-cerium conversion coating showed better self-healing properties as compared to conventional chromate conversion coating. Zand et al. (2015) investigated the self-healing performance of silane hybrid coating containing activated ceria-zirconia nanoparticles. Shchukin et al. (2006) demonstrated the self-healing and anti-corrosive properties of composite based on ZrO$_2$/SiO$_2$ coating loaded with benzotriazolenano reservoirs. Similarly, Lvov et al. (2008) suggested some new designs of self-healing, anti-corrosion coating with the addition of nanocontainers equipped with corrosion inhibitors. This system was tested by loading halloysite nanotubes with the inhibitor benzotriazole, incorporated it into a ZrO$_2$-SiO$_2$ sol-gel coating and then deposited it onto aluminum alloy. It was demonstrated that after exposure to saline medium, corrosion protection efficiency was found to increase, as well as self-healing ability, was also achieved. Grigoriev et al. (2017) reviewed the recent achievements in the development of nanocontainers for self-healing corrosion protection coatings. The functionality and design of layer-by-layer assembled, polymer and inorganic nanocontainers were evaluated in the coatings for protection of steel and aluminum alloys. The release of the corrosion inhibitors from nanocontainers occurred only when triggered by local pH changes or other internal/external stimuli, which inhibited leakage of the corrosion inhibitor out of the coating and enhanced the coating durability. This led to the self-healing ability of the coating and terminated the rate of corrosion.

Kowalski et al. (2010) evaluated the self-healing ability of an intrinsically conducting polymer coating. Development of conducting polymer nanocomposites to be used as protection of metals against corrosion having hydrophobic, self-healing and brilliant mechanical properties are expected to inspire a major revolution in the field of corrosion. Huang et al. (2011) and Mahmood et al. (2016) reported the preparation and characterization of conductive polyaniline/zirconia nanocomposites. They investigated the effect of zirconia nanoparticles on thermal, electrical, morphological and crystalline behavior of polyaniline and results showed that all these properties were found to improve with the incorporation of zirconia nanoparticles in polyaniline matrix. Similarly, Ozkazanc (2016) reported the charge transport mechanism of poly-thiophene and its nanocomposites having various weights of zirconium dioxide nanoparticles.

Recently, hydrophobic nanocomposite based coatings have increased the life of materials prone to corrosion and this approach has led to better saving. Such type of coatings technologies has the market potential for an extensive range of applications like pipeline, marine, aerospace, chemical and construction industries. Hydrophobic substrates have the ability to eliminate the dust particles, water droplets and other contaminants from the surface. Yang et al. (2012) demonstrated the anti-corrosive and hydrophobic properties of the epoxy resin cured by electroactive amine-capped aniline trimer. The water repellant properties of hydrophobic electroactive epoxy acted as a barrier and prevented the entrance of water or other species, and thus protected the metal surface from corrosion. Stankiewicz et al. (2013) reported that the fluoropolymers are characterized by good thermal and chemical stability and the attractive property of such coatings is water repellency, which plays a vital part in corrosion protection.

Yabuki et al. (2007) reported that the self-healing ability of the fluorine-based coating could be improved by the incorporation of some metal powder. Yabuki et al. (2009) also investigated that the coating containing fluoropolymer have self-healing ability,by forming a barrier film on the damaged surface. Weng et al. (2015) synthesized advanced anti-corrosion coating materials using F-PANI/ silica composites which have synergistic effect of super-hydrophobicity and redox catalytic capability. These PANI-based super-hydrophobic surfaces were achieved by lowering surface energy by adding strongly hydrophobic substituent group (i.e., fluoro-moieties) into polyaniline chains. Weng et al. observed that three-dimensional microstructures/nanostructures of polyaniline were self-assembled by template-free method, combined with interfacial polymerization in the presence of per-fluoro-sebacic acid as a dopant, and the 3D microstructures/nanostructures of polyaniline showed both electrical conductivity and super-hydrophobicity.

Therefore, the present chapter describes the development of highly durable, inexpensive, hydrophobic and self-healing coating based on epoxy resin modified with conducting copolymer nanocomposite such as poly(aniline-co-pentafluoro aniline)/ZrO_2 and explains their physico-mechanical properties as well as anti-corrosion performance. This chapter also examines the influence of comonomer (pentafluoroaniline) and ZrO_2 nanoparticles on the surface morphology, anti-corrosion behavior and mechanical properties of the epoxy coating.

4.2 Synthesis of zirconia (ZrO_2) nanoparticles

Several methods have been employed to synthesize zirconia nanoparticles; for example, the microwave plasma method (Vollath and Sickafus 1992), sol–gel method (Stocker and Baiker 1998), precipitation method (Chen et al. 2001), hydrothermal method (Padovini et al. 2014), chemical vapor method (Srdic and Winterer 2006), etc. Sundram et al. (2007) described the synthesis of ZrO_2 nanoparticles of size 60 nm using microwave assisted sol-gel method. Zirconium oxide was obtained by drying the sample at 60°C for 3 hours and calcined at 400°C for 6 hours. Shukla et al. (2002) had prepared nanosized zirconia powder using a sol-gel technique which involved hydrolysis and condensation of zirconium (IV) n-propoxide in an alcohol solution and hydroxypropyl cellulose polymer as a steric stabilizer.

Dorigato et al. (2010) also synthesized zirconia nanoparticles using sol-gel route, and dispersed into an epoxy base for structural adhesives. Zirconia nanoparticles were used as synthesized or after its calcination. The authors reported that calcined zirconia nanoparticles showed better mechanical properties.

Kim et al. (2007) demonstrated the synthesis of zirconia nanopowder by the glycol-thermal process. Roy (2007) described the synthesized tetragonal zirconia using poly-acryl amide as a gel and matrix, and the average particle size of tetragonal zirconia was found to be about ~ 20 nm. Both the methods for preparing the zirconia nanoparticles used a synthetic template material, which was not economical and was relatively more expensive. Septawendara et al. (2011) synthesized zirconia using a precursor calcination process, involving hydrolysis of zirconium (IV) iso-propoxide and utilizing Oryza Sativa pulp (Merang Pulp) as the economical natural template. Apart from the above given methods, controlled precipitation method was one of the most employed methods that had been used to synthesize zirconia nanoparticles. This method was eco-friendly, cost-effective, rapid and convenient (Rezaei et al. 2007; Chen et al. 2001). In this method, zirconium oxychloride hydrate [($ZrOCl_2.8H_2O$)] and ammonia solution (30% w/w) was used as a precursor and solutions were prepared in double distilled water. The reaction was initiated by drop wise addition of ammonia solution to zirconium oxychloride solution at room temperature, and the reaction mixture was stirred

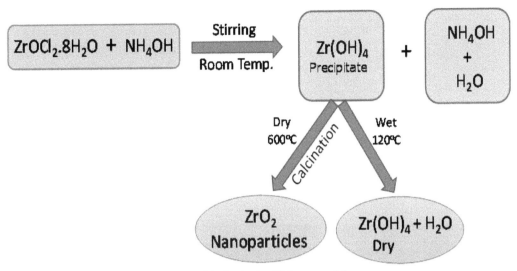

Scheme 4.1: Synthesis of ZrO₂ nanoparticles

vigorously till the pH of the solution was 10.0. White precipitates of $Zr(OH)_4$ was obtained which were then centrifuged at 4,000 rpm followed by washing and filtration to obtained $Zr(OH)_4$. The paste of $Zr(OH)_4$ was calcinated at about 600°C which resulted in the formation of ZrO_2 powder. The formation of ZrO_2 nanoparticles was confirmed by FTIR spectroscopy, SEM and TEM analysis. The chemical reactions involved in the synthesis of zirconia nanoparticles using precipitation method are shown in Scheme 4.1.

4.3 Preparation of poly(AN-co-PFA)/ZrO₂ nanocomposites

Various methods have been employed to synthesize conducting polymer/metal oxide nanocomposite. *In situ* oxidative polymerization method involves the polymerization of a solution containing monomer, dopant and the metal nanoparticles in presence of an oxidizing agent. Although, some physical deposition, thermal, electrochemical and photochemical methods have been developed for synthesizing various conducting polymer-metal oxide nanocomposites. They generally yield comparatively small quantities of the product and are not eco-friendly (Cho et al. 2003). Prasanna et al. (2016) synthesized polyaniline/zirconium oxide nanocomposites by oxidative interfacial polymerization of aniline in the presence of ZrO_2 nanoparticles using an ammonium persulphate (APS) as an oxidant in acidic medium.

In this chapter, synthesis of conducting copolymer/zirconia nanocomposites using *in situ* chemical oxidative polymerization is explained, which involves the incorporation of ZrO_2 nanoparticles during copolymerization of aniline (AN) and 2,3,4,5,6 pentafluoro aniline (PFA) in presence of ammonium persulfate (APS) as an oxidant and orthophosphoric acid as a dopant.

For the preparation of poly(AN-co-PFA)/ZrO_2 nanocomposite, aqueous mixture of aniline (0.1 M), PFA (0.01 M), orthophosphoric acid (0.2 M) and ZrO_2 (20 g) was homogenized using high speed blender followed by transferring the reaction mixture to double-walled glass reactor under constant stirring. The copolymerization was initiated by the dropwise addition of aqueous solution of APS (0.1 M) at low temperature (< 5°C) with continuous stirring for 4–6 hours.

Synthesized copolymer nanocomposite was isolated from the reaction mixture by filtration and followed by washing to eliminate oligomers and oxidant. The paste of copolymer nanocomposites was dried at about 60°C. Synthesis of zirconia nanoparticles and its nanocomposites with conducting copolymer is shown with the help of flow diagram (Figure 4.1). Scheme 4.2 displays the schematic diagram for the preparation of poly(AN-co-PFA)/ZrO_2 nanocomposite.

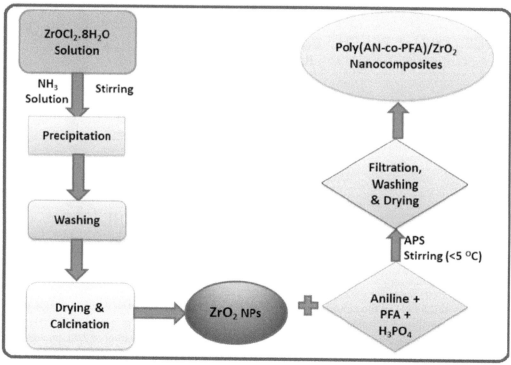

Figure 4.1: Flow diagram of preparation of poly(AN-co-PFA)/ZrO$_2$ nanocomposites

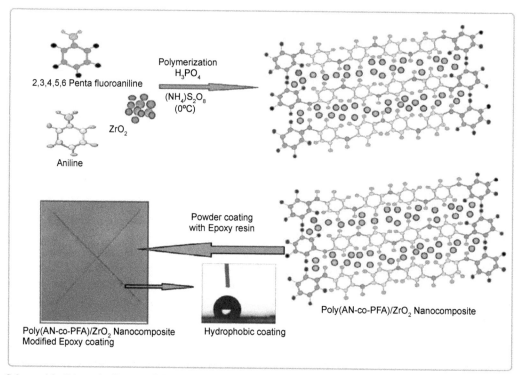

Scheme 4.2: Schematic diagram of the formation of poly(AN-co-PFA)/ZrO$_2$ nanocomposite (Bisht et al. Current Smart Materials, DOI: 10.2174/2405465802666170707163408)

Color version at the end of the book

4.4 Development of epoxy formulated poly(AN-co-PFA)/ZrO₂ nanocomposites coating on mild steel

Mild steel panels were coated with different wt. ratio of the poly(AN-co-PFA)/ZrO₂ formulated with the epoxy resin used for corrosion studies. Before applying these formulations on the mild steel surface, surface treatment of the panels was carried out. This involved the cleaning using suitable solvents such as acetone and trichloroethylene to remove extra impurities on the metal surface. The powder copolymer nanocomposites were mixed with epoxy resin in various loading like 1.0 wt.%, 2.0 wt.%, 3.0 wt.% and 4.0 wt.% followed by blending with epoxy resin using a laboratory ball mill and then applied on mild steel panels using an electrostatic spray gun. Powder coating technique is one of the best methods to improve the adherence ability of the coating on the metal surface; therefore, powder coating method is performed in the present chapter.

The powder coated mild steel panels were cured at 140–150°C for 30 minutes. Mild steel panels coated with epoxy resin is designated as EC, epoxy coatings with different wt.% loadings of poly(AN-co-PFA) are designated as PF2 (2.0 wt.%), PF4 (4.0 wt.%) and epoxy coatings with different wt.% loadings of poly(AN-co-PFA)/ZrO₂ nanocomposites are designated as PFZ1 (1.0 wt.%), PFZ2 (2.0 wt.%), PFZ3 (3.0 wt.%) and PFZ4 (4.0 wt.%).

4.5 Characterization of copolymer nanocomposite and epoxy modified copolymer nanocomposite coated substrate

4.5.1 Fourier Transform Infrared Spectroscopy (FTIR)

Formation of zirconia powder and their inclusion in poly(AN-co-PFA) matrix can be determined by FTIR spectroscopy. Figure 4.2 shows the FTIR spectra of ZrO₂, poly(AN-co-PFA) and poly(AN-co-PFA)/ZrO₂. The FTIR spectra of zirconia showed a characteristic absorption band at 462 cm⁻¹ which was assigned to the vibration of the Zr-O bond in ZrO₂ (Santos et al. 2008).

Ozkazanc (2016) studied the FTIR spectra of zirconia nanoparticles. The characteristic adsorption bands of zirconia nanoparticles were observed at 3,458, 1,634, 1,052, 746 and 519 cm⁻¹. The bands at

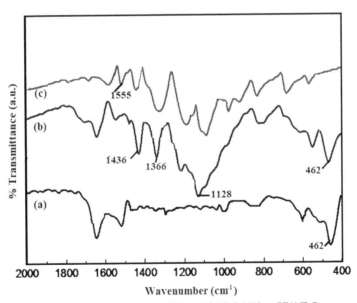

Figure 4.2: FTIR spectra of (a) Zirconia (b) Poly(AN-co-PFA) and (c) Poly(AN-co-PFA)/ZrO₂ nanocomposite (Bisht et al. Current Smart Materials, DOI: 10.2174/2405465802666170707163408)

3,458 and 1,634 cm^{-1} were related to OH bending and stretching vibration representing the presence of molecular water. The peak observed at 1,052 cm^{-1} was due to Zr–O$_2$–Zr asymmetric mode. The peaks at 746 and 519 cm^{-1} were related to characteristic monoclinic and tetragonal phases of zirconia nanoparticles.

These peaks were also observed in poly(AN-co-PFA)/ZrO$_2$ nanocomposite which indicated interaction of zirconia nanoparticles in the copolymer chain. The main characteristics bands at 1,555 and 1,436–1,440 cm^{-1} were attributed to the stretching mode of C=N and C=C respectively. The bands at 1,251 cm^{-1} indicated the C–N stretching mode of benzenoid ring attributed to the polymerization of aniline. Poly(AN-co-PFA) and poly(AN-co-PFA)/ZrO$_2$ showed the characteristic strong peaks at 1,128 cm^{-1} and 1,366 cm^{-1} due to C-F stretching mode which indicated the interaction of PFA and aniline moieties in the polymer chain. Furthermore, a characteristic broad absorption band at 1,620 cm^{-1} was related to vibrations of the absorbed water molecules.

FTIR spectra of polyaniline-zirconia nanocomposite had also been described in the literature (Prasanna et al. 2016). It was observed that the characteristic peaks at about 743 and 508 cm^{-1}, which were also present in polyaniline-zirconia nanocomposites, revealed that interaction of zirconia nanoparticles in polyaniline matrix.

Hu et al. (2015) also compared the FTIR spectra of PANI/ZrO$_2$, poly(o-toluidine) and poly(o-toluidine)/ZrO$_2$ nanocomposites. It was observed that the FTIR spectra of PANI and poly(o-toluidine) were almost similar, except a characteristic peak around at 2,913 cm^{-1} in poly(o-toluidine) which was due to the C–H stretching of the methyl group. On comparing the FTIR spectra of Poly(o-toluidine) and poly(o-toluidine)/ZrO$_2$ nanocomposite, it was observed that the characteristic peaks of poly(o-toluidine)/ZrO$_2$ were found to shift to the lower wave numbers. This might be attributed to the incorporation of ZrO$_2$ nanoparticles into the poly(o-toluidine) matrix as a result of this certain physico-chemical interaction which occurred between Poly(o-toluidine) and ZrO$_2$ nanoparticles due to the hydrogen bond that developed by an interaction between the hydroxyl on the surface of zirconia nanoparticles and the imines group of poly(o-toluidine).

4.5.2 X-Ray diffraction analysis

Figure 4.3 shows the X-ray diffractogram pattern of zirconia, poly(AN-co-PFA) and poly(AN-co-PFA)/ZrO$_2$ nanocomposites. Besides FTIR spectroscopy, the powder X-ray diffraction (XRD) pattern of poly(AN-co-PFA)/ZrO$_2$ nanocomposites further confirmed the incorporation of zirconia nanoparticles into the copolymer matrix. Comparatively sharp diffraction peak at respective diffraction angles were shown by zirconia nanoparticles which indicated its crystalline nature. The powder X-ray diffraction technique was used to recognize the crystalline phases of the samples using a mono-chromatized Cu-K$_\alpha$ (1.5056 Å).

Figure 4.3(a) shows the XRD pattern of the crystalline zirconia powder, there are four strong diffraction peaks at 2θ = 29.87°, 34.40°, 50° and 59.69°, which is attributed to the formation of the zirconia nanoparticles (Ranjbar et al. 2012). These peaks are also observed in copolymer-zirconia nanocomposites, which indicate the crystalline nature of nanocomposites due to incorporation zirconia nanoparticles in conducting copolymer matrix. Meanwhile amorphous nature of copolymer based on poly(AN-co-PFA) is evident from the appearance of weak reflections peaks centered at 2θ values of approximately 49.2° 56.2°, 65.1° and 78°. The average size of the zirconia particles has also been calculated from the full width at half maximum (FWHM) values of the diffraction peaks using the Scherrer equation.

Crystalline size = K.λ/W.cos θ

where K = 0.9, the shape factor, λ = the wavelength of the X-ray used and W = (W$_b$ – W$_s$), the difference of the broadened profile width of the experimental sample and the standard width of the reference silicon sample. The average crystallite size of the synthesized zirconia particles is estimated to be 12 nm.

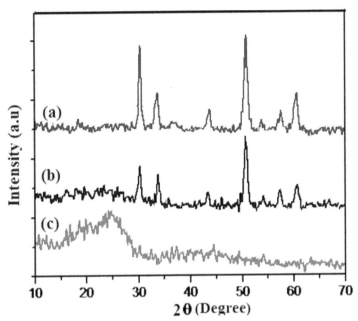

Figure 4.3: XRD curves of (a) ZrO₂, (b) Poly(AN-co-PFA)/ZrO₂ nanocomposite and (c) Poly(AN-co-PFA) (Bisht et al. Current Smart Materials, DOI: 10.2174/2405465802666170707163408)

Dorigato et al. (2010) studied the XRD pattern of zirconia nanoparticles, which clearly showed that the non-calcined zirconia nanoparticles were an amorphous. Meanwhile, after the calcination step, the powders were crystallized and showed mainly tetragonal zirconia, showing the monoclinic zirconia peaks with very low intensity. This showed that the increase of the density values was induced by the thermal treatment. Prasanna et al. (2016) reported the XRD analysis of polyaniline-zirconia nanocomposites. It was observed that the crystalline nature of polyaniline was due to the presence of zirconia nanoparticles in the polymer matrix. They also calculated the crystalline size using Scherrer's formula, which was found to be in the range of 23–25 nm. Hu et al. (2015) also studied the XRD pattern of ZrO₂ nanoparticles, poly(o-toluidine) and poly(o-toluidine)/ZrO₂ nanocomposite.

They explained that broad diffraction peaks occur between 2θ = 15°–25° was owing to the parallel and perpendicular periodicity of the polyaniline chains. XRD pattern of polyaniline showed that it existed due to the crystallization performance to a certain extent. According to Shi et al. (2009), low crystallinity of the polymers was due to the repetition of benzenoid and quinoid rings in polyaniline chain. Authors also noticed that the poly(o-toluidine) showed a single broad diffraction peak at 2θ = 20° due to the effects of the existence of the methyl substituent on the regularity of the whole polymer, which led to partial crystallization property to be covered. Furthermore, XRD pattern of ZrO₂ particles and poly(o-toluidine)/ZrO₂ nanocomposite were found to be similar. The broad diffraction peak of poly(o-toluidine) became sharp due to the interaction of ZrO₂ particles and the degree of crystallinity of poly(o-toluidine)/ZrO₂ increased.

4.5.3 Thermogravimetric analysis (TGA)

TGA is used to evaluate the thermal stability of copolymer nanocomposite in a nitrogen atmosphere at the heating rate of 10°C/minutes between 20°C to 700°C. Figure 4.4 shows the TGA curves of zirconia, conducting copolymer based on poly(AN-co-PFA) and poly(AN-co-PFA)/ZrO₂ nanocomposites. ZrO₂ nanoparticles show an excellent thermal stability up to 700°C and only 0.1% weight loss was calculated in zirconia powder. The TGA curve of poly(AN-co-PFA) and poly(AN-co-PFA)/ZrO₂

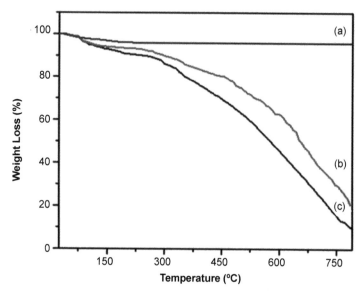

Figure 4.4: Thermo-gravimetric curves of (a) Zirconia (b) Poly(AN-co-PFA) and (c) Poly(AN-co-PFA)/ZrO$_2$ nanocomposite (Bisht et al. Current Smart Materials, DOI: 10.2174/2405465802666170707163408)

nanocomposite signify the first weight loss at 100–110°C is attributed to the loss of water molecules and other volatile species (Huang et al. 2011; Mahmood et al. 2016).

The weight loss in the second step at about 195°C indicates the loss of dopant ions or low molecular weight fragments as well as the onset of degradation of copolymer and copolymer nanocomposites. If the TGA traces of poly(AN-co-PFA) and poly(AN-co-PFA)/ZrO$_2$ nanocomposites are compared, it is observed that the first stage of weight loss is similar whereas the difference is noticed in the second weight loss. Figure 4.4b indicates that the poly(AN-co-PFA) is thermally stable up to 195°C, whereas thermal stability of poly(AN-co-PFA)/ZrO$_2$ nanocomposites is about 240°C. TGA studies indicate that the decomposition of copolymer nanocomposites has reduced and it shows greater thermal stability as compared to copolymer due to strong interaction of ZrO$_2$ nanoparticles in the copolymer matrix (Mostafaeia and Zolriasatein 2012; Shi et al. 2009).

According to the literature, the size of the nanoparticle influences the thermal stability of the polymer matrix. Gomez et al. (2015) developed a conducting polymer nanocomposites based on polypyrrole/silica nanocomposite with 20–100 nm diameter. The authors concluded that the nanoparticles improved the thermal stability of polypyrrole through the adsorption of volatile compounds on the surface where the smaller nanoparticles revealed better stabilization.

4.5.4 Morphological analysis

Scanning electron microscopy is performed on powdered materials at different magnifications. Figure 4.5 shows the SEM micrograph of poly(AN-co-PFA) and poly(AN-co-PFA)/ZrO$_2$ nanocomposite. The SEM micrograph ZrO$_2$ particles that was synthesized by precipitation method showed spherical morphology (Srdic and Winterer 2006), while poly(AN-co-PFA) showed cauliflower like morphology. SEM images of copolymer-zirconia nanocomposites indicated that ZrO$_2$ particles were found to be well-dispersed in the copolymer matrix which proposed good compatibility and strong interaction between zirconia and copolymer matrix as also shown in the TEM image of copolymer-zirconia nanocomposite. SEM image of polyaniline-zirconia nanocomposites also indicated the well-dispersed zirconia nanoparticles in polyaniline matrix as reported by Prasanna et al. (2016).

TEM and HRTEM image of zirconia showed that these particles were uniform and the average particle size calculated was about 10–12 nm as shown in Figure 4.6. When these nanoparticles were

Figure 4.5: SEM micrographs and EDS analysis of copolymer and its composite with zirconia nanoparticles (Bisht et al. Current Smart Materials, DOI: 10.2174/2405465802666170707163408)

incorporated in the conducting copolymer matrix, they showed agglomerated morphology and the dimension of copolymer-zirconia nanocomposite calculated was about 20–24 nm. Thus, it was confirmed from both XRD and TEM analysis that the crystallite size of zirconia and its composites with copolymer showed the nano range.

Figure 4.5 shows the EDS analysis of zirconia nanoparticles and its composite with the copolymer. EDS analysis of poly(AN-co-PFA)/ZrO₂ composite shows the peaks of carbon, oxygen, nitrogen, fluorine, and zirconium which confirms the presence of pentafluoroaniline and zirconia moieties in copolymer chain.

SEM analysis of cross-sectional view of the epoxy formulated coating is performed to demonstrate the thickness and surface topography of the coated surface. Figure 4.7 displays the SEM micrograph of the cross-sectional view of the epoxy coating formulated with 4.0 wt.% loading of poly(AN-co-PFA), the thickness of the coating is ~ 90 μm, as well as topography of the coating, exhibits the appearance of a homogenous surface (Figure 4.7).

4.5.5 Wettability test (contact angle measurement)

This test is used to evaluate the wettability of coated surfaces. Figure 4.8 shows the images obtained by the contact angle measurements using the sessile drop method. Water is used as a test liquid to evaluate whether a coated surface is hydrophilic (angle < 90°) or hydrophobic (angle > 90°) in nature. A low contact angle value (< 90°) indicates the hydrophilic nature of the coating. Hydrophobic surfaces

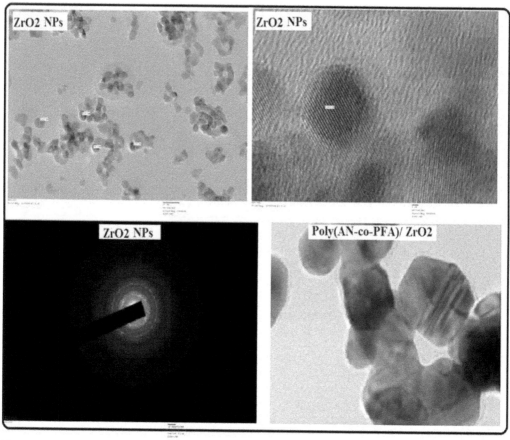

Figure 4.6: TEM and HRTEM images of zirconia nanoparticles and its composite with poly(AN-co-PFA) (Bisht et al. Current Smart Materials, DOI: 10.2174/2405465802666170707163408)

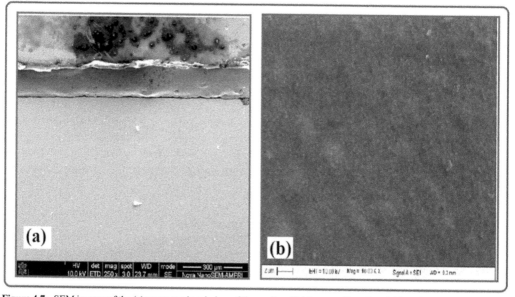

Figure 4.7: SEM images of the (a) cross-sectional view of the coating (thickness ~ 90 μm) and (b) surface topography of the coated surface (Bisht et al, Current Smart Materials, DOI: 10.2174/2405465802666170707163408)

are considered to be anti-corrosives nature since they repel water and reduce the rate of corrosion. Superhydrophobic surfaces are characterized by an apparent contact angle above 150°.

Xu et al. (2016) synthesized super-hydrophobic polyaniline by modification with different surfactants such as sodium dodecylbenzenesulfonate, polyethylene glycol, and cetyltrimethylammonium bromide, and the modified polyaniline was used as corrosion protection coatings. The authors observed that the unmodified polyaniline showed hydrophilic nature due to low contact angle value, i.e., 80°, while all modified PANI samples with water contact angles of 164°, 160° and 157° for polyethylene glycol modified PANI, cetyltrimethylammonium bromide modified PANI and sodium dodecylbenzenesulfonate modified PANI respectively. This behavior of conducting polymer indicated that the modification of surfactants had changed polyaniline matrix from hydrophilic to super-hydrophobic. Therefore, the authors illustrated that the surfactants have long chain alkyl group which showed hydrophobicity. When these molecules were linked polyaniline chain, the hydrophobic alkyl chain of surfactant protruded and reduced the surface energy of polyaniline. The larger the hydrophobic alkyl chain, the greater was the contact angle.

Dorigato et al. (2010) demonstrated the contact angle measurement of the epoxy coating containing zirconia nanoparticles. It was observed that in the pure epoxy sample an equilibrium contact angle of 82.1° was evaluated, while for the epoxy resin having 0.5 volume % loading of calcined zirconia nanoparticle, a mean contact angle of 71.5° was observed. At higher concentration of zirconia nanoparticles, the vibrated contact angle was slightly increased. The authors concluded that the increase of contact angle for relatively high filler contents was associated with the worsening of the dispersion degree of the nanofiller. Hence, the introduction of zirconia nanoparticles led to better interfacial wettability and chemical compatibility between the adhesive and the substrate. Therefore, in the case of poly(AN-co-PFA) and poly(AN-co-PFA)/ZrO$_2$ nanocomposite interaction of pentafluoroaniline comonomer contributed to increasing the surface hydrophobicity of coating.

Dhawan et al. (2015) demonstrated the hydrophobic and anti-corrosive properties of perfluoro-octanoic acid doped polyaniline and polyaniline-SiO$_2$ composites. The results showed that the contact angle values of perfluoro-octanoic acid doped polyaniline and polyaniline-SiO$_2$ composites film on the electrode surface was found to be in the range of 126° and 115° respectively. Recently, Adhikari et al. (2017) reported the corrosion resistant hydrophobic coating using modified conducting polyaniline. Modified polyaniline was prepared by two different techniques, conventional and rapid mixing method, in the presence of phenyl phosphonic acid as a dopant. The prepared polymers were dispersed in an epoxy resin and coated on mild steel samples. The authors observed that the polyaniline based epoxy coating prepared by rapid mixing method showed a remarkable improvement in hydrophobicity as compared to the epoxy coating containing polyaniline prepared by the conventional method. It was observed that the contact angle value for the epoxy coating was 67.7°, whereas the contact angle value was increased to 88.4° when polyaniline prepared by the conventional method was incorporated in the epoxy resin. A remarkable increase in contact angle upto 146.9° was observed in case of the epoxy coating containing polyaniline prepared by a rapid mixing method. In order to prepare nanosized conducting polymer and thus to improve the dispersibility of the conducting polymer in the coating, the authors had employed rapid mixing method to synthesize polyaniline. In addition, polyaniline which was prepared by rapid mixing methods showed enhanced contact angle due to its nanostructural features and better dispersibility in epoxy resin.

The contact angle of epoxy formulated polyaniline and polyaniline-zirconia nanocomposites less than 90° showed higher wettability (hydrophilic behavior). For the coating to be hydrophobic, the contact angle must be equal to or greater than 90° (Yang et al. 2012). It was observed that the surface contact angle of epoxy coating with 4.0 wt.% poly(AN-co-PFA) and poly(AN-co-PFA)/ZrO$_2$ nanocomposite were found to be 102° displaying its hydrophobic nature as shown in Figure 4.8. While the contact angle of the epoxy coating with water droplets was 78°.

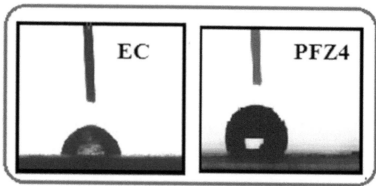

Figure 4.8: Image of the contact angle of epoxy and epoxy formulated copolymer-zirconia nanocomposites (Bisht et al. (2017) Current Smart Materials, DOI: 10.2174/2405465802666170707163408)

4.6 Physico-mechanical properties

Mirabedini et al. (2012) reported the mechanical properties of epoxy-based amino-propyltrimethoxy silane (APS) treated zirconia nanocomposites. The results of mechanical testing showed that the incorporation of ZrO_2 nanoparticles slightly increased the tensile strength and Young modulus, thus reducing elongation at break compared to the neat epoxy coating. ZrO_2 nanoparticles were rigid and had a higher modulus than that of neat epoxy coating. The authors reported that in the addition of 1 wt.% loading of ZrO_2 nanoparticles, the tensile strength was increased from 39.9 MPa to 43.6 MPa and the elongation decreased from 3.9% to 3.8% as compared to neat epoxy coating. Furthermore, with addition of 3 wt.% loading of ZrO_2 nanoparticles in epoxy resin the tensile strength and Young modulus gradually decreased. Hence, the authors explained on the basis of literature (Zhang et al. 2006) that the observed variations in the mechanical properties at higher loading of ZrO_2 nanoparticles in epoxy resin might be due to an undesirable dispersion of the nanoparticles and in turn to the presence of aggregates, air pockets as well as discontinuity within the film. Physico-mechanical properties of epoxy modified poly(AN-co-PFA) and poly(AN-co-PFA)/ZrO_2 nanocomposites are as follows:

4.6.1 Cross-cut tape test

Figure 4.9A showed the images of cross-cut tape test which was carried out on mild steel panels with epoxy and epoxy modified poly(AN-co-PFA) and poly(AN-co-PFA)/ZrO_2 nanocomposites. It was found that the small flakes of coating were detached along edges and at the intersection of cuts in epoxy and epoxy modified poly(AN-co-PFA) coated mild steel. The area affected in epoxy coated panel was noticed to be about 5% (Grade 4B). Meanwhile, epoxy coating modified with different loading of copolymer-zirconia nanocomposites exhibited absolutely smooth edges of cuts and none of the squares of the lattice was detached. The adhesion noticed refers to the 5B scale (Grade 5B), which was the representative of a perfect adherence. Results indicated that the interaction of zirconia nanoparticles in the copolymer matrix strengthened the coating on the metal substrate.

4.6.2 Taber abrasion and scratch resistance test

Taber abrasion test is carried out to investigate the abrasion/wear resistance of coatings. Abrasion resistance is calculated as a loss in weight at a specified number of abrasion cycles or a number of cycles to remove a unit amount of coating thickness. Figure 4.9B and Table 4.1 shows the Taber abrasion test results of epoxy coating and epoxy modified different loading of copolymer

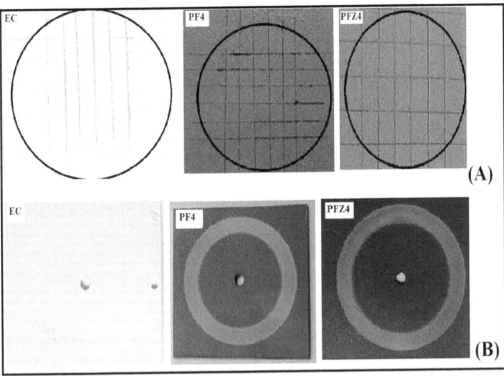

Figure 4.9: Pictures of epoxy and epoxy modified copolymer-zirconia nanocomposites coated samples after performing (A) Cross-cut adhesion test and (B) Taber Abrasion test (Bisht et al. Current Smart Materials, DOI: 10.2174/24054658026 66170707163408)

Table 4.1: Comparative physico-mechanical test results of epoxy and epoxy formulated coatings

Test samples	Cross-cut adhesion test	Mandrel bend test (observation)	Scratch resistance test (observation)	Taber abrasion test results coating loss in (mg/1000 cycle)
EC	Grade 4B	Minor cracks	Minor scratch	60.0
PF2	Grade 4B	Minor cracks	Minor scratch	58.5
PF4	Grade 4B	Minor cracks	Minor scratch	54.9
PFZ1	Grade 5B	No cracks	Scratch-free coating	50.0
PFZ2	Grade 5B	No cracks	Scratch-free coating	42.1
PFZ3	Grade 5B	No cracks	Scratch-free coating	36.8
PFZ4	Grade 5B	No cracks	Scratch-free coating	30.0

nanocomposite coating. The coating loss in the epoxy coating is observed to be about 60.0 mg/1,000 cycles. Epoxy coating modified with 2 wt.% and 4 wt.% loading of copolymers does not exhibit any remarkable reduction in coating loss as shown in Table 4.1. It is calculated to be 58.5 and 54.9 mg/1,000 cycles respectively. While epoxy coating that has 3 wt.% and 4 wt.% loading of copolymer-zirconia nanocomposites revealed remarkably less coating loss per 1,000 cycle. The coating loss is reduced from 60.0 mg/1,000 cycle for epoxy coating to 30.0 mg/1,000 cycles for 4.0 wt.% loading of copolymer-zirconia nanocomposite based epoxy coating. Hence, epoxy modified copolymer-zirconia nanocomposites, i.e., poly(AN-co-PFA)/ZrO$_2$ coatings exhibits greater wear resistance as compared to epoxy coating and epoxy formulated poly(AN-co-PFA) coating.

Similarly, scratch resistance of epoxy coating modified with copolymer-zirconia nanocomposite coating was found to be superior to epoxy and epoxy modified copolymer, i.e., poly(AN-co-PFA)

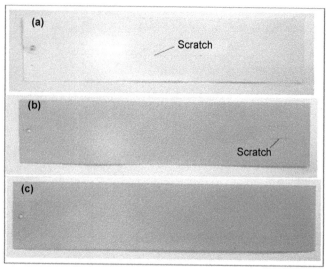

Figure 4.10: Pictures of coated substrates after performing scratch resistance test for (a) Epoxy coating, (b) epoxy modified with Poly(AN-co-PFA) coating and (c) epoxy modified with Poly(AN-co-PFA)/ZrO₂ coating (Bisht et al. Current Smart Materials, DOI: 10.2174/2405465802666170707163408)

coatings as indicated by Figure 4.10. It was also observed that the mechanical properties of the epoxy coatings were found to be significantly enhanced due to the loading of copolymer-zirconia, i.e., poly(AN-co-PFA)/ZrO₂ nanocomposite in an epoxy coating.

4.6.3 Mandrel bend test

Figure 4.11 shows the results of the Mandrel Bend Test of the epoxy coating and epoxy modified copolymer nanocomposite coating. It can be noticed from the Figure 4.11(a) that the epoxy coated shows minor cracks, this is owing to detachment of coating from the metal surface and such cracks does not observe on the epoxy modified poly(AN-co-PFA) and poly(AN-co-PFA)/ZrO₂ nanocomposites coated panels as shown in Figure 4.11(b) and 4.11(c). This indicates that the binding and adhesion ability of coating improved with the loading of copolymers and copolymer-zirconia nanocomposites in epoxy resin.

4.7 Corrosion protection performance of the coating

4.7.1 Salt spray test

Salt spray test is performed on epoxy coated, epoxy formulated with the copolymer and copolymer-zirconia nanocomposite coated mild steel panels in 5.0 wt.% NaCl medium (pH of the solution 6.5–7.2) for 180 days as per ASTM B117 method. In order to carry out this test, all the coated mild steel panels are provided with a scribe mark across the panel and placed in the salt spray chamber as per ASTM B117 standard. Figure 4.16 shows the images of epoxy coated and epoxy formulated 4.0 wt.% loading of copolymer and copolymer nanocomposite (i.e., PFZ1, PFZ2, PFZ3 and PFZ4) coated mild steel panels after their exposure to salt spray fog of 5.0 wt.% NaCl solution. The results show that the occurrence of the spread of corrosion along the scribe mark for epoxy coated mild steel is observed only on 60 days of exposure to salt spray fog. Epoxy coated panels show the fast degradation with rust formation. The occurrence of rusting indicates the loss of the epoxy coating from the metal surface during prolonged exposure to salt spray fog. The epoxy coating also contains

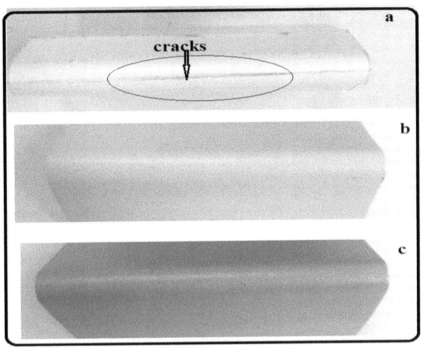

Figure 4.11: Image of coated substrates after forming Mandrel Bend Test with (a) epoxy resin (b) epoxy having copolymer (PF4) and (c) epoxy having copolymer-zirconia nanocomposites (PFZ4) (Bisht et al. Current Smart Materials, DOI: 10.217 4/2405465802666170707163408)

severe blistering and pinholes as shown in Figure 4.12. While epoxy coating modified with 4.0 wt.% of copolymer and copolymer nanocomposite (PFZ1, PFZ2, PFZ3 and PFZ4) exhibit the better corrosion resistance properties after 180 days of exposure. This can be clearly observed from the photographs (Figure 4.12). The epoxy coated specimen with 4.0 wt.% loading of the copolymer (PF4) shows less extended corrosion along the scribe. The same behavior is observed in samples coated with PFZ1 and PFZ2. Further, no sign of blistering, delamination, pinholes and rust formation in the scratched region is noticed for the samples coated with PFZ3 and PFZ4 (Figure 4.12) and the sample still retains its originally shiny surface after the prolonged exposure. The salt spray results clearly indicate that the coating based on poly(AN-co-PFA)/ZrO₂ nanocomposite reduce the spread of corrosion more effectively near the scribe mark. It is assumed that the copolymer; poly(AN-co-PFA) and its nanocomposite with might have supported the adhesion of the epoxy coating to the metal substrate and enhanced the corrosion resistance performance under accelerated test conditions. Moreover, the results obtained from the salt spray test reveals that the inorganic nanofiller, i.e., zirconia nanoparticles also play a significant role in improving the anti-corrosion performance of the coating.

Therefore, coating based on copolymer, i.e., poly(AN-co-PFA) and copolymer-zirconia nanocomposites, i.e., poly(AN-co-PFA)/ZrO₂ provided an efficient protective function against delamination around scribed areas and ensured the intense galvanic function of the primer. The corrosion resistance performance of the coating without defect depends on mainly its barrier properties, which indicate its ability to reduce the penetration of corrosive ions through the film (Radhakrishnan et al. 2009).

Mirabedini et al. (2012) had performed the salt spray test of the epoxy coating containing various wt.% loading of zirconia nanoparticles. The results showed that the epoxy coated sample suffered from corrosion after exposure for 72 hours while the rust formation in the epoxy coating containing zirconia nanoparticles was detected after 480 hours. Hu et al. (2015) reported the results of immersion test of the epoxy coating containing poly(o-Toluidine)/ZrO₂ nanocomposites after immersing the coated

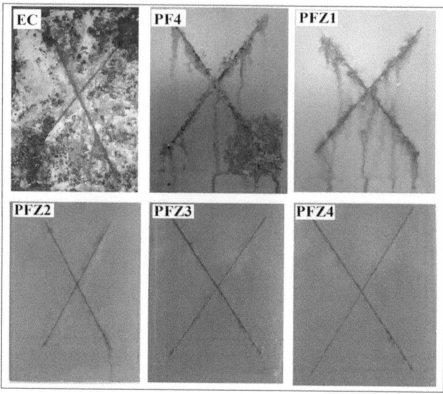

Figure 4.12: Images of (a) epoxy coated and epoxy modified with the copolymer and different loading of copolymer-zirconia nanocomposite coated mild steel panels exposed to salt spray fog for 180 days Bisht et al. (2017) Current Smart Materials, DOI: 10.2174/2405465802666170707163408)

Color version at the end of the book

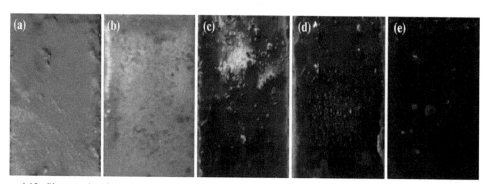

Figure 4.13: Photographs of (a) uncoated steel, (b) pure epoxy coated steel, (c) polyaniline/epoxy coated steel, (d) poly(o-toluidine)/epoxy coated steel and (e) poly(o-Toluidine)/ZrO$_2$ coated steel after immersion in 3.5% NaCl solution for 120 days (Hu et al. Journal of Inorganic and Organometallic Polymers and Materials, DOI: 10.1007/s10904-014-0158-1)

samples in 3.5% NaCl solution for 120 days. Figure 4.13 shows the photographs of uncoated and coated steel coupons after immersion test in 3.5% NaCl solution for 120 days. It was observed that the red rust layer was developed on the entire surface of uncoated steel samples after the immersion test. The rust formation (less than uncoated metal) was also observed in case of epoxy coated metal as shown in the figure.

Epoxy coating containing polyaniline and poly(o-Toluidine) showed some surface bubbled and localized corrosion, but the surface of the epoxy coating having polyaniline suffered much severe corrosion as compared to poly(o-Toluidine). Furthermore, the authors observed that in steel panel coated with epoxy formulated poly(o-Toluidine)/ZrO₂ nanocomposites coating film surface did not show any bubbles or rust, which was found to be entirely intact with no peeling off the coating, and showed good appearance and gloss even after immersing in such high corrosion conditions for a long time.

Hence, it was confirmed that the epoxy coatings containing poly(o-toluidine)/ZrO₂ nanocomposites showed much better corrosion protection property than the single component systems such as polyaniline and poly(o-toluidine). Therefore, their studies showed that the use of poly(o-toluidine) as fillers can enhance the corrosion protection performance of epoxy coating, and ZrO₂ nanoparticles are able to decrease the porosity of composite coating, leading to excellent barrier property toward corrosive ions.

4.7.2 Electrochemical studies of the coating

4.7.2.1 Open Circuit Potential (OCP) versus time measurement

Anti-corrosive performance of epoxy coated and epoxy modified copolymers and copolymer-zirconia nanocomposites coated mild steel electrode were investigated by OCP verus time measurements. The samples were immersed in 3.5 wt.% NaCl solution, and the OCP was recorded versus time at room temperature. The protection time was illustrated by the time during which the OCP of the covered electrode remains in the passive state of mild steel before it fell to the corrosion potential of unprotected mild steel. The OCP versus time curves of bare and coated mild steel electrode in 3.5% NaCl solution are exhibited in Figure 4.14.

The OCP variation for bare mild steel shows a continuous shifting of OCP towards negative potential. This is owing to the occurrence of general corrosion on the metal surface in the corrosive medium as shown in Figure 4.14a. The OCP curves of epoxy, epoxy modified PF2 and PF4 coated mild steel is initially high at about 0.01 V, 0.1 V and 0.4 V respectively, which indicates the passivation of these coated mild steel electrodes at the initial time of immersion. The epoxy coating shows a sharp shift of OCP towards negative potential at a few minutes of immersion is mainly due to the

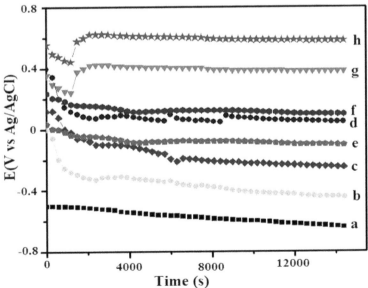

Figure 4.14: OCP versustime curves of (a) uncoated mild steel and mild steel coated with, (b) epoxy, (c) PF2, (d) PF4, (e) PFZ1, (f) PFZ2, (g) PFZ3 and (h) PFZ4 exposed to 3.5 wt.% NaCl solution at room temperature 25 ± 2°C (Bisht et al. Current Smart Materials, DOI: 10.2174/2405465802666170707163408)

diffusive nature of the electrolyte. The shift of OCP towards negative potential continues until the end of the immersion time as the surface cannot attain the equilibrium state (stable potential region). The observed trend of epoxy coating is somewhat similar to the bare mild steel surface. Therefore, the progress of corrosion process can be speculated at the coating/metal interface. This also gives a preliminary idea about the weak barrier property of the epoxy coating under freely corroding condition.

The OCP curves of epoxy modified copolymers (PF2 and PF4) also exhibited a similar pattern of the variation of potential with time (Figure 4.14c and 4.14d). However, the steady state OCP was observed to be 219 mV and 415 mV nobler than the epoxy coating. The OCP curves of epoxy coating modified with poly(AN-co-PFA)/ZrO_2 exhibited different curve trend (Figure 4.14e, 4.14f, 4.14g and 4.14h). The OCP value of the mild steel electrode coated with PFZ1, PFZ2 maintained the stable surface potential with respect to immersion time. The coated surface attained the equilibrium state within a few minutes of immersion in 3.5% NaCl solution. A completely different trend of OCP versus time curves was noticed for the coatings PFZ3 and PFZ4.

The initial shift of potential toward negative potential is evident from the curves. However, a significant shift of OCP toward positive potential occurs and this positive potential is maintained until the end of immersion. It can be interpreted from the trend of the curves that the surface passivation occurs with 3.0% and 4.0% loading of poly(AN-co-PFA)/ZrO_2 in epoxy. Literature reports that the shifting of OCP towards positive direction indicates the passive state of underlying metal because of the good corrosion protection ability of the surface film (Kowalski et al. 2010).

Hence, the high positive value of OCP measured for PFZ3 and PFZ4 is attributed to the effective barrier property of copolymer nanocomposites toward the ingression of corrosive ions into the metal/coating interface. The corrosion protection offered by copolymer nanocomposite is attributed to the combined effect of the barrier property of the coating and the formation of a passive oxide layer due to redox reaction at the metal/polymer interface. The barrier property of the coating works as long as the coating remains undamaged, and it also inhibits the ingression of chloride ions into the metal surface. With the passage of time, corrosive ions penetrate into the metal surface through the pores and cause deterioration of metal. This is due to the weakening of the barrier property of the coating during that period. However, it is assumed that the presence of aniline and pentafluoroaniline moieties in the copolymer chain offer immediate repassivation due to the occurrence of redox reaction at the metal-coating interface. This is evident from the remarkable shifting of potential toward positive direction. The inclusion of ZrO_2 NPs as a reinforcing material in copolymer matrix results in the enhancement of the mechanical strength of the coatings in a corrosive medium.

A high positive value of OCP is also observed by Hu et al. (2015) in steel coated with an epoxy resin containing poly(o-toluidine)/ZrO_2 nanocomposite in 3.5% NaCl solution. The results of the OCP curves are shown in Figure 4.15. It is observed that the initial OCP of the epoxy coating containing poly(o-toluidine)/ZrO_2 nanocomposite is found to be more positive than poly(o-toluidine)/epoxy, polyaniline/epoxy, pure epoxy coated steel and uncoated steel. With the increase in immersion time, the OCPs of all steel samples is seen to shift to a negative direction and then return back to positive direction. While in the case of epoxy containing poly(o-toluidine)/ZrO_2 nanocomposite coated steel sample, the final OCP is found to increase to a higher potential as compared to other coatings as shown in Figure 4.15e, indicating the formation of protective passive layer on the steel surface. Therefore, the results indicates that the presence of substituted conducting polymer, i.e., poly(o-toluidine) and nanofiller such as ZrO_2 in epoxy resin had improved the barrier properties of the epoxy coating.

Kowalski et al. (2010) used the OCP to evaluate the self-healing ability of an intrinsically, conducting polymer based coating, polypyrrole (PPy) doped with molybdate, on a carbon steel to repair defects in the coating and restore the passive state of the steel. The self-healing proficiency of the coating was examined by making the defects on coating surface while measuring the OCP in a 3.5 wt.% NaCl solution. The OCP versus time shows a speedy decrease in the potential was observed when the defect was formed and the potential steadily returned to the passivation level which indicates that the coating is undergoing self-healing process.

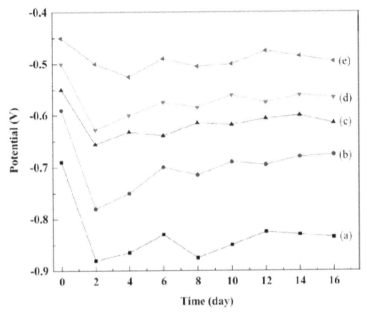

Figure 4.15: OCP versus time curves of (a) uncoated steel, (b) epoxy coated steel, (c) Polyaniline/epoxy coated steel, (d) poly(o-toluidine) coated steel and (e) poly(o-toluidine)/ZrO₂ nanocomposite coated steel (Hu et al. (2015) Journal of Inorganic and Organometallic Polymers and Materials, DOI: 10.1007/s10904-014-0158-1)

4.7.2.2 Tafel extrapolation measurement

Tafel polarization studies are carried out to analyze the trend of corrosion current density of the coated surfaces under potentiodynamic conditions. The Tafel parameters like corrosion potential (E_{corr}), corrosion current density (i_{corr}), anodic (β_a) and cathodic (β_c) Tafel constants provide sufficient explanation about the barrier property of the surface coatings. The Tafel polarization curves of the epoxy coated and epoxy modified copolymer nanocomposite coated mild steel sample (PFZ1, PFZ2, PFZ3 and PFZ4) are drawn after immersing the samples in 3.5 wt.% NaCl solution for 24 hours (see Figure 4.16). The cross-sectional view of the coated surface (Figure 4.11) evidences the presence of thick coating (thickness ~ 90 μm). The thick surface coatings offer high corrosion resistance during initial period of immersion. However, the possibilities of weakening of the barrier property of the coatings increases with lapse of immersion time. To evaluate the corrosion resistance of the coatings with time, Tafel polarization curves are drawn after 30 days of immersion in NaCl (Figure 4.16). The different Tafel parameters measured after 24 hours and 30 days of immersion are mentioned in Tables 4.2 and 4.3 respectively.

Figure 4.16a exhibits remarkably lower i_{corr} values and more positive corrosion potential (E_{corr}) for the mild steel electrodes coated with copolymers (PF2 and PF4) and copolymer nanocomposites as compared to epoxy coated electrode.

The presence of defect free passive layer on the substrate exhibits low i_{corr} values. The i_{corr} value of epoxy coated sample is measured to be 89.0 μA/cm² on 24 hours of exposure in NaCl medium, which further reduces to 19.0 μA/cm² for the sample PF4. Hence 78% protection efficiency is calculated for PF4 coated mild steel. The i_{corr} values of the coated samples PFZ1, PFZ2, PFZ3 and PFZ4 are found to 26 μA/cm², 21 μA/cm², 10 μA/cm² and 0.8 μA/cm² respectively for the same immersion time.

The i_{corr} value is found to decrease on increasing the loading of poly(AN-co-PFA)/ZrO₂ nanocomposite in epoxy coating. The corrosion protection efficiency (% P.E.), as calculated from equation 4.1 is 89% and 99% for the samples PFZ3 and PFZ4, respectively (Table 4.2).

$$P.E.\ (\%) = \frac{i^E_{corr} - i^P_{corr}}{i^E_{corr}} \times 100$$

[4.1]

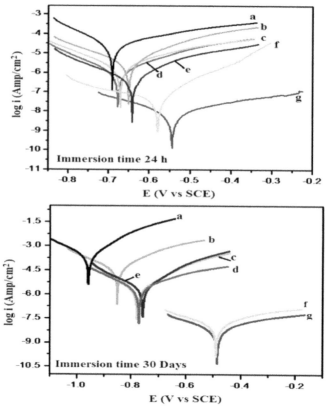

Figure 4.16: Tafel plots of (a) epoxy, (b) PF2, (c) PFZ1, (d) PFZ2, (e) PF4, (f) PFZ3 and (g) PFZ4 coated mild steel at 24 hours and 30 days of exposure to 3.5 wt.% NaCl solution (Bisht et al. Current Smart Materials, DOI: 10.2174/2405465802 666170707163408)

Table 4.2: Comparative corrosion test results obtained from Tafel extrapolation method for the coated samples exposed to 3.5% NaCl solution for 24 hours

Test sample	Tafel test results				
	i_{corr} (μA/cm²)	E_{corr} (mV)	$β_a$ (mV/dec)	$β_c$ (mV/dec)	P.E. (%)
EC	89	−693	265.1	99.2	--
PF2	32	−650	269.0	108.5	64
PF4	19	−645	261.5	208.5	78
PFZ1	26	−666	268.1	219.5	70
PFZ2	21	−667	312.2	288.1	76
PFZ3	10	−580	208.0	261.0	89
PFZ4	0.8	−541	408.0	389.2	99

Results indicate that the addition of ZrO_2 in poly(AN-co-PFA) improves the corrosion protection efficiency of the epoxy coatings. Among all the coated test samples, PFZ4 exhibits least i_{corr} and most noble E_{corr} values. The value of anodic Tafel constant ($β_a$) is also found to be high as compared to other coated samples (Table 4.2).

The high $β_a$ represents anodically controlled corrosion process. The Tafel polarization curves drawn after 30 days of immersion in NaCl solution exhibited significant changes in the values of E_{corr}, i_{corr}, $β_a$ and $β_c$ (Table 4.3). The E_{corr} of the electrode coated with epoxy coating shifted significantly

Table 4.3: Comparative corrosion test results obtained from Tafel extrapolation method for the coated samples exposed to 3.5% NaCl solution for 30 days

Test sample	Tafel test results				
	i_{corr} (µA/cm²)	E_{corr} (mV)	β_a (mV/dec)	β_c (mV/dec)	P.E. (%)
EC	108	−963	89.1	195.9	--
PF2	55	−850	205.0	215.0	49
PF4	42	−759	215.3	238.0	61
PFZ1	35	−775	298.0	281.5	68
PFZ2	30	−769	205.0	218.0	72
PFZ3	20	−491	308.0	305.9	81
PFZ4	2	−489	417.0	418.2	98

toward more negative potential and a drastic increase in the i_{corr} was also noticed after 30 days of immersion.

Table 4.3 mentions a low β_a value (89.1 mV/decade), indicating the progress of active corrosion process at coating/metal interface. It can be inferred that the barrier property of the epoxy coating has severely compromised during the exposure period of 30 days. The hydrogen bonding between epoxy resin and the surface hydroxyl groups of metals are the basis of superior adhesion of epoxy coatings to the metal substrate. However, these hydrogen bonds are replaced by water molecules and diffusive ions (Cl⁻ ions) when exposed to corrosive electrolyte for longer durations. The condition causes severe degradation to the adhesion of epoxy coatings to its substrate. On the contrary, the i_{corr} and E_{corr} values of the samples PFZ1 and PFZ2 changed almost negligibly (Table 4.3). This indicates that the barrier property of the surface coatings is not compromised much. Further, the samples PFZ3 and PFZ4 exhibits remarkably good test results as the i_{corr} is maintained significantly low and E_{corr} shifted toward positive potential as shown in Table 4.3. The β_a values of both the samples are also measured to be high, indicating the corrosion process is anodically controlled. The protection efficiency (P.E.) of PFZ4 coated sample is measured to as high as 98%, even after 30 days of immersion in 3.5% NaCl solution. The behavior indicates that the barrier property of the epoxy coatings modified with copolymer nanocomposite is not affected much with time. The synergistic combination of the properties of conjugated polymer and filler acts as efficient physical barrier against the diffusion of corrosive ions and passivates the underlying metal surface.

Similar action of conducting polymer/ZrO$_2$ nanocomposite based epoxy coating against corrosion are depicted by Hu et al. (2015). The authors have studied the Tafel plots of uncoated, epoxy coated, and epoxy formulated with polyaniline, poly(o-toluidine) and poly(o-toluidine)/ZrO$_2$ nanocomposites. It is observed that among the all steel samples, poly(o-toluidine)/ZrO$_2$ nanocomposites based epoxy coating showed best corrosion resistance properties in 3.5 wt.% NaCl medium as indicated by its highest E_{corr} and lowest i_{corr} values as compared to other samples.

In addition, their studies also confirms that the corrosion protection efficiency of substituted polyaniline such as poly(o-toluidine) based epoxy coating is found to be better as compared to polyaniline based epoxy coating in similar corrosive conditions. This is due to the presence of electron donating substituent, i.e., methyl group in poly(o-toluidine) chain which have longer molecular size as compared to polyaniline that leads to greater adsorption on the metal surface and reduces the effective area for corrosion reaction by blocking the reaction sites. Furthermore, the authors also explain the mechanism of the enhanced corrosion protective efficiency shown by poly(o-toluidine)/ZrO$_2$ nanocomposites based epoxy coating. According to the authors, presence of well-dispersed ZrO$_2$ nano particles in poly(o-toluidine) matrix decreases the small pores nature of polymer and synergistic effect of three different materials in coating such as poly(o-toluidine), zirconia nanoparticles and epoxy resin that protects the underlying metal effectively in aggressive condition.

4.7.2.3 Electrochemical Impedance Spectroscopy (EIS)

Electrochemical Impedance Spectroscopy is a technique in which the impedance of an electrochemical system is studied as a function of the frequency of an applied A.C. wave. This is a nondestructive method to evaluate the quality and corrosion protection performance of organic coatings (Mueller et al. 2004). Impedance experiment is performed by immersing the test specimens in 3.5 wt.% NaCl solution at OCP condition for 24 hours to 30 days at room temperature. Sathiyanarayanan et al. (2008) reported the electrochemical impedance data of epoxy coating formulated phosphate doped polyaniline. They confirmed that the phosphate doped polyaniline containing acrylic and epoxy coatings were able to protect the aluminum alloy from corrosion due to the passivation ability of polyaniline.

The corrosion kinetics of the epoxy coated mild steel (EC) and epoxy modified with different wt.% loading of copolymer nanocomposite (PFZ1, PFZ2, PFZ3 and PFZ4) coated electrode are shown in Nyquist (Figure 4.17) and Bode (Figure 4.19) plots. Figure 4.18 shows the electrochemical equivalent circuits used to extract the EIS parameters like pore resistance (R_{pore}), coating capacitance (C_c), constant phase element (Y_0) and Warburg impedance (W). Figure 4.18a is the simple Randles, which is used to extract EIS data for intact coating. In Figure 4.18b, an additional circuit element, Warburg impedance (W) is added in series with the resistor to explain the occurrence of diffusion controlled corrosion processes at coating/electrolyte interface. Constant Phase Element (CPE), Y_0 is also introduced in place of C_c in this equivalent circuit.

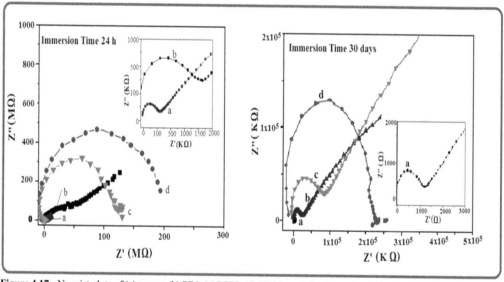

Figure 4.17: Nyquist plots of (a) epoxy, (b) PF4, (c) PFZ3, (d) PFZ4, coated mild steel at 24 hours and 30 days of exposure to 3.5 wt.% NaCl solution (Bisht et al. Current Smart Materials, DOI: 10.2174/2405465802666170707163408)

The values of EIS studies were carried out to calculate the values of pore resistance (R_p), constant phase element, CPE (Y_0) and Warburg impedance (W) and coating capacitance (C_c). Here, constant phase element (CPE) was used as a circuit element in place of coating capacitance (C_c) when obtained semicircle was slightly depressed. This was done to take into account the deviation of the coated surfaces from the pure capacitive behavior and was related to C_c as follows:

$$C_c = Y_0(\omega'')^{n-1}$$

[4.2]

Here, ω'' is the angular frequency at which Z'' is maximum and n is the CPE exponent. The measurements are carried out in a frequency range of 100 kHz to 0.1 Hz at OCP by applying a sine

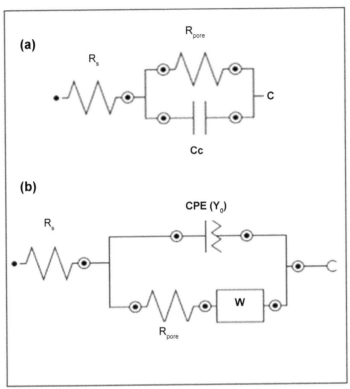

Figure 4.18: Electrical equivalent circuits of (a) normal intact coating, (b) coating showing diffusion of electrolyte. The circuit elements are solution resistance (R_s), Pore resistance (R_{pore}), Coating capacitance (C_c), Warburg impedance (W) and Constant Phase Element (Y_0) (Current Smart Materials, DOI: 10.2174/2405465802666170707163408)

Table 4.4: Comparative corrosion test results obtained from EIS studies for the coated samples exposed to 3.5% NaCl solution for 24 h.

Test sample	EIS test results		
	C_c (F/cm²)	R_{pore} (Ω)	W (Mho)
EC	8.5×10^{-7}	8.1×10^{4}	2.3×10^{-4}
PF2	1.1×10^{-8}	5.9×10^{5}	2.5×10^{-5}
PF4	2.1×10^{-9}	1.0×10^{6}	4.1×10^{-6}
PFZ1	2.9×10^{-8}	6.8×10^{5}	5.5×10^{-5}
PFZ2	7.1×10^{-9}	9.2×10^{6}	1.1×10^{-7}
PFZ3	9.8×10^{-9}	8.9×10^{7}	--
PFZ4	1.1×10^{-10}	1.9×10^{8}	--

potential signal of 10.0 mV. Each experiment is repeated thrice to check the reproducibility of the test results. The reproducibility of the EIS test results is found to be 2–3%.

The EIS parameters measured for 24 hours and 30 days are summarized in Tables 4.4 and 4.5 respectively. The Nyquist plots obtained after 24 hours immersion showed two different types of corrosion mechanism. For samples EC and PFZ1 evidenced high frequency capacitive behavior followed by low frequency diffusion controlled corrosion process (low frequency diffusion tail). The appearance of low frequency diffusion tail demonstrated occurrence of diffusion of electrolyte at coating/electrolyte interface (Xu et al. 2014).

However, the high frequency capacitive arc of PFZ1 is much bigger (inset of Figure 4.17a) than that of sample EC, showing comparatively high Z' vs Z" impedance. The respective Nyquist plots

Table 4.5: Comparative corrosion test results obtained from EIS studies for the coated samples exposed to 3.5% NaCl solution for 30 days

Test sample	EIS test results		
	$CPE(Y_0 \text{ F/cm}^2)$	$R_{pore}(\Omega)$	$W(Mho)$
EC	9.1×10^{-3}	9.9×10^{2}	3.3×10^{-3}
PF2	5.1×10^{-5}	6.1×10^{4}	1.6×10^{-4}
PF4	2.1×10^{-6}	1.9×10^{5}	2.5×10^{-5}
PFZ1	8.2×10^{-5}	9.1×10^{5}	5.1×10^{-5}
PFZ2	8.9×10^{-7}	1.0×10^{6}	8.2×10^{-6}
PFZ3	9.1×10^{-8}	4.1×10^{7}	9.2×10^{-7}
PFZ4	$7.8 \times 10^{-9} (C_c)$	1.7×10^{8}	--

of the samples PFZ3 and PFZ4 demonstrates pure capacitive behavior, equivalent to an undamaged coating. The plots appear like a perfect semicircle (Figure 4.17a). Since, these samples have shown one time constant, simple Randles circuit is applied to extract the values of R_{pore} and C_c (Table 4.4). For samples EC and PFZ1, a modified Randles circuit with Warburg impedance and CPE (Y_0) (Figure 4.18b) is applied.

The electrical resistances offered by the surface coating, i.e., R_{pore} and the water uptake tendency of the coating, i.e., C_c are the decisive EIS parameters to evaluate corrosion protection property of the coatings. The measurement of low R_{pore} (8.1×10^4 Ω) and high C_c (8.5×10^{-7} F/cm²) for epoxy coating as compared to the epoxy modified Poly(AN-co-PFA)/ZrO$_2$ based coatings (PF4, PFZ3, PFZ4) explains the weak barrier nature of the neat epoxy coating. Literature reports that the R_{pore} is maintained high for a normal intact coating (coatings without defects) (Peebre et al. 1989). The comparatively low R_{pore} and high C_c of the epoxy coating is explained by the assumption that the diffusion of electrolyte into the coating increases the coating conductivity. This results in the change of the behavior of the coating from purely capacitive to capacitive at high frequency region and diffusive at low frequency region. The respective Bode plot of the epoxy coating evidences low impedance (|Z|) in the low frequency region (Figure 4.19a).

The epoxy modified with Poly(AN-co-PFA)/ZrO$_2$ coatings (PFZ1, PFZ2, PFZ3 and PFZ4) have shown comparatively high R_{pore} and low C_c values (Table 4.4). Especially for samples PFZ3 and PFZ4 the R_{pore} is measured as high as 8.9×10^7 Ω and 1.9×10^8 Ω, respectively. Accordingly, low coating capacitance value, i.e., 9.8×10^{-9} F/cm² for PFZ3 and 1.1×10^{-10} F/cm² for PFZ4 is observed. The Bode plots of PFZ3 and PFZ4 in the middle frequency region maintains the slop of the curve close to −1 and the low frequency impedance |Z| values are measured to be appreciably high. It can be concluded from above observations that the epoxy modified coatings have shown superior corrosion resistance for short immersion periods.

The corrosion kinetics shows that the surface of the organic coatings becomes heterogeneous with the lapse of immersion time. The condition allows easy penetration/diffusion of water, ions and oxygen through the defects/pores (Feliu et al. 1990). These ions reach the metal/coating interface and initiate corrosion process. Therefore, evaluation of corrosion resistance of the coatings after prolonged immersion period is required. Figure 4.17b and 4.19b present the Nyquist and Bode plots for the coated samples immersed in 3.5% NaCl solution for 30 days. The corrosion mechanism of epoxy coating remains unchanged as the capacitive-diffusive behavior can be easily noticed (Figure 4.17b inset).

The capacitive semicircle appeared to slightly depress and the R_{pore} value of epoxy coated mild steel (9.9×10^2 Ω) further reduced with immersion time and the Y_0 measured as (9.1×10^{-3} F/cm²), exhibiting its higher affinity toward electrolyte uptake (Table 4.5) (Barranco et al. 2004). The poor corrosion resistance of epoxy coating was supported by the low |Z| value in the lower frequency region (Figure 4.19b). The corrosion mechanism of the epoxy modified coating, PFZ3 changed with the lapse of immersion time as the low frequency diffusion tail could be noticed from the Figure

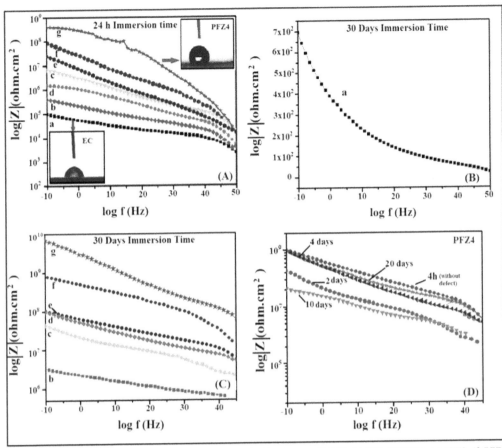

Figure 4.19: (A), (B) and (C) Bode plots of mild steel coated with (a) epoxy (b) PF2, (c) PF4, (d) PFZ1, (e) PFZ2, (f) PFZ3 and (g) PFZ4 exposed to 3.5 wt.% NaCl solution for 24 hours and 30 days at room temperature 25 ± 2°C. (D) Bode plots of PFZ4 before and after the defect at different interval. Inset image indicate the image of contact angle of coated metal (Bisht et al. Current Smart Materials, DOI: 10.2174/2405465802666170707163408)

4.17b. However, the epoxy modified coating, PFZ4 maintained the purely capacitive behavior even after prolonged immersion period of 30 days.

Table 4.5 mentions the remarkably high R$_{pore}$ (1.7×10^8 Ω) and coating capacitance as low as 7.8×10^{-9} F/cm², for sample PFZ4. The respective Bode plot exhibits remarkably high |Z| in the lower frequency region (Figure 4.19c). The superior corrosion resistant properties of epoxy modified with Poly(AN-co-PFA)/ZrO$_2$ coatings, specially PFZ4, is due to the synergistic effect of the combined properties of conducting copolymer and ZrO$_2$ nanoparticles. Additionally, the water contact angle of PFZ4 as shown in Figure 4.19a is 102°, which shows its low affinity toward water. The hydrophobic surface nature of the sample PFZ4 has reinforced the exhibition of superior corrosion resistance in 3.5% NaCl solution.

In order to evaluate the self-healing property of coating, we have collected the impedance data of PFZ4 coated mild steel sample at different intervals such as 4 hours, 2 days, 4 days, 10 days and 20 days of immersion in 3.5% NaCl medium. Figure 4.17d shows the impedance spectra for the PFZ4 based coating before and after the defect formation as a function of immersion time. The impedance value at low frequencies for the sample without the defect is close to 1.1×10^9 Ωcm² during the first 4 hours of the immersion. After 4 hours of immersion, an artificial defect is made on the coated sample. At further check of the pore resistance value at 2 days, a small drop of the resistance value at low frequency is observed at 2.6×10^8 Ωcm². This can be linked with the water uptake through the

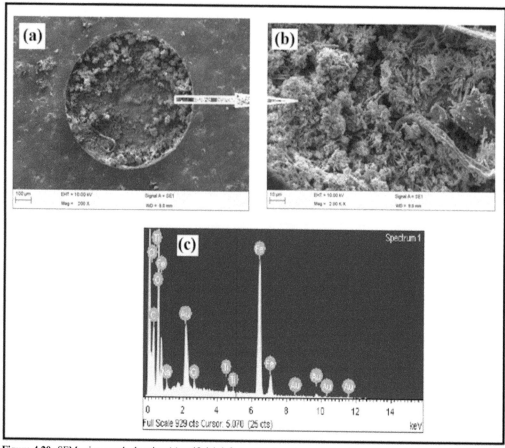

Figure 4.20: SEM micrograph showing (a) artificial defect created on the coated surface, (b) The high magnification SEM micrograph showing the oxide layer present in the defect. (c) The EDS spectrum of the defect area (Bisht et al. Current Smart Materials, DOI: 10.2174/2405465802666170707163408)

defects in the coating (Molera et al. 2003) for a longer immersion time, i.e., at 4 days; coating is further maintained its high impedance value at low frequency which is found to be 1.0×10^9 Ωcm^2. At the 10 day of immersion, the resistance value is further reduced to 4.1×10^7 Ωcm^2 due to the weakening of barrier performance of coating. The impedance value is found to significantly increase after 10 days, and maintains it high values up to 1.0×10^9 Ωcm^2 at the 20th day. This increase of impedance value at longer immersion time indicates the passivated form of steel surface.

SEM micrograph of the defect (diameter ~ 0.7 mm) shows presence of oxide layer in the defect region (Figure 4.20a). The high magnification SEM micrograph (Figure 4.20b) of the defect region reveals the needle like structure of the oxide. The EDS spectrum (Figure 4.18c) of the defect region confirms presence of elements, Fe (51.3%), O (21.7%), C (11.7%), Na (0.5%), Ti (0.8%) and Cl (0.6%).

4.8 Role of poly(AN-co-PFA)/zirconia nanocomposites on metal protection

In this chapter, the use of epoxy coating modified with poly(Aniline-co-Pentafluoroanililne) and its nanocomposites with zirconia were demonstrated. Combining these results with the previous chapter, we could also presume about the protection mechanism reported by these copolymer nanocomposites.

The anti-corrosion mechanism was clearly explained by Wessling (1994); Lu et al. (1995); Sathiyanarayan et al. (2008); Schauer et al. (1998) and Nguyen et al. (2004) in their respective work.

Wessling and Posdorfer (1999) highlighted that the passivation of the metal surface responsible for the protective activity of polyaniline, which mainly resulted the formation of Fe_3O_4 and γ-Fe_2O_3 as precursors of a passive layer to protect the underlying steel.

Jain et al. (1986) explained the corrosion protection mechanism of conducting polymers. They emphasized that the protection of corrosion by conducting polymers occurred by generation of electric field in polymers, which restricts the flow of electrons from the metal to an outside oxidizing species.

Chen and Liu (2011) have confirmed that the polyaniline based epoxy coating had higher adhesion and mechanical effects than the pure epoxy coating.

According to the literature studied by Bhanvase and Sonawane (2010), the amino group and the aromatic ring of aniline are in the same plane, which renders polyaniline fillers dispersed in the epoxy coating with a great capacity to form homogeneous films that leads to better adhesion to the metal surface. In the case of ortho substituted polyaniline which have longer molecular size as compared to polyaniline shows better adsorption on the metal surface, and reduces the effective area for corrosion reaction by blocking the reaction sites.

Hu et al. (2015) concluded that epoxy coating formulated with substituted polyaniline/zirconia, i.e., poly(o-toluidine)/ZrO_2 nanocomposites showed higher corrosion protection property than that of poly(o-toluidine)/epoxy composite coated steel which is attributed to the enhanced mechanical and barrier properties shown by ZrO_2 nanoparticles in polymer matrix. The authors confirmed that the water absorption ability of poly(o-toluidine)/ZrO_2 nanocomposites based coating was found to reduce due to presence of ZrO_2 nanoparticles, which acted as a barrier structure that restricted penetration of water molecules through coating film.

Similar protection behavior can also be explained when we employ a system composed by epoxy coating modified with conducting copolymer-zirconia nanocomposites. The epoxy coating having copolymer without metal oxide also present the oxide formation but does not protect the metal in the same condition than the epoxy having copolymer-zirconia coating does. These copolymer-zirconia nanocomposites delay the corrosion process in greater time than epoxy coating and epoxy having copolymer coating does. Shi et al. (2009) demonstrates the role of inorganic nanofiller such as SiO_2, Zn, Fe_2O_3 and halloysite clay in improving the mechanical properties and anti-corrosion performance of epoxy coating. The possible mechanisms of corrosion protection of PANI on steel are essentially of three types, i.e., barrier properties (involving physical action), corrosion inhibition (involving chemical action) and anodic protection (involving electrochemical action). The barrier performance of epoxy coatings can be enhanced by the incorporation of a second phase that is miscible with the epoxy polymer, by decreasing the porosity and zig-zagging the transmission path for damaging species. For example, the inorganic nanofiller particles as well as their nanocomposites with conducting polymer matrix can be formulated with epoxy resin, which offers environmentally gentle solutions to enhancing the mechanical integrity and durability of the coatings; as the nanosized polymer composites dispersed in coatings are able to fill cavities in epoxy coating and cause crack bridging, crack deflection and crack bowing. In addition, epoxy coatings containing polymer nanocomposites provide substantial barrier properties for corrosion protection, and reduce the trend for the coating to blister or delaminate in aggressive medium (Huong 2006; Becker et al. 2002; Yang et al. 2005; Umaka et al. 2007). In addition, better barrier properties, durability, self-healing ability and hydrophobic nature of conducting copolymer-zirconia nanocomposites are able to make epoxy resin an excellent anti-corrosion material.

4.9 Conclusions

Epoxy coating modified with conductive copolymer nanocomposites based on poly(AN-co-PFA)-zirconia can be an excellent corrosion resistant material under harsh chlorine ion environment, which may also disclose a new opportunity in various technological applications for the marine engineering materials that require very high salt resistance ability. The corrosion test study of epoxy coating based on poly(AN-co-PFA)/ZrO$_2$ showed superior corrosion protection performance as compared to poly(AN-co-PFA) based coating. Impedance spectra of poly(AN-co-PFA)/ZrO$_2$ nanocomposite modified epoxy coating at different time interval also indicate its self-healing property. Tafel parameters of coated mild steel samples showed a significant reduction of corrosion current density for coatings containing 3.0 and 4.0 wt.% loading of poly(AN-co-PFA)/ZrO$_2$ nanocomposite at prolonged exposure to 3.5 wt.% NaCl solution. The results of mechanical testing such as abrasion resistance, scratch resistance, bend test and cross-cut tape test for poly(AN-co-PFA)/ZrO$_2$ modified epoxy coating were found to be superior than epoxy and poly(AN-co-PFA) modified epoxy coating. Apart from the prevention of corrosion, it was observed that the (AN-co-PFA)-zirconia nanocomposites modified epoxy coating showed better mechanical properties as compared to epoxy and epoxy with (AN-co-PFA) coating. Mechanical properties of poly(AN-co-PFA)-zirconia modified epoxy coatings were found to significantly improve with the loading of copolymer nanocomposites in coating due to the presence of zirconia nanoparticles in polymer matrix, which reinforced the polymer by providing addition mechanical strength. The synergistic interaction between poly(AN-co-PFA) and zirconia nanoparticles in epoxy coating resulted in the better thermal, mechanical, hydrophobic and anti-corrosive properties of coating. Hence, such system could be used as an effective anti-corrosive coating on mild steel where the probability of corrosion is greater.

References

Adhikari, A., Karpoormath, R., Radha, S., Singh, S. K., Mutthukannan, R., Bharate, G. and Vijayan, M. (2017). Corrosion resistant hydrophobic coating using modified conducting polyaniline. High Performance Polymers, 30(2): 181–191.

Anwer, T., Ansari, M. O. and Mohammad, F. (2013). Dodecylbenzene sulfonic acid micelles assisted *in situ* preparation and enhanced thermoelectric performance of semiconducting polyaniline–zirconium oxide nanocomposites. Journal of Industrial and Engineering Chemistry, 19: 1653.

Arul JeevaNijanthan, J., Bhagyanathan, C. and Karuppuswamy, P. (2017). Study of nano coating on zirconium by sol-gel process. International Journal of ChemTech Research, 10(14): 131–142.

Barranco, V., Feliu, S. and Feliu, Jr, S. (2004). EIS study of the corrosion behavior of zinc-based coatings on steel in quiescent 3% NACL solution. part. 1: directly exposed coatings. Corrosion Science, 46(9): 2203–2220.

Becker, O., Varley, R. and Simon, G. (2002). Morphology, thermal relaxations and mechanical properties of layered silicate nanocomposites based upon high-functionality epoxy resins. Polymer, 43(16): 4365.

Behzadnasab, M., Mirabedini, S. M., Kabiri, K. and Jamali, S. (2011). Corrosion performance of epoxy coatings containing silane treated ZrO$_2$ nanoparticles on mild steel in 3.5% NaCl solution. Corrosion Science, 53: 89–98.

Behzadnasab, M., Mirabedini, S. M. and Esfandeh, M. (2013). Corrosion protection of steel by epoxy nanocomposite coatings containing various combinations of clay and nanoparticulate zirconia. Corrosion Science, 75: 134–141.

Bhanvase, B. A. and Sonawane, S. H. (2010). New approach for simultaneous enhancement of anticorrosive and mechanical properties of coatings: Application of water repellent nano CaCO$_3$–PANI emulsion nanocomposite in alkyd resin. Chemical Engineering Journal, 156: 177.

Bhattacharya, A., Ganguly, K. M., De, A. and Sarkar, S. (1996). A new conducting nanocomposite—PPy-zirconium (IV) oxide. Materials Research Bulletin, 31(5): 527–530.

Brusciotti, F., Batan, A., Graeve, I. De., Wenkin, M., Biessemans, M., Willem, R., Reniers, F., Pireaux, J. J., Piens, M., Vereecken, J. and Terryn, H. (2010). Characterization of thin water-based silane pre-treatments on aluminium with the incorporation of nano-dispersed CeO$_2$ particles. Surface and Coatings Technology, 205(2): 603–613.

Brzezińska-Miecznik, J., Macherzyńska, B., Lach, R. and Nowak, B. (2014). The effect of calcination and zirconia addition on HAP hot pressed materials. Ceramics International, 40: 15815–15819.

Chen, K. L., Chiang, A. S. T. and Tsao, H. K. (2001). Preparation of zirconia nanocrystals from concentrated zirconium aqueous solutions. Journal of Nanoparticle Research, 3: 119–126.

Chen, F. and Liu, P. (2011). Conducting polyaniline nanoparticles and their dispersion for waterborne corrosion protection coatings. ACS Applied Materials & Interfaces, 3(7): 2694–2702.

Cho, M. S., Cho, Y. H., Choi, H. J. and John, M. S. (2003). Synthesis and electrorheological characteristics of polyaniline-coated Poly(methyl methacrylate) microsphere: Size effect. Langmuir, 19(14): 5875–5881.

Cullity, B. D. and Stock, S. R. (2001). Elements of X-ray Diffraction, Prentice Hall, Englewood Cliffs.

Deflorian, F., Fedel, M., Rossi, S. and Kamarchik, P. (2012). Study of the effect of mechanically treated CeO_2 and SiO_2 pigments on the corrosion protection of painted galvanized steel. Progress in Organic Coatings, 74(1): 36–42.

Dhawan, S. K., Kumar, A., Bhandari, H., Bisht, B. M. S. and Khatoon, F. (2015). Development of highly hydrophobic and anticorrosive conducting polymer composite coating for corrosion protection in marine environment. American Journal of Polymer Science, 5(1A): 7–17.

DoğruMert, B. (2016). Corrosion protection of aluminum by electrochemically synthesized composite organic coating. Corrosion Science, 103: 88–94.

Dorigato, A., Pegoretti, A., Bondioli, F. and Messori, M. (2010). Improving epoxy adhesives with zirconia nanoparticles. Composite Interfaces, 17(9): 873–892.

Feliu, S., Galvan, J. C. and Morcillo, M. (1990). The charge transfer reaction in Nyquist diagrams of painted steel. Corrosion Science, 30(10): 989–998.

Figueroa-Lara, J. J., T-Rodríguez, M., Gutiérrez-Arzaluz, I. M. and R-Romo, M. (2017). Effect of zirconia nanoparticles in epoxy-silica hybrid adhesives to join aluminum substrates. Materials, 10(10): 1135.

Gnedenkov, S. V., Sinebryukhov, S. L., Mashtalyar, D. V., Imshinetskiy, I. M., Samokhin, A. V. and Tsvetkov, Y. V. (2015). Fabrication of coatings on the surface of magnesium alloy by plasma electrolytic oxidation using ZrO_2 and SiO_2 nanoparticles. Journal of Nanomaterials, 16(1): 196.

Grigoriev, D., Shchukina, E. and Shchukin, D. G. (2017). Nanocontainers for self-healing coatings. Advanced Materials Interfaces, 4(1): 1600318.

Gomez, M., Bracho, D., Palza, H. and Quijada, R. (2015). Effect of morphology on the permeability, mechanical and thermal properties of polypropylene/SiO_2 nanocomposites. Polymer International, 64(9): 1245–1251.

Govindaraj, Y., Premkumar, K. P. and Balaraju, J. N. (2015). Evaluation of corrosion resistance and self-healing behavior of zirconium-cerium conversion coating developed on AA2024 alloy. Surface and Coatings Technology, 270: 249–258.

Hawthorne, H. M., Neville, A., Troczynski, T., Hu, X., Thammachart, M., Xie, Y., Fu J. and Yang, Q. (2004). Characterization of chemically bonded composite sol–gel based alumina coatings on steel substrates. Surface and Coatings Technology, 176(2): 243–252.

Hu, C., Zheng, Y., Qing, Y., Wang, F., Mo, C. and Mo, Q. (2015). Preparation of Poly(o-toluidine)/nano zirconium dioxide (ZrO_2)/epoxy composite coating and its corrosion resistance. Journal of Inorganic and Organometallic Polymers and Materials, 25(3): 583–592.

Huang, H., Guo, Z. C., Zhu, W. and Li F. C. (2011). Preparation and characterization of conductive polyaniline/zirconia nanoparticles composites. In Advanced Materials Research, 221: 302–307.

Huong, N. (2006). Improvement of bearing strength of laminated composites by nanoclay and Z-pin reinforcement. PhD. Dissertation. University of New South Wales, Australia.

Jain, F. C., Rosato, J. J., Kalonia, K. S. and Agarwala, V. S. (1986). Formation of an active electronic barrier at Al/semiconductor interfaces: a novel approach in corrosion prevention. Corrosion, 42(12): 700–707.

Kim, J. -S., Lee, D. -H., Kang, S., Bae, D. -S., Park, H. -Y. and Na, M. -K. (2009). Synthesis and microstructure of zirconia nanopowders by glycothermal processing. Transactions of Nonferrous Metals Society of China, 19: s88–s91.

Kowalski, D., Ueda, M. and Ohtsuka, T. J. (2010). Self-healing ion-permselective conducting polymer coating. Journal of Material Chemistry, 20: 7630–7633.

Kumar, A., Bhandari, H. and Dhawan, S. K. (2013). A new smart coating of Polyaniline-SiO_2 composite for protection of mild steel against corrosion in strong acidic medium. Polymer International, 62(8): 1192–1201.

Lenz, D. M., Ferreira, C. A. and Delamar, M. (2002). Distribution analysis of TiO_2 and commercial zinc phosphate in polypyrrole matrix by XPS. Synthetic Metals, 126(2-3): 179–182.

Lu, W. -K., R. L. E. and Wessling, B. (1995). Corrosion protection of mild steel by coatings containing polyaniline. Synthetic Metals, 71(1-3): 2163–2166.

Lvov, Y. M., Shchukin, D. G., Mohwald, H. and Price, R. R. (2008). Halloysite clay nanotubes for controlled release of protective agents. ACS Nano, 2(5): 814–820.

Mahmood, W. K. and Azarian, M. H. (2016). Sol-gel synthesis of polyaniline/zirconia composite conducting materials. Journal of Polymer Research, 23(5): 88.

Mansour, S. F., El-dek, S. I. and Ahmed, M. K. (2017). Physico-mechanical and morphological features of zirconia substituted hydroxyapatite nano crystals. Scientific Reports, 7: 43202.

Masim, F. C. P., Tsai, C. -H., Lin, Yi-F., Fu, M.-Lai., Liu, M., Kang, F. and Wang, Ya-F. (2017). Synergistic effect of PANI–ZrO_2 composite as antibacterial, anti-corrosion, and phosphate adsorbent material: synthesis, characterization and applications. Environmental Technology, 40(2): 226–238.

Medina, R., Haupert, F. and Schlarb, A. K. (2008). Improvement of tensile properties and toughness of and epoxy resin by nanozirconium-dioxide reinforcement. Journal of Materials Science, 43(9): 3245–3252.

Mirabedini, S. M., Behzadnasab, M. and Kabiri, K. (2012). Effect of various combinations of zirconia and organoclay nanoparticles on mechanical and thermal properties of an epoxy nanocomposite coating. Composites Part A: Applied Science and Manufacturing, 43(11): 2095–2106.

Milŏsev and Frankel, G. S. (2018). Review-conversion coatings based on zirconium and/or titanium. Journal of The Electrochemical Society, 165(3): C127–C144.

Montemor, M. F. and Ferreira, M. G. S. (2007). Cerium salt activated nanoparticles as fillers for silane films: Evaluation of the corrosion inhibition performance on galvanised steel substrates. Electrochimica Acta, 52(24): 6976–6987.

Molera, P., Montoya, J. and Valle, M. Del. (2003). Zinc phosphate as corrosion inhibitor in epoxy paints. Corrosion Reviews, 21(4): 349–358.

Mostafaeia, A. and Zolriasatein, A. (2012). Synthesis and characterization of conducting polyaniline nanocomposites containing ZnO nanorods. Materials International, 22(4): 273–280.

Mueller, R., Jossen, R. and Pratsinis, S. E. (2004). Zirconia nanoparticles made in spray flames at high production rates. Journal of the American Ceramic Society, 87(2): 197–202.

Nguyen, T. D., Nguyen, T. A., Pham, M. C., Piro, B., Normand, B. and Takenouti, H. (2004). Mechanism for protection of iron corrosion by an intrinsically electronic conducting polymer. Journal of Electroanalytical Chemistry, 572(2): 225–234.

Ozkazanc, H. (2016). Novel nanocomposites based on polythiophene and zirconium dioxide. Materials Research Bulletin, 73: 226–232.

Padovini, D. S. S., Pontes, D. S. L., Dalmaschio, C. J., Pontes, F. M. and Longo, E. (2014). Facile synthesis and characterization of ZrO_2 nanoparticles prepared by the AOP/hydrothermal route. RSC Advances, 4(73): 38484–38490.

Pareja, R., Ibáñez, R., Martín, R. L., Ramos-Barrado, F. and Leinen, J. R. D. (2006). Corrosion behaviour of zirconia barrier coatings on galvanized steel. Surface and Coatings Technology, 200(22-23): 6606–6610.

Peebre, N., Picaud, T., Duprat, M. and Dabosi, F. (1989). Evaluation of corrosion performance of coated steel by the impedance technique. Corrosion Science, 29(9): 1073–1086.

Piconi, C. and Maccauro, G. (1991). Zirconia as a ceramic biomaterial. Biomaterials, 20(1): 1–25.

Prasanna, B. P., Avadhani, D. N., Muralidhara, H. B., Chaitra, K., Thomas, V. R., Revanasiddappa, M. and Kathyayini, N. (2016). Synthesis of polyaniline/ZrO_2 nanocomposites and their performance in AC conductivity and electrochemical supercapacitance. Bulletin of Materials Science, 39(3): 667–675.

Radhakrishnan, S., Sonawane, N. and Siju, C. R. (2009). Epoxy powder coatings containing polyaniline for enhanced corrosion protection. Progress in Organic Coatings, 64(4): 383–386.

Ranjbar, M., Yousefi, M., Lahooti, M. and Malekzadeh, A. (2012). Preparation and characterization of tetragonal zirconium oxide nanocrystals from isophthalic acid-zirconium(IV) nanocomposite as a new precursor. International Journal of Nanoscience and Nanotechnology, 8(4): 191–196.

Rezaei, M. S., Alavi, M., Sahebdelfar, S., Xinmei, L. and Yan, Z. F. (2007). Synthesis of mesoporous nanocrystalline zirconia with tetragonal crystallite phase by using ethylene diamine as precipitation agent. Journal of Materials Science, 42(17): 7086–7092.

Roy, S. (2007). Nanocrystalline undoped tetragonal and cubic zirconia synthesized using poly-acrylamide as gel and matrix. Journal of Sol-Gel Science and Technology, 44(3): 227–233.

Santos, V., Zeni, M., Bergmann, C. P. and Hohemberger, J. M. (2008). Correlation between thermal treatment and tetragonal/monoclinic nanostructured zirconia powder obtained by sol gel process. Reviews on Advanced Materials Science, 17(1/2): 62–70.

Sathiyanarayanan, S., Syed Azim, S. and Venkatachari, G. (2008). Performance studies of phosphate-doped polyaniline containing paint coating for corrosion protection of aluminium alloy. Journal of Applied Polymer Science, 107(4): 2224–2230.

Sathiyanarayanan, S., Azim, S. and Venkatachari, G. (2008). Corrosion protection coating containing polyaniline glass flake composite for steel. Electrochimica Acta, 53(5): 2087–2094.

Schauer, T., Joos, A., Dulog, L. and Eisenbach, C. D. (1998). Protection of iron against corrosion with polyaniline primers. Progress in Organic Coatings, 33(1): 20–27.

Shchukin, D. G., Zheludkevich, M., Yasakau, K., Lamaka, S., Ferreira, M. G. S. and Möhwald, H. (2006). Layer-by-layer assembled nanocontainers for self-healing corrosion protection. Advanced Materials, 18(13): 1672–1678.

Shi, X., Nguyen, T. A., Suo, Z., Liu, Y. and Avci, R. (2009). Effect of nanoparticles on the anticorrosion and mechanical properties of epoxy coating. Surface and Coatings Technology, 204(3): 237–245.

Shi, L., Wang, X., Lu, L., Yang, X. and Wu, X. (2009). Preparation of TiO_2/polyaniline nanocomposite from a lyotropic liquid crystalline solution. Synthetic Metals, 159(23-24): 2525–2529.

Shukla, S., Seal, S., Vij, R. and Bandyopadhyay, S. (2002). Effect of HPC and water concentration on the evolution of size, aggregation and crystallization of sol-gel nano zirconia. Journal of Nanoparticle Research, 4(6): 553–559.

Septawendar, R., Purwasasmita, B. S., Suhanda, Nurdiwijayanto, L. and Edwin., F. (2011). Nanocrystalline ZrO$_2$ powder preparation using natural cellulosic material. Journal of Ceramic Processing Research, 12(1):110–113.

Srdic, V. V. and Winterer, M. (2006). Comparison of nanosized zirconia synthesized by gas and liquid phase methods. Journal of the European Ceramic Society, 26(15): 3145–3151.

Stankiewicz, A., Szczygieł, I. and Szczygieł, B. (2013). Self-healing coatings in anti-corrosion applications. Journal of Materials Science, 48(23): 8041–8051.

Stocker, C. and Baiker, A. (1998). Zirconia aerogels: effect of acid-to-alkoxide ratio, alcoholic solvent and supercritical drying method on structural properties. Journal of Non-crystalline Solids, 223(3): 165–178.

Sundaram, N. T., Vasudevan, T. and Subramania, A. (2007). Synthesis of nanoparticles in microwave hydrolysis of Zr (IV) salt solutions—Ionic conductivity of PVdF-co-HFP-based polymer electrolyte by the inclusion of nanoparticles. Journal of Physics and Chemistry of Solids, 68(2): 264–271.

Lamaka, S. V., Zheludkevich, M. L., Yasakau, K. A., Serra, R., Poznyak, S. K. and Ferreira, M. G. S. (2007). Nanoporous titania interlayer as reservoir of corrosion inhibitors for coatings with self-healing ability. Progress in Organic Coatings, 58(2-3): 127–135.

Vollath, D. and Sickafus, K. E. (1992). Synthesis of nanosized ceramic oxide powders by microwave plasma reactions. Nanostructured Materials, 1(5): 427–437.

Wang, L., Ding, C. M., Zhu, Y., Wan, M. X. and Jiang, L. (2012). Superhydrophobic conducting polyaniline prepared via interfacial polymerization. Chemical Journal of Chinese Universities, 33(06): 1355–1359.

Weng, C. J., Chang, C. H., Lin, L. I., Yeh, M. J., Wei, Y., Hsu, L. C. and Chen, H. P. (2012). Advanced anticorrosion coating materials prepared from fluoroaniline silica composites with syngeric effects. Surface and Coatings Technology, 207: 42–49.

Wessling, B. (1994). Passivation of metals by coating with polyaniline: Corrosion potential shift and morphological changes. Advanced Materials, 6(3): 226–228.

Wessling, B. and Posdorfer, J. (1999). Corrosion prevention with an organic metal (polyaniline): corrosion test results. Electrochimica Acta, 44(12): 2139–2147.

Xu, A., Zhang, F., Jin, F., Zhang, R., Luo, B. and Zhang, T. (2014). The evaluation of coating performance by analyzing the intersection of bode plots. International Journal of Electrochemical Science, 9: 5116–5125.

Xu, H., Liu, J., Chen, Y., Tang, J. and Zhao, Z. (2016). Facile fabrication of superhydrophobic polyaniline structures and their anticorrosive properties. Journal of Applied Polymer Science, 133(47).

Yabuki, A., Yamagami, H. and Noishiki, K. (2007). Barrier and self healing abilities of corrosion protective polymer coatings and metal powders for aluminium alloys. Materials and Corrosion, 58(7): 497–501.

Yabuki, A. and Kaneda, R. (2009). Barrier and self-healing coating with fluoro-organic compound for zinc. Mater Corrosion, 60(6): 444–449.

Yang, L. H., Liu, F. C. and Han, E. H. (2005). Effects of P/B on the properties of anticorrosive coatings with different particle size. Progress in Organic Coatings, 53(2): 91–98.

Yang, T. -I., Peng, C. -W., Lin, Y. L., Weng, C. -J., Edgington, G., Mylonakis, A., Huang, T. -C., Hsu, C. -H., Yeh, J. -M. and Wei, Y. (2012). Synergistic effect of electroactivity and hydrophobicity on the anticorrosion property of room-temperature-cured epoxy coatings with multi-scale structures mimicking the surface of Xanthosoma sagittifolium leaf. Journal of Materials Chemistry, 22(31): 15845–15852.

Zand, R. Z., Flexer, V., Kim, M. De. Keersmaecker and Adriaens, V. (2015). Effects of activated ceria and zirconia nanoparticles on the protective behaviour of silane coatingsin chloride solutions. International Journal of Electrochemical Science, 10(1): 997–1014.

Zhang, Q., Shen, J., Wang, J., Wu, G. and Chen, L. (2000). Sol–gel derived ZrO$_2$–SiO$_2$ highly reflective coatings. International Journal of Inorganic Materials, 2(4): 319–323.

Zheludkevich, M. L., Serra, R., Montemor, M. F., Yasakau, K. A., Salvado, I. M. M. and Ferreira, M. G. S. (2005). Nanostructured sol–gel coatings doped with cerium nitrate as pre-treatments for AA2024-T3: Corrosion protection performance. Electrochimica Acta, 51(2): 208–217.

Zhong, X., Wu, X., Jia, Y. and Liu, Y. (2013). Self-repairing vanadium–zirconium composite conversion coating for aluminum alloys. Applied Surface Science, 280: 489–493.

5

Polypyrrole-Based Composite Coatings

5.1 Introduction

Corrosion has always been a matter of great concern, especially when metals and alloys are in contact with a corrosive environment. Nearly 88% of the corrosion control strategies are based on the application of organic coatings. Normally, corrosion preventive coating schemes have three coats, viz. primer coat, intermediate coat and topcoat. Conventional coating formulations can prevent corrosion either by barrier type operation or by use of inhibitive pigments that have the following disadvantages:

1. Corrosion at pin holes or scratches
2. Use of highly hazardous chromates
3. Use of heavy metal ions

Recently, conducting polymer-based paints/coatings have found immense application to fight against corrosion. These coatings overcome the drawbacks of conventional anti-corrosive coatings and offer following advantages like self-healing ability which helps in passivating the regions such as scratches, pores, etc., are environmentally friendly/based on green technology (free from heavy metal ions and hazardous chromates), have long service life, are economically feasible and also have additional anti-static properties.

Among conducting polymers, polypyrrole (PPy) is widely used conducting polymer for commercial applications because of its good conductivity, good environmental stability and its potential to perform as a better corrosion preventive material. In the present chapter, we have reported an approach to synthesize polypyrrole composite via *in situ* emulsion polymerization for corrosion preventive application.

Various approaches to the synthesis of polypyrrole (PPy) have been reported. Electrochemical polymerization and chemical polymerization are very frequently used techniques for obtaining conducting polymers. In 1979, Diaz et al. prepared the polymer in the form of flexible films by electrolysis of an aqueous solution of pyrrole (Diaz and Kanazawa 1979). This work gave a start to the extensive use of electrochemical synthesis of PPy and other conducting polymers. Electrochemical polymerization is performed in a one-compartment cell with three electrodes. In the electrochemical oxidation method, a pyrrole and an electrolyte salt are dissolved in a suitable solvent and then the solution is subjected to oxidation which results in the deposition of a conducting PPy film on the inert anodic working electrode like Pt, Au, glassy carbon or stainless steel. The advantages of electrochemical deposition of polypyrrole are that films can be prepared simply with the one-step

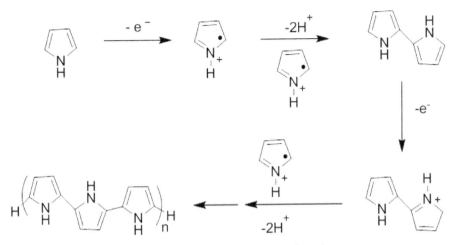

Figure 5.1: Oxidative polymerization of pyrrole

procedure and exact control of thickness. The properties of electronically conducting polymers strongly depend on their synthesis condition, such as the film growth rate, electro-polymerization potential, current density, the nature of the solvent and the types and concentrations of dopant anions. The electrochemical polymerization is generally preferred as cleaner polymers are produced, and it provides a better film thickness and morphology control when compared to chemical oxidation. However, the bulk quantity of polypyrrole can be obtained by using chemical polymerization of the monomer by selecting transition metal ions in water or various other solvents. Chemical polymerization is a simple and fast process as compared to the electrochemical polymerization process as there is no need for any special instrument. Aqueous and anhydrous ferric chloride is the most widely used oxidant for the chemical oxidative polymerization. The yield and conductivity of the chemically produced polypyrrole depend on various factors among which are the choice of solvent and oxidant, monomer/oxidant ratio, time and temperature of the reaction. The main strategies for all synthesis methods are to synthesize new and novel structures, increasing the order of the polymer backbone (and also the conductivity), good processing ability, easier synthesis, more defined three-dimensional structure, stability in certain solvents and many other unique properties. The mechanism of formation of polypyrrole from pyrrole is shown in Figure 5.1.

In the polymerization of pyrrole, the first step consists of the oxidation of the monomer to its radical cation. The second step involves the coupling of two radical cations to yield diradical dication that leads to a dimer after losing two protons and initiates reorganization. The dimer has less oxidation potential than the monomer, undergoes further coupling with a monomeric radical that leads to the formation of termer, tetramer and subsequently polymer. A radical cation is formed by the removal of one electron from the π-conjugated system of conducting polymer which is not completely delocalized. The radical cation is partially delocalized over a segment of the polymer, and it is stabilized through the polarization of the surrounding medium. This partially delocalized radical cation is known as polaron. With further oxidation loss of electron occurred that can be obtained from different part of the polymeric chain resulting in an independent polaron or from a polaron level (removal of an unpaired electron) to create a dication known as bipolaron which is shown in Figure 5.2.

Oxidation of pyrrole to conducting polypyrrole by ferric salts represents a versatile alternative to the usual anodic oxidation. In another study by Dhawan et al. (1992), polypyrrole coating on insulating surfaces was obtained by the oxidation of pyrrole with actively generated $FeCl^{4-}$ which provided a novel method of getting a highly conductive film. The electrochemical stoichiometry of the pyrrole polymerization suggests a mechanism which involves a combination step of two radical cations. In the present case, these cations of intermediate stability get adsorbed on the surface of the insulating polymeric surface, and these cations, in a further radical coupling reaction, lead to the formation

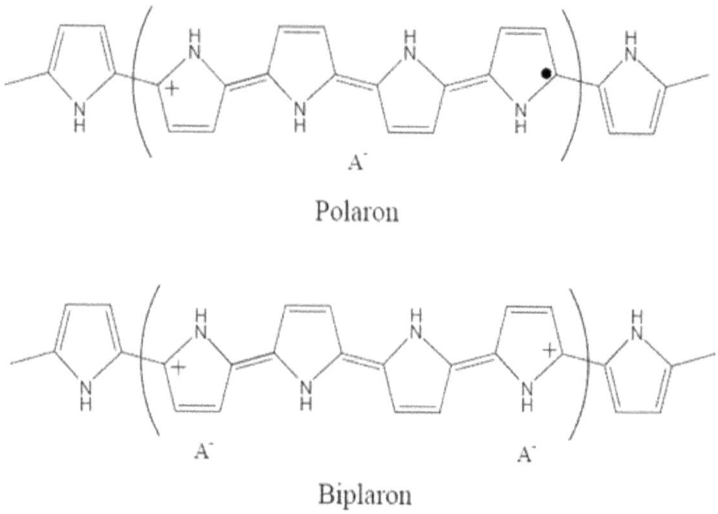

Polaron

Biplaron

Figure 5.2: Polaronic and bipolaronic states of polypyrrole

of a stable and adherent electrically conductive film (1–10 μ). In the present study, it has also been observed that the rate of formation of conducting polypyrrole, PPy, on the insulating surface depends not only on the concentration of the monomer but also on the activity of the freshly prepared $FeCl^{4+}$. On the $FeCl^{4+}$ impregnated surfaces, the polymerization is completed within 30 minutes to yield a bluish to blackish PPy film covered surfaces.

The corrosion protective properties of conducting polymers like polypyrrole, polyaniline, polythiophene, etc., are well documented in literature by numerous researchers (Troch-Nagels et al. 1992; Ferreira et al. 1999; Hien et al. 2005; Xu et al. 2005; Van Schaftinghen et al. 2006; Sharifirad et al. 2010; Gergely et al. 2011; Yan et al. 2013; Yi et al. 2013). The proposed mechanisms of corrosion protection are barrier protection as a corrosion inhibitor, anodic protection and self-healing property. The initial work on the corrosion protection of steel surface by conducting polymers was reported by David DeBerry in 1985. He pointed out that the surface of stainless steel covered with polyaniline (PANI) coating remains passivated in the solution of sulfuric acid (DeBerry 1985). The corrosion inhibition properties of conducting polymers are due to the formation of a passive oxide layer at the metal/coating interface (anodic protection). Another important explanation was given by Wessling et al. (1994) who reported that the conducting polymer-based coatings exhibited self-healing properties. They explained that the oxidative nature of PANI and PPy caused spontaneous healing of the micro-defects/flaws which develop in the coatings during prolonged exposure to the corrosive electrolyte. Apart from PANI, polypyrrole (PPy) is also an important conducting polymer that is used for corrosion protection applications. PPy-based coatings have been reported to have superior corrosion resistance (Janssen and Beck 1989; Hülser and Beck 1990; Grgur et al. 1998; Herrasti et al. 2005; Paliwoda-Porebska et al. 2005; Zhang and Zeng 2005; Koene et al. 2006; Hosseini et al. 2007). Some more advantages of PPy coatings are their environmental stability, good electrical conductivity, thermal stability, ease of synthesis/deposition on the metal surface, etc. (Redondo and Breslin 2007; Mollahosseini and Noroozian 2009). Polypyrrole-based coatings applied on metal surfaces (ferrous/non-ferrous) are low cost, non-toxic alternative to the chrome and phosphate-based surface treatments. Lehr et al. reported that the electrodeposited PPy bilayer coatings on SAE 4140 steel are superior in corrosion protection as compared to the single-layer coatings (Lehr and Saidman 2014). The presence of molybdate ions (as corrosion inhibitor), inside the inner layer and AOT (as dopant), in the outer layer reduces the film porosity and inhibits the ingress of water through it. In another reported work, Kowalski et al. designed PPy coatings on a steel surface with an inner layer of PPy doped with phosphomolybdate ions and outer layer doped with dodecylsulfate (Kowalski et al. 2007; Kowalski et al. 2008). The phosphomolybdate helps in the formation of a passive layer at the coating/metal interface, and the dodecylsulfate inhibits the diffusion of anions through the coating layers. These reported works claimed PPy-based coatings as a potential candidate to mitigate corrosion in active metals. However, the adhesion of PPy coatings to the substrate, mechanical integrity and corrosion resistance properties are compromised under harsh climatic conditions. Apart from this, porosities develop in the coatings during long periods of exposure to the corrosive electrolytes. Specifically, in the presence of chloride ions, localized corrosion is initiated by the formation of tiny galvanic cells. To overcome these shortcomings and to further improve the corrosion characteristics of these coatings, designing composite coating systems by adding certain fillers that can reinforce the mechanical integrity, barrier property and impart self-healing ability to the PPy coatings are extremely important. In recent years, the use of polymer composites has gained momentum to design protective/decorative/ intelligent coatings. The addition of various organic/inorganic fillers in the conventional polymers or conducting polymeric matrices have paved the way for the design of composite systems with superior properties and has added value to the end product. The added fillers have marked influence on the physical, mechanical and chemical properties of the polymers and also alter their degradation/failure mechanism. However, the addition of fillers may affect the processibility of the polymers. Composite coating systems usually have a polymer matrix comprising of conducting polymer in which micro/ nano-sized fillers are dispersed homogenously. Fillers play a crucial role in enhancing the properties of the polymer because the shape, size and volume fraction of fillers affect the properties of the polymer. Some of the common fillers used in composite coating formulations are SiO_2/TiO_2/clay/ ZnO/Al_2O_3/graphene/CNTs, etc. Conducting polymer-based composite coatings are known to have superior corrosion resistance and excellent mechanical integrity as compared to the neat conducting polymer coatings (Hosseini et al. 2007; Montoya et al. 2014). The added advantage of the composite coatings over neat coatings is the synergistic interaction between the polymer matrix and the fillers to achieve the best possible combination of properties. The fillers provide mechanical integrity to the coating and improve the thermal resistance and corrosion resistance of the coating layer. The fillers are the interlocking points in the cured coatings that inhibit the propagation of corrosion at the metal/coating interface. Polypyrrole-based composite coatings of superior corrosion resistance are reported by various researchers (Ciric-Marjanovic et al. 2009; Al-Dulaimi et al. 2011; Kumar et al. 2013; Ashassi-Sorkhabi et al. 2016). These studies have suggested that PPy-based composite coatings are designed to achieve advanced corrosion resistance properties for metals/alloys.

As mentioned above, there are several reported literature on the corrosion protection of ferrous and non-ferrous metals by mono/bi-layer polypyrrole composite coatings. In this chapter, our efforts will be to discuss the corrosion studies of polypyrrole-based (PPy) composite coatings on various metallic substrates. This chapter will review the role of fillers (SiO_2/TiO_2/clay/ZnO/Al_2O_3 nano-particles, etc.) in improving the properties of the designed coatings. The chapter is sub-divided into different synthesis routes of the polypyrrole-based composites, development of composite coatings, micro-structural analysis, compositional analysis and thermal studies of the composites and cross-sectional analysis of the coatings. The electrochemical studies section includes Open Circuit Potential (OCP) versus time study, potentiodynamic polarization studies and Electrochemical Impedance Spectroscopy (EIS). The detailed discussion on the corrosion-resistant properties of the composite coatings is included in the chapter. The self-healing property of the composite coatings is explained with the help of creating artificial defects in the coatings. The mechanism of healing of the defect is also discussed in detail in this chapter.

5.2 Polypyrrole composites and their synthesis routes

Polypyrrole-based (PPy) composites are designed using facile steps that are carried out under ambient atmospheric conditions. The synthesis process includes *in situ* polymerization of pyrrole monomer in the presence of different filler materials. The polymerization of pyrrole is an important step during the synthesis process of composites. There are two different routes of polymerization of pyrrole: the chemical and the electrochemical polymerization routes. In the chemical oxidative polymerization route, pyrrole is subjected to oxidation in presence of oxidizing agents like ferric chloride (FeCl$_3$), ferric sulfate (Fe$_2$(SO$_4$)$_3$), ferric p-toluene sulfonate (FePTS), ammonium persulfate (APS), potassium persulfate (KPS), quinines, etc. The oxidation process is carried out in presence of suitable dopants/surfactant (sodium lauryl sulfate (SLS), dodecyl benzene sulfonic acid (DBSA), dioctyl sodium sulfosuccinate (AOT), etc.), dissolved in aqueous/non-aqueous solvents. The surfactants play a crucial role in improving the conductivity of the polymer and increasing its environmental stability and solubility in organic solvents (Kudoh 1996; Kudoh et al. 1998; Omastová et al. 2003; Omastova et al. 2003; Stejskal et al. 2003). In addition to this, anionic surfactants like sodium dodecyl benzene sulfonate (DBSNa) also bring morphological changes to the polymer matrix (Omastova et al. 2003; Stejskal et al. 2003). The chemical route of synthesis of polypyrrole composites is more appropriate for the bulk production of the composite and has an advantage of tailoring the structural features and conductivity of the synthesized composites by optimizing various experimental parameters. The other synthesis route, i.e., the electrochemical oxidative polymerization is carried out by dissolving the monomer, surfactant/dopant in a suitable solvent, followed by oxidation of the solution by applying certain potential/current. This results in the growth of conducting film of polypyrrole on the working electrode. Some of the early reported works discussed the electrochemically synthesized doped PPy films with good mechanical properties and high conductivity (Diaz et al. 1979; Nogami et al. 1994). Zhou et al. (1999) have carried out electrochemical polymerization experiments with pyrrole monomer in the presence of acetonitrile at a very low potential. The synthesis of polypyrrole carried out potentiodynamically and galvanostatically at extremely low reaction rates exhibited an additional oxidation wave at (–0.23 V) which was more negative than the normal oxidation wave of polypyrrole. Pei et al. (1992) discussed the formation of a smooth, uniform film of PPy by electro-polymerization of pyrrole in aqueous Kolthoff buffer solutions. They concluded that the ideal pH range for the growth of PPy films is between 2 to 3.5 because polymerization takes place smoothly in this pH range. The pH > 4 produces films of non-uniform thickness having a structure different from that of the conventional PPy film. In comparison to the chemical oxidative polymerization, the electrochemical polymerization is less time consuming and high conductivity films can be obtained with ease. The most common electrodes used for electrochemical deposition of PPy films are stainless steel, platinum, gold and glassy carbon as polymerization takes place easily on the surface of these inert electrodes. However, one of the difficulties of this process arises when the film growth is carried out on the surface of active metals, like low carbon steel, aluminum, etc. With these metals, oxidation of metal takes place in preference to the electrochemical polymerization of the monomer. The basic reason is that the oxidation potentials of these metals are comparatively more negative than the potential required in oxidizing the pyrrole monomer. Numerous research works have been carried out to overcome this practical difficulty (Su and Iroh 1997; Wencheng and Iroh 1998; Iroh and Su 1999; Su and Iroh 1999). The most common approach is to carry out the polymerization process in oxalic acid at low current density. In this system, the surface of the active metal (steel) is passivated by iron oxalate interlayer, which protects the metal surface, and the oxidation of pyrrole then takes place with ease. In another reported work, Petitjean et al. discussed the growth of PPy film on low carbon steel using an aqueous solution of sodium salicilate in the one-step electrochemical synthesis (Petitjean et al. 1999). In this system, sodium salicilate metal complex is formed which inhibits the metal dissolution without inhibiting the oxidation of pyrrole.

The above explanations are focused on the different polymerization routes of the polypyrrole. In the composites, fillers are an integral part of the system and influence the overall properties of the composites. Polypyrrole composites are synthesized in such a way that the polymerization step is carried out *in situ* in a reaction medium containing fillers/dopants in a definite molar ratio. The desired filler materials present in the reaction medium either function as a template on the surface of which polymerization of pyrrole takes place (Tallman et al. 2008) or the fillers are simply embedded in the polymer matrix. The most common fillers used with PPy are SiO$_2$/TiO$_2$/clay/ZnO/Al$_2$O$_3$/CNTs/graphene, etc. (Lenz et al. 2003; Riaz et al. 2007; Ioniţă and Prună 2011; Yan et al. 2013; Mondal et al. 2014; Ruhi et al. 2015; Valença et al. 2015; Ashassi-Sorkhabi et al. 2016; Jiang et al. 2016; Jadhav et al. 2017; Liu et al. 2017; Yan et al. 2017). Among the various PPy-based composites, PPy/carbon nano-tube (PPy/CNT) composites need a special mention because of their unique combination of properties (Fan et al. 1999; Chen et al. 2000; Chen et al. 2001). Due to the high aspect ratio and remarkable electrical and mechanical properties of CNTs, they are an ideal reinforcing material for the polypyrrole matrix. During the synthesis process, CNTs act as a template for the polymerization of pyrrole. Apart from this, the anionic nature of CNTs helps them to act as a strong conducting dopant, and the growth of PPy coating is reported to accelerate with the increase of CNT concentration (Chen et al. 2001). Further, PPy/γ-Fe$_2$O$_3$ composites have excellent optical, electrical and superparamagnetic properties and are synthesized by *in situ* polymerization in water/oil micro-emulsion (Sunderland et al. 2004). In this work, micro-emulsion was formed in the presence of a surfactant (sodium dodecylbenzene sulfonate) and a co-surfactant (1-butanol). The polymerization of freshly distilled pyrrole was carried out in the oil phase as well as in the aqueous phase (containing γ-Fe$_2$O$_3$) by stirring the reaction mixture overnight at room temperature. The residual surfactants were removed from the system by washing the precipitates with de-ionized water and acetone. The final product, i.e., the PPy/γ-Fe$_2$O$_3$ nano-composites were obtained by centrifugation followed by vacuum drying in an oven. Recently, polypyrrole/carbon composites were synthesized by chemical oxidative polymerization in the presence of FeCl$_3$. Here, carbon nano-tubes (CNTs) and carbon-black were used as fillers in the composite systems which were reported to have applications in capacitors (Lota et al. 2015; Wang et al. 2017).

Among the various applications of the polypyrrole, designing of corrosion-resistant coatings for ferrous and non-ferrous metals (Fenelon and Breslin 2002; Fenelon and Breslin 2003) have a lot of potentials. Anna et al. have electro-polymerized pyrrole on a copper electrode in a neutral sodium oxalate solution to form a uniformly adherent film of polypyrrole (Fenelon and Breslin 2002). The copper electrode surface was pretreated (oxidized) in the oxalate solution to facilitate the formation of copper oxalate passive layer. This layer firmly inhibits further dissolution of the copper electrode and is found to be conducive in enabling the electro-polymerization of pyrrole monomer. The obtained polypyrrole film was stable and conducting and had offered good corrosion resistance under acidic conditions and in neutral NaCl solution. Polypyrrole coatings were improvised by the incorporation of various fillers to form composite systems. Numerous reported works investigated the synthesis routes, micro-structural studies and the corrosion tests of polypyrrole-based composite coating systems. The summarized PPy composites coatings along with their fillers and their synthesis routes are mentioned in Table 5.1.

Apart from the above-mentioned composite coatings, PPy/SiO$_2$ and PPy/flyash composites have been studied to make a significant improvement in the corrosion resistance of the mild steel substrates (Ruhi et al. 2014; Ruhi 2015). The reason for using flyash for designing composites with polypyrrole was because flyash is a solid by-product that is generated in huge quantity during the combustion of coal in thermal power stations. The mass scale dumping of flyash has led to a severe environmental threat by contaminating the surrounding air and land. Therefore, utilization of flyash in various applications is required to avoid these problems. Flyash, being rich in metal oxides like SiO$_2$, Al$_2$O$_3$, Fe$_2$O$_3$ and TiO$_2$, has tremendous potential in corrosion protection purposes. The ternary blends of flyash cement and silica fume is reported to improve its resistance to chloride ion penetration and reduce corrosion significantly. Further, the fineness of flyash improves the pore refinement and minimizes the access to deteriorating agents even at accelerated corrosion process. Flyash as filler is important from both economic and commercial point of view. The objective of this work is to design polypyrrole-flyash (PPy-flyash) composite coatings for corrosion protection of mild steel substrate. The corrosion inhibition property of polypyrrole and reinforcing ability of flyash are utilized to design coatings with superior corrosion resistance for saline conditions. Ruhi et al. (2015) have synthesized the composites by chemical oxidative emulsion polymerization of pyrrole in the presence of ferric chloride (an oxidant) and sodium lauryl sulfate (SLS as surfactant/homogenizing agent). In the synthesis steps, the dropwise addition of ferric chloride in the suspension containing the fillers

Table 5.1: Polypyrrole-based different composites

PPy composite	Fillers/Dopants	Synthesis route	Substrate	Reference
PPy/Al flake	Al flakes/molybdate Phosphate/vandate	Aqueous chemical oxidative emulsion polymerization	Aluminum	(Yan et al. 2013)
PPy/DGEBA	Diglycidyl ether of bisphenol A (DGEBA)	Non-aqueous chemical oxidative polymerization	Mild steel	(Riaz et al. 2007)
PPy/PW12	Phosphotungstic acid	Electropolymerization using cyclic voltametry	Carbon steel	(Liu et al. 2017)
PPy/ND/DBSA	Nano-diamond powder	Electro-polymerization	St-12 steel	(Bagheri et al. 2016)
PPy/flyash	Flyash	Aqueous chemical oxidative emulsion polymerization	mild steel	(Ruhi et al. 2015)
PPy/Al$_2$O$_3$	Al$_2$O$_3$	Electropolymerization	316 stainless steel	(Yan et al. 2017)
PPy/TiO$_2$	TiO$_2$	Electrochemical synthesis	AISI 1010 steel	(Lenz et al. 2003)
PPy/Mica	Mica/molybdate/DBSA	Aqueous chemical oxidative polymerization	Cold rolled steel	(Jadhav et al. 2017)
PPy/GO	Graphene oxide	Electropolymerization	304 stainless steel	(Marandi et al. 2014)
PPy/G	Graphene	Aqueous chemical oxidative polymerization	316 stainless steel	(Jiang et al. 2016)
PPy/ZnO	ZnO nano-particles	Aqueous chemical oxidative emulsion polymerization	1020 carbon steel	(Valença et al. 2015)
PPy/PO$_4^{3-}$	Phosphate	Cyclic voltametry	Mild steel	(Montoya et al. 2014)
PPy/CNTs	Carbon nano-tubes	Electropolymerization	Carbon steel	(Ioniţă and Prună 2011)

such as flyash, monomer and SLS commences the polymerization process. The appearance of black precipitate shows the formation of the composites. The composite was filtered, washed and dried under vacuum. The various steps of the synthesis of PPy/flyash composite and the coating development are presented in Figure 5.3. Coatings were developed on low carbon mild steel. The steel sheets of dimension 1 cm × 4 cm × 0.2 cm for corrosion studies and 15 cm × 10 cm × 0.2 cm for salt spray tests were cut and polished metallographically. The final coating formulations were prepared by blending the polymer composites with epoxy powder resin in different weight percent. These coating formulations were spray coated onto the mild steel substrate by employing an electrostatic spray gun. The coating step is completed by curing the spray-coated mild steel samples at 150°C in an oven. The synthesized polymer composite was blended with epoxy powder coating formulation in various wt% loadings (1.0, 2.0, 3.0 and 4.0%) using a laboratory ball mill. The composition of epoxy powder coating formulation is as follows: resin [epoxy (bisphenol A + polyester) (70%), flow agent (D-88) (2.3%), degassing agent (benzoin) (0.7%), fillers (TiO$_2$ and BaSO$_4$) (27%)]. A homogeneously mixed polymer composite in epoxy was applied on mild steel specimens using an electrostatic spray gun held at 67.4 KV potential. The powder-coated steel specimens were baked in the oven at 180°C for 30 minutes. The epoxy coating developed on steel substrate is designated as EC and epoxy coatings with different wt% loading of polymer composite are designated as PF1 (1.0%), PF2 (2.0%), PF3 (3.0%) and PF4 (4.0%).

Even polypyrrole/SiO$_2$ composite was synthesized by chemical oxidative polymerization using ferric chloride as an oxidant and sodium lauryl sulfate as the medium as shown in Figure 5.4. In this study, even nano-silica was synthesized by hydrolysis of tetraethyl orthosilicate in ethanol medium using ammonia as a catalyst.

Another interesting work on polypyrrole/silicon nitride composite is reported by Ashassi-Sorkhabi et al. (Ashassi-Sorkhabi and Bagheri 2014). The work highlights a sonoelectrochemical synthesis route to design PPy/SiN coatings with superior corrosion properties. The Taguchi method is applied to design the experiment and various experimental parameters like current density, time taken during

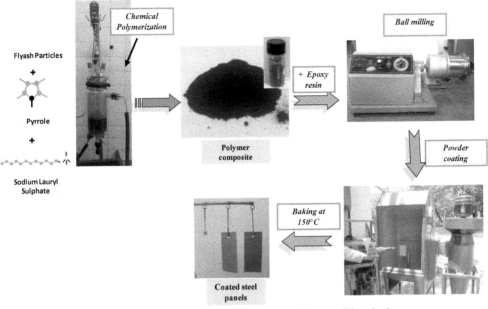

Figure 5.3: Development of PPy/Flyash composite coatings on mild steel substrate

Color version at the end of the book

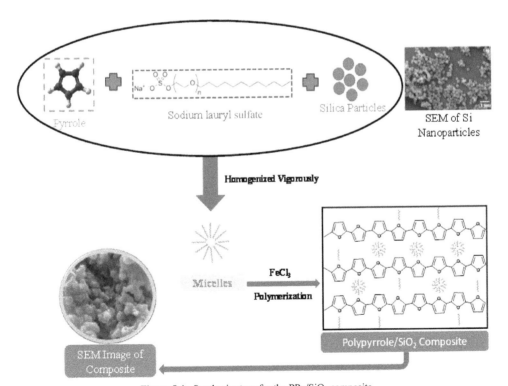

Figure 5.4: Synthesis steps for the PPy/SiO$_2$ composite

Color version at the end of the book

synthesis, the number of SiN nano-particles used, etc., are optimized. Electrochemical deposition of the polypyrrole and PPy/SiN films was carried out using a constant current density in the presence of the sono-irradiation power. A conventional three-electrode system consisting of saturated calomel electrode (as reference electrode), platinum foil (as a counter electrode) and the uncoated/coated St-12 steel specimens (as working electrodes). The electrodeposited films were found to be dense, compact and homogenous. This work reports that the SiN nano-particles improve the corrosion properties of the coatings and coating micro-structure. In another works by Ashassi-Sorkhabi et al. (Ashassi-Sorkhabi et al. 2014), sonoelectrochemically synthesized PPy/Au nano-composite coatings were developed on St-12 steel electrodes using the galvanostatic technique. The designing of the experimental steps were done by applying the Taguchi method in terms of optimum synthesis time, the concentration of $HAuCl_4$ and applied current density to obtain the minimum average size of Au nano-particles. The corrosion tests of the designed coatings were performed by employing Open Circuit Potential (OCP) time plots, potentiodynamic polarization technique and Electrochemical Impedance Spectroscopy (EIS) in a NaCl 3.5% solution. A similar synthesis route is adopted to design PPy/MWCNT/chitosan composite coatings on St-12 steel samples (Ashassi-Sorkhabi et al. 2015). The research work reports the synergistic effect of nano-material (MWCNT) and organic polymer (chitosan) when incorporated in conducting polymer matrix. *In situ* polymerization of pyrrole was adopted in the solution of oxalic acid (0.1 mol/L) and dodecylbenzen sulfonic acid (100 mg/L) dispersed with a certain amount of chitosan and MWCNT. The polymerization step is carried out under ultrasonic irradiation using a standard three-electrode electrochemical setup.

5.3 Compositional, thermal and micro-structural studies of the PPy-based composites

5.3.1 FTIR spectroscopy

Fourier Transform Infrared (FTIR) spectroscopy is a strong tool to study the chemical composition of the polymers and their composites, like the presence of different functional groups, the interaction between the filler and the bulk polymer, etc. IR spectra provide valuable information needed to help enhance our understanding of polymer composites. The use of such powerful technique in understanding the chemistry of the material has allowed designing of new polymer composites. The interaction between polymers and fillers in the composite is studied by FTIR plots in terms of the appearance of IR peaks at new frequencies, shifting of IR peaks, appearance/broadening of the absorption bands, etc. The IR spectra are recorded in the wave number range of 4,000–600 cm^{-1}. The molecules present in the composites selectively absorb radiation of the specific wavelength and the absorption spectrum is obtained. For example, different fillers like SiO_2, α-Al_2O_3, TiO_2, ZnO, etc., show their characteristic stretching vibrations of Si-O (at 1,108 and 817 cm^{-1}) (Ruhi et al. 2014), Ti-O (at 1,383 cm^{-1}), Zn-O (at 480 cm^{-1}) (Valença et al. 2015) and Al-O (at 459, 595 and 656 cm^{-1}) (Djebaili et al. 2015). The typical FTIR absorption spectrum of polypyrrole shows characteristic peaks of pyrrole ring at 1,559 cm^{-1} (Radhakrishnan et al. 2009). The absorption peaks due to =CH in-plane vibrations appear at 929 cm^{-1} and the peak due to =CH out of plane vibrations at 789 cm^{-1} (Lei et al. 1992; Saravanan et al. 2007; Fu et al. 2012). The IR absorption peak of N–H in-plane deformation in the polypyrrole molecule is present at 1,038 cm^{-1}. In polypyrrole composites, like PPy/SiO_2 composites, the absorption peak of C-N stretching of polypyrrole at 1,186 cm^{-1} is overlapped with a band of Si-O (1,108 cm^{-1}) (Cheah et al. 1998). The other IR absorption peaks of PPy/SiO_2 composite, like polypyrrole ring, is shifted to 1,554 cm^{-1}, =CH vibrations in and out of the plane at 1,474 and 771 cm^{-1}, respectively (Dai et al. 2007). For the PPy/SiO_2 composite, Ruhi et al. have discussed the shifting of FTIR peak of polypyrrole due to the presence of SiO_2 nano-particles in the polymer matrix. The =CH out of plane vibration at 789 cm^{-1} in neat polypyrrole is shifted toward lower wave number (771 cm^{-1}) in the PPy/SiO_2 composite as shown in Figure 5.5. In addition to

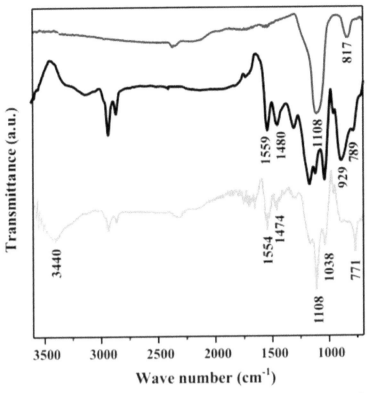

Figure 5.5: FTIR spectra of synthesized SiO$_2$ nano-particle (a), polypyrrole (b) and PPy-SiO$_2$ composite (c) (Ruhi et al., Progress in Organic Coatings DOI: 10.1016/j.progcoat. 2014. 04.013)

this, the presence of a broadband at 3,440 cm^{-1} for the PPy/SiO$_2$ composite is due to the hydrogen bonding between SiO$_2$ nano-particles and the polypyrrole.

Konwer et al. (Konwer et al. 2011) have compared the FTIR spectra of PPy/graphene oxide, graphene oxide (GO) and PPy. The PPy/GO composite evidenced shifting of peaks toward higher and explained that the peak corresponding to C=O stretching in GO at 1,730 cm^{-1} is downshifted to 1,728 cm^{-1}, which is possibly due to the π-π interaction of GO layers and the polypyrrole rings. The FTIR spectrum obtained for polypyrrole (PPy)/MWCNT composites synthesized by *in situ* chemical oxidation polymerization in the presence of various concentrations of cationic surfactant cetyl trimethyl ammonium bromide (CTAB) showed the presence of interfacial interaction between polypyrrole and MWCNT (Wu and Lin 2006).

5.3.2 X-Ray diffraction studies

X-ray diffraction (XRD) is one of the advanced analytical methods of characterization that evaluates the structure and chemical composition of the materials. XRD is a non-destructive technique to evaluate the amount of crystalline content present in the material. The assessment of the crystalline content of the material is carried out by exposing the sample to the high energy X-ray beam. The X-ray diffraction pattern evaluates the degree of crystallinity of the material by measuring the width of the peaks on the diffractogram. Since polymers and their composites are semi-crystalline in nature, XRD is a pioneer tool for analysis, evaluation and characterization. The powder diffraction technique gives crucial information regarding the crystalline/semi-crystalline/amorphous nature of the polymer composites. The XRD pattern of the polymer composites also assesses the presence of crystalline phases in the composites. The typical XRD pattern of the inorganic oxide fillers Al$_2$O$_3$,

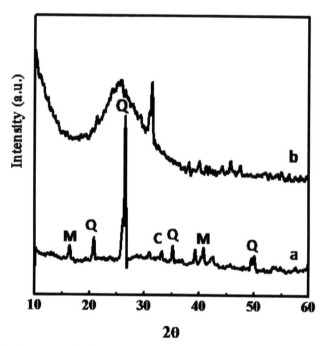

Figure 5.6: X-ray diffraction pattern of (a) flyash and (b) PPy/flyash composite (Ruhi et al., American Journal of Polymer Science, DOI: 10.5923/s.ajps.201501.03)

TiO_2, SiO_2, flyash, ZnO, etc., show sharp well-defined peaks depicting the crystalline nature of the different phases of these filler particles. Interesting X-ray diffraction patterns are obtained for the polymer composites. For example, the XRD pattern of polypyrrole/ZnO nano-composites (Tang et al. 2000) shows the presence of broad diffraction pattern at $2\theta = 20°$ due to the polypyrrole matrix. Furthermore, all the peaks of the pure ZnO nano-particles do not appear in the diffraction pattern but only the peak at $2\theta = 31.8°$ (along 100 planes). Tang et al. (Tang et al. 2000) have given a possible explanation of the above observations as ZnO nano-particles are well embedded in the polymer matrix and probably enveloped with the polymer. The reported work of Ruhi et al. mentioned the XRD diffraction pattern of flyash and the PPy/flyash composite (Ruhi et al. 2015). The X-ray diffraction pattern of flyash shows well-defined peaks of mullite (M) at 2θ of 16.29° and 40.02°, quartz (Q) at 2θ of 20.86°, 26.58°, 35.13° and 49.96° and calcite (C) at 2θ of 33.4°, depicting its crystalline nature, whereas PPy/flyash composite exhibits a broadband at 2θ value of 25.7° showing the amorphous nature of the composite (Figure 5.6).

In another work of Ruhi et al. (Ruhi et al. 2014) they elaborated on the XRD patterns of SiO_2 and PPy/SiO_2 composites. Figure 5.7 showed semi-crystalline in nature of the synthesized SiO_2 particles with a broad diffraction peak at $2\theta = 24.3°$, whereas for the PPy/SiO_2 composite the peak is observed to be shifted at $2\theta = 22.2°$.

Su Li et al. (2012) reported that the XRD pattern of polypyrrole/ASPB nano-composite (here ASPB stands for anionic spherical polyelectrolyte brushes) has shown a shifting of the XRD peak of polypyrrole from 2θ, 24.38° to 22.2°. The reported work mentions that the full width at half maximum (FWHM) of the peak of PPy/ASPB composite positioned at $2\theta = 10.1°$ is higher as compared to that of the neat PPy at 7.6°. The inference is drawn that the crystallinity in the composite is decreased as a result of the synergistic interaction between the filler and the polymer matrix. Selvaraj et al. confirm the amorphous nature of the neat PPy and MnO_2 particles through XRD analysis in their reported work (Selvaraj et al. 2010). Furthermore, discussing the XRD spectrum of PPy-MnO_2 nano-composite, they explained that the nano-composites are amorphous, having average particle size ~ 50 nm (calculated by applying Scherrer's formula). The reported research work of Lingappan et al.

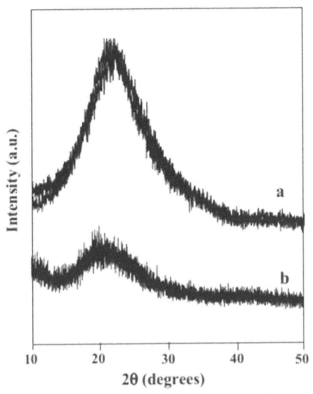

Figure 5.7: XRD diffractogram of SiO$_2$ and polypyrrole/SiO$_2$ composite (Ruhi et al., Progress in Organic Coatings DOI: 10.1016/j.progcoat.2014.04.013)

have discussed the XRD pattern of reduced graphene oxide (RGO)/PPy composite (Lingappan et al. 2013). In this work, the comparative X-ray diffraction patterns of pristine graphite, graphene oxide (GO), reduced graphene oxide (RGO), reduced graphene oxide/polypyrrole (RGO/PPy) composites and neat PPy are shown. Reported work evidenced strong XRD peak (at 2θ = 26.4°) for pristine graphite, whereas GO exhibits broad diffraction peak at 2θ = 10.45° as a result of the complete oxidation of the pristine graphite. In the RGO sample, the original GO peak is replaced by new broadband at 2θ = 26.3° corresponding to the graphitic nano-structure. The XRD pattern confirms the amorphous nature of PPy. However, the RGO/PPy composite contains the broad diffraction peak of PPy but the corresponding peak of RGO is missing in the composite. The authors conclude that a thin layer of PPy is formed on the surface of RGO during the polymerization process.

5.3.3 Thermogravimetric Analysis (TGA)

Thermogravimetric Analysis (TGA) is a technique to study the changes in physical and chemical properties of a material with respect to the increasing temperature (under a constant heating rate) in a controlled atmosphere. During the course of analysis, the studied material can be kept at a constant temperature and the analysis is carried out as a function of time. The weight gain/loss in the material with the increase of temperature provides crucial information regarding the physical/chemical changes, e.g., oxidation, reduction, phase transition, chemisorptions, thermal decomposition, etc., occurring in the material. This technique is very useful to study the thermal behavior of polymeric materials, including thermosets, thermoplastics, composites, fibers, paints and films. The TGA of polymer composites provides valuable data regarding the improvement in its thermal stability, compositional analysis by assessing the filler content, decomposition kinetics of the composite system, assessment of moisture and volatile content, etc. The comparative TGA thermograms of neat polymers and polymers with different wt% loadings of fillers particles give a clear explanation about the effective role of the fillers in increasing the thermal stability of the composites. The enhanced thermal stability of the composites is due to the high thermal stability of the fillers like SiO_2, flyash, clay, Al_2O_3, ZnO, ZrO_2, SiC, etc. The presence of fillers in the composite restricts the thermal motion of the polymeric chains and shields the thermal degradation of the polymer (Zhang et al. 2012). The amount of fillers has a significant influence on the thermal expansion, stiffness, etc., of the composite system. However, the overall thermal stability of the polymer composite also depends on the thermal stability of the parent polymer as well as the homogenous dispersion of the fillers in the polymer matrix. M Ramesan has studied the thermal behavior of PPy/CuS nano-composites (Ramesan 2013). The author reported that the neat polypyrrole evidenced weight loss below 100°C mainly due to the removal of residual water molecules and oligomers from the polymer matrix. However, PPy/CuS composite evidenced negligibly small weight loss in this temperature range, exhibiting better thermal stability over the neat polymer. The decomposition of polypyrrole occurred at 204°C, whereas the incorporation of CuS in the PPy matrix delayed the decomposition of the polymer by almost 22°C. These nano-composites have improved the thermal stability, and this property increases with the increase of wt% loading of CuS in the PPy matrix. The possible explanation is that the nano-particles play a crucial role in changing the thermal behavior of polymer as the nano-particles inculcate an ordered arrangement of polypyrrole chains around them. The reported work also emphasized that the thermal stability of the composite also depends on the way the fillers are dispersed in the polymer matrices. Furthermore, literature reports that the thermal stability of polypyrrole (PPy) is due to the delocalized π-electrons over the heterocyclic ring (Batool et al. 2012). During the composite formation, the incorporation of thermally stable ZnO particles sharply increases the thermal stability of the composite. The synergistic interaction between the polymer and the filler reduces the mobility of the polymer chains when bound onto the filler particles, are the reasons behind the increased thermal stability of the composite. The existence of a coordination bond between the filler (ZnO particles) and the polymer (PPy) is also mentioned in the reported work (Batool et al. 2012). In another interesting work, Han et al. studied the thermal behavior of graphite oxide (GO)/PPy composites (GPs) and 1,5-naphthalene disulfonic acid (1,5-NDA) doped GPs (1,5-NGPs) under nitrogen atmosphere (Han and Lu 2007). The TGA curve of GO evidenced sharp weight loss at 290°C mainly due to the deflagration of GO. The comparative weight loss of GPs and 1,5-NGPs in this temperature range is found to be less than that of GO. The authors gave an explanation that the surface modification of GO by the PPy has prevented the deflagration of GO. A strong interaction among the PPy, GO and the dopant (1,5-naphthalene disulfonic acid) is suggested to be responsible for the improved thermal stability of the composite. The GO layers have negatively charged oxygen atoms and SO_3^- groups (from 1,5-naphthalene disulfonic acid) which provide active sites for the polymerization of pyrrole and the chain growth on GO. The authors proposed that the polymeric chains are arranged in a compact and absolute ordered way on the surface of GO layers. This has restricted heat transport, hence prevented the deflagration of GO. For the purpose of comparison, the TGA of GO-PPy blends were also monitored. The obtained TGA curve evidenced a sharp weight loss at 260°C. The observation suggested no interaction between PPy and GO. Another reported work demonstrated the thermal degradation behavior of PPy/MWCNT composites (Yuvaraj et al. 2010). The presence of thermally stable MWCNT was found to be responsible for the improvement of the thermal stability of PPy.

The reported research works of Ruhi et al. 2014; Ruhi et al. 2015 have given a comparative TGA data for PPy/SiO_2 and PPy/flyash composites with neat polypyrrole. The thermal characteristic of the mentioned composites is explained in detail between the temperature ranges of 25–700°C as shown in Figure 5.8. The fillers (SiO_2 and flyash) showed excellent thermal stability as almost no weight loss is noticed throughout the temperature range. The first derivative curves of the PPy/SiO_2 composites mentioned in the reported work clearly evidenced the delay of thermal decomposition of the composite by 80°C as compared to the neat polypyrrole. The authors have explained that the fillers are interlocking points in the polymer matrix that causes the compact structure of the composite.

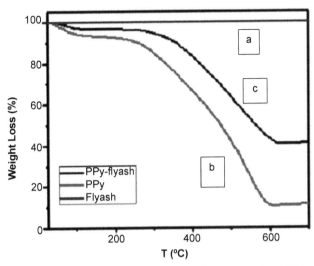

Figure 5.8: Thermograms of flyash (a), polypyrrole (b) and polypyrrole/flyash composite (c) (Ruhi et al., American Journal of Polymer Science, doi:10.5923/s.ajps.201501.03)

5.3.4 Micro-structural studies

Polymer composites are specially designed materials having superior physical, chemical and mechanical properties. For these types of materials, it is important to know the morphology of polymer, the size of the filler material, the distribution of filler in the polymer matrix, the presence of contaminants and inclusions, etc. The micro-structural analysis of the composites is carried out using a Scanning Electron Microscope (SEM) equipped with Energy Dispersive X-Ray Analyser (EDX) and Transmission Electron Microscope (TEM). The above-mentioned techniques provide detailed high-resolution images of the composites by bombarding electron beam over the surface of the test specimens and detecting the backscattered or secondary electron signal, whereas EDX is used to provide information about the presence of different elements and quantitative compositional analysis. The SEM micro-structure of the neat polypyrrole shows cauliflower-like morphology constituted of micro-spherical globules (Eisazadeh 2007; Sanches et al. 2015). However, the polypyrrole-based composites have demonstrated interesting variations in the morphology and size of the polymer. For PPy/mica composites, the presence of mica influence controls the morphology of the polypyrrole (PPy) (Jadhav et al. 2017). The surface of mica was found to be evenly covered with PPy. The PPy/ZnO nano-composites also evidenced nano-sized rod shapes ZnO core coated with polypyrrole particles (Valença et al. 2015). Gao et al. (2014) have reported the micro-structural features of germanium/polypyrrole (Ge/PPy) composite (Gao et al. 2014). The SEM micrographs revealed PPy particles of an average size of 200 nm, whereas the pristine Ge particles are extremely small-sized and clustered to form large agglomerates. The micrographs of the composite showed uniform covering of Ge particles on the surface of PPy which eventually gave spherical morphology (400 nm diameter) of the final composite. The high-resolution SEM micrographs clearly presented the rough surface having very small-sized particles attached on the surface of the composite.

The authors also proposed that PPy acted as a barrier to reduce the extent agglomeration of Ge nano-particles in the composite. The Energy Dispersive Spectroscopy (EDS) mapping confirmed the uniform distribution of elements, like nitrogen (N) and germanium (Ge) in the composite. An elaborate explanation of the micro-structural features of the composite was given by employing High Resolution Transmission Electron Microscopy (HRTEM) with SAED patterns. It was observed that the pristine Ge nano-particles and the Ge nanoparticles in the composite are amorphous. The pristine amorphous Ge nano-particles were observed to have a diameter around 5–20 nm but were clustered into large agglomerates. TEM images also confirmed the uniform covering of amorphous Ge nano-particles on polypyrrole.

Lota et al. (2015) have studied the micro-structural features of carbon/polypyrrole composites. In the composite, the carbon black and carbon nano-tube form the skeleton of the composite having covered with the aggregates of polypyrrole particles. Varshney et al. have discussed the SEM and TEM images of ferrofluid-based nano-architectured polypyrrole composites (Varshney et al. 2014). In this reported work, the SEM study revealed a change in morphology from porous nano-spheres to coral-like morphology with the increase in the amount of ferrofluid in the composite. An important observation carried out by the authors is that the morphology of the composite is influenced by the structure/molar ratio of the precursor materials (monomer, dopant, oxidant) and the experimental conditions (temperature, time of stirring, etc.). The reported work concluded that the change in morphology of the composite is mainly due to the different concentration of ferrofluid in the feed system that helps the formation of ferrofluide/pyrrole complexes in the aqueous phase. The TEM images exhibit irregularly shaped Fe_3O_4 nano-particles being uniformly dispersed in the PPy matrix. The authors have given a creative explanation to their composite as a bunch of cotton balls with entangled black seeds. The TEM results indicated the amorphous nature of PPy, whereas Fe_3O_4 nano-particles, d-spacing of 0.25 nm (311 planes) is found to be present in the polymer matrix. The micro-structural features of PPy/SiO_2 composites are discussed by Ruhi et al. Figure 5.9a and b show the SEM micrographs of the SiO_2 particles and the PPy/SiO_2 composite. The typical cauliflower = like morphology of PPy is shown in the micrograph. The composite evidenced embedded SiO_2 particles in the PPy matrix. Furthermore, the TEM micrograph of the PPy/SiO_2 is shown in Figure 5.9c. The micrographs reveal the filler (SiO_2 particles) dispersed homogenously in the polymer matrix.

Figure 5.10 shows the SEM micrograph of flyash and flyash embedded with PPy. The micrograph reveals the spherical morphology of flyash (Figure 5.10a) SEM micrograph of PPy/flyash composite shows a regular cauliflower-like morphology in which flyash particles are encapsulated in the polypyrrole matrix.

In another work, Ahmed et al. (2016) discussed that the introduction of TiO_2 particles in the polypyrrole matrix which makes the composite denser and causes the porosities to reduce as compared to the pure polypyrrole. SEM analysis also provides important information about the surface morphology of the coated surface and cross-sectional views of the coatings. Yan et al. (2017) have presented the SEM micrographs of the PPy/Al_2O_3 coatings on 316 stainless steel surface. The micro-structural features of the coated surface evidenced homogenously covered surface without any detectable cracks/flaws. The reported work on DGEBA/polypyrrole composite coatings (DGEBA is Diglycidyl ether of bisphenol A) have demonstrated the surface morphology of corroded coated/uncoated steel surfaces after 480 hours of exposure in 5% HCl, 5% NaOH and 5% NaCl solutions (Riaz et al. 2007). The SEM micrographs displayed the appearance of blisters/pin holes on the corroded steel surface, whereas the DGEBA/PPy coatings showed superior passivation to the metal surface in these corrosive media. The micro-structural features of the PPy/nano-diamond (PPy/ND) composite coatings are found to be different from that of the pure PPy coatings (Ashassi-Sorkhabi et al. 2016). The composite coating was observed to have fine, dense and tortuous morphology, having nano-diamond particles embedded in the polypyrrole matrix. Apart from the morphological features, the cross-sectional views of the composite coatings also reveal some interesting morphology and compositional information. Ruhi et al. have presented the FESEM micrographs of the cross-sectional views of the epoxy coatings with different wt% loadings of PPy/SiO_2 composite (Ruhi et al. 2014). The results obtained after the EDS analysis of the coatings confirmed the presence of elements like carbon, oxygen, silicon, nitrogen and titanium. The reported work of Zeybek et al. have discussed the comparative studies of the cross-sectional views of the poly(N-methylpyrrole)-dodecylsulfate (PNMPy-DS) coatings and poly(N-methylpyrrole)-dodecylsulfate/multi-walled carbon nano-tubes composite PNMPy-DS/MWCNTs coatings on the stainless steel substrate (Zeybek et al. 2015). The SEM micrographs of PNMPy-DS coatings evidenced the typical cauliflower-like structure of substituted polypyrrole formed by micro-spherical grains on the stainless steel substrate. The cross-

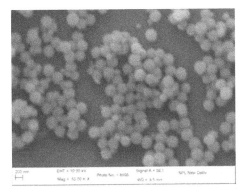

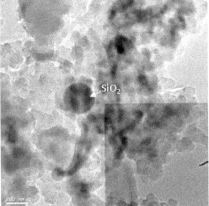

Figure 5.9: SEM micrographs showing morphology of (a) SiO$_2$ particles, (b) PPy/SiO$_2$ composite and (c) TEM micrograph of PPy/SiO$_2$ composite showing distribution of SiO$_2$ particles in the composite (Ruhi et al., Progress in Organic Coatings DOI: 10.1016/j.progcoat.2014.04.013)

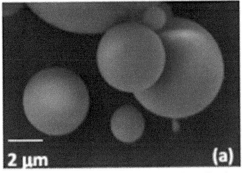

Figure 5.10: SEM micrograph of flyash and PPy/flyash composite (Ruhi et al., American Journal of Polymer Science, DOI: 10.5923/s.ajps.201501.03)

sectional views of the PNMPy-DS/MWCNTs composite coatings exhibit thick coatings with an additional layer of MWCNTs (Zeybek et al. 2015).

5.4 Electrochemical studies to evaluate corrosion resistance

The electrochemical studies are important to evaluate the corrosion resistance of the surface coatings. The corrosion resistance behavior of the coatings is evaluated by employing methods like Open

Circuit Potential (OCP) versus time, potentiodynamic polarization, Electrochemical Impedance Spectroscopy (EIS), weight loss, etc. In OCP versus time method, the surface potential of the metal surface (coated/uncoated) is analyzed under freely corroding conditions without the application of an external current. The variation of OCP with time provides information regarding the various phenomena (corrosion, diffusion of electrolyte, surface passivation, etc.) occurring on the exposed metal surface. The potentiodynamic polarization (Tafel analysis) is a conventional DC technique which facilitates the measurement of cathodic and anodic current densities by polarizing the cathodic and anodic half-cells, respectively. In this measurement, the applied potential is increased with time and the current is constantly observed. Thereafter, the current or current density is plotted against the potential. However, the current-potential curve (Tafel plot) is not linear, and the logarithmic plot is considered. The Tafel plot shows the cathodic curve for reduction reaction and anodic curve for the oxidation reaction. Each curve has a small linear portion which on extrapolating gives the values of various Tafel parameters like corrosion potential (E_{corr}) and corrosion current density (i_{corr}). The slopes of the respective anodic and cathodic curves give the values of Tafel coefficients like cathodic (β_c) and anodic (β_a) Tafel constants. Although this method of analysis is easy to perform, it fails when the corrosion rate is low, the electrolyte has low conductivity or the coating is highly insulating in nature. The corroding metal surface (uncoated/coated) of the electrochemical cell is a part of an equivalent electrical circuit. In this electrical circuit various contributions are from: the solution resistance (R_s) (the polarization resistance that is related to the corrosion rate), the charge transfer resistance (R_{ct}) (that is due to the exchange of charge at coating/metal interface) and the double-layer capacitance (C_{dl}) (that is due to the accumulation of ions at the surface of the electrode). Therefore, in order to obtain a more precise measurement of corrosion kinetics, Electrochemical Impedance Spectroscopy (EIS) is employed. The technique uses AC current rather than DC current. The EIS parameters are obtained by fitting the data in the suitable equivalent electrical circuit model. This technique has unique importance in studying the degradation mechanism of paints/coatings. The paints/coatings pass through a number of degradation stages when exposed to any corrosive electrolyte. These stages are identified and quantified by various equivalent electrical circuit modeling. EIS provides elaborate insight about the mechanism of corrosion phenomena by introducing various parameters like the Warburg diffusion coefficient (Z_w) to explain the occurrence of diffusion of electrolyte through the coating, etc. The EIS data are presented in terms of Nyqiust and Bode plots.

The mentioned methods are established and powerful techniques to study corrosion kinetics. For carrying out the measurements, a potentiostat/galvanostat having three electrodes assembly, namely, the working electrode (uncoated/coated steel specimens 1 cm² area exposed), a platinum counter electrode and an Ag/AgCl reference electrode are used (Figure 5.11). The electrical connection of the electrodes is maintained by connecting it to the potentiostat. The working, counter and reference electrodes are kept immersed in an electrolyte that closely resembles the actual corrosive environment.

5.4.1 Open Circuit Potential (OCP) versus time

The electrode potential that develops at the metal/electrolyte interface when metal is immersed in a solution is called Open Circuit Potential (OCP). In other words, OCP is the potential of the working electrode relative to the reference electrode under a situation when no current or potential is applied to the corroding system. The OCP certainly depends on the tendency of a particular electrode (coated/uncoated) to participate in the electrochemical reactions of corrosion when exposed to a particularly corrosive environment. Thus, the electrodes with nobler (positive) OCP are thermodynamically more stable than the electrodes with lower OCP values. The value of OCP of an electrode exposed to the corrosive medium changes with the lapse of time. Specifically, in the case of the corrosion resistance of a surface coating, watching the OCP trend is interesting. Thus, the variation of OCP with time when the test specimens are immersed in an electrolyte gives the preliminary idea about the barrier property of the surface coating/film under freely corroding condition. The shifting of potential (OCP) toward more positive values generally denotes the passivation behavior of the surface film, whereas

Autolab Potentiostat/ Galvanostat

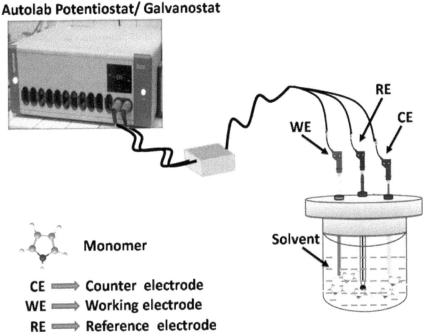

CE ⟹ Counter electrode
WE ⟹ Working electrode
RE ⟹ Reference electrode

Figure 5.11: Diagram of the Auto-lab Potentiostat consisting of conventional three electrodes (working electrode, counter electrode and reference electrode) electrochemical cell.

the negative shift of the potential denotes the commencement of the diffusion of electrolyte/ions through the coating. The surface coatings are defect-free during the initial period of immersion; however, the defects appear in the coatings in the form of pores, fissures, pinholes, etc., as a result of a continuous attack of the corrosive species at coating/electrolyte interface. Over time coatings lose their structural integrity and protective properties. The defects in the coatings create easy pathways for the corrosive species to reach the substrate. So, we can say that OCP versus immersion time is a real method to rank coating performance. The surface coatings show a positive shift of the OCP as compared to the bare substrate, but this value dips gradually over a period of hours/days/months of immersion. Finally, the OCP of the coatings attains a value equal to the bare substrate, indicating the area available for the anodic reactions through pores/defects become somewhat equivalent in terms of the reaction rate intensity to the area of the substrate before coating. Among the various corrosive environments, the chloride-containing seawater environment is found to be highly aggressive in nature and the OCP versus time data can provide important information regarding the surface condition of the coated/uncoated metal surface. Furthermore, this study also gives information about the point where the transition between different states (from active to passive and vice versa) occurs. Here, the passive state indicates the protective nature of the surface coating, and the active state signifies corroding uncoated/coated metal surface.

In saline water, the OCP versus time curve of the bare metal shows a gradual shift toward negative potential. This is basically due to the occurrence of general corrosion on the metallic surface which is exposed to the electrolyte. However, the metal surface coated with films/coatings exhibits interesting OCP trends with the lapse of immersion time. Initially, the potential shifts toward negative values showing the initiation of the diffusion of electrolyte/ions across the coating. However, gradually the potential shifts toward more positive values because of the passivation offered by the surface coating (Li et al. 2016). The steady-state OCP (when the exposed surface reaches equilibrium state) of a protective coating remains more positive than the bare metal surface as the metal remains passivated/protected under the coating layer. The positive shift of the OCP is also associated with the recuperation of the micro-defects/pores in the coatings. Under these conditions, it is noticed that the coatings show self-healing behavior. Conducting polymers (polypyrrole, polyaniline, polythiophene) based organic coatings show self-healing properties because of their unique redox properties. These coatings maintained the OCP of the metal in the noble region exhibiting the passivated state of the underlying metal surface. The incorporation of certain additives/fillers in the polymer matrix makes the composite coating system robust and shifts the OCP toward more noble potential. The fillers trapped in the polymer matrices are perfect interlocking points in the baked coatings which discourage the degradation of coatings under corrosive conditions. The OCP versus time trends of various reported works of polypyrrole (PPy) based composite coatings are discussed in this section. Niratiwongkorn et al. (2016) have discussed the self-healing behavior of polyvinyl butyral (PVB) based organic coating formulations having polypyrrole-carbon black (PPyCB) composite as an inhibiting pigment. A comparative study of OCP of the bare steel, PVB coated and PVB/PPyCB (polyvinyl butyral/polypyrrole/carbon black) composite coated steel are carried out by exposing the test specimens in 4.0% NaCl solution. The OCP values of the initial period of immersion are found to be more positive than the bare steel which shows the superior barrier properties of the coatings. However, the PVB coating demonstrated an abrupt shift of potential toward negative value, and it reached almost equal to the bare steel. The observed trend clearly indicated the failure of the protective nature of PVB coating during prolong immersion time. Interesting OCP trend is observed for the PVB/PPyCB composite coatings. The composite coatings with 20 wt% loading of PPy/CB exhibited an initial negative shift due to the initiation of ingress of electrolyte through the coating. However, a gradual positive shift (noble direction) of potential is noticed after 10 hours of immersion. This is basically due to the surface passivation of the metal by a protective surface coating. The proposed mechanism was that the composite coatings provide dual protection, passivation of the underlying metal surface and the re-passivation of the defects that appear in the coatings during prolong immersion period.

An interesting affirmation of the self-healing property of bilayered PPy-PMo/PPy-DoS composite coating (here, PMo is phosphomolybdate ions and DoS is dodecylsulfate) is given by Kowalski et al. (2010) and Ohtsuka et al. (2006). A small defect is created on the coated surface and is exposed to 3.5% NaCl solution. The OCP shifted toward negative potential in the initial hours of exposure. However, the potential rose up and maintained in the passive region, confirming the passivation of the defect. The Raman Spectroscopy of the defect site evidenced the presence of molybdate salt. A possible mechanism was proposed as the PMo passivated the steel surface by stabilizing the potential in the passive region, whereas the DoS restricted the ingress of anions through the coating. The reported research works of Ruhi et al. demonstrated interesting OCP versus times trends of the PPy/SiO$_2$ and PPy/flyash composite coatings (Figure 5.12a and b) on mild steel substrate. Both the composite coating systems exhibit an initial negative shift of the potential followed by gradual positive shift. The more positive steady-state OCP values of the composite coatings as compared to the neat epoxy coatings clearly evidenced the well-protected metal surface underneath the composite coatings. Authors have specifically mentioned that the redox behavior of the polypyrrole passivates the underlying metal surface, and the fillers (SiO$_2$ and flyash particles) provide mechanical integrity to the coatings even for prolonged immersion period. Montoya et al. have discussed the OCP versus time curves of carbon steel coated with polypyrrole (PPy), PPy and magnetite (PPy-Fe$_3$O$_4$), Glycidoxypropyltrimethoxysilane-Tetraethoxysilane-γ-aminopropyl-silane (GPTMS), PPy/GPTMS-TEOS-APS hybrid bi-layer (PPy/GPTMS) and PPy-Fe$_3$O$_4$/GPTMS in natural aerated aqueous solution at room temperature (Montoya et al. 2014). The OCP curves were drawn for 5,000 seconds. The respective curve of bare steel shows a gradual decrease of potential that is typical for an uncoated steel surface under the active dissolution process. The steel surface coated with GPTMS also evidenced a drop in the potential indicating the penetration of the electrolyte through the coating. Authors observed interesting OCP trends for the steel surface coated with polypyrrole (PPy) coating with an initial passivation action followed by decrease potential showing loss of barrier property. The steady-state potential was measured to be 400 mV which is more positive than the bare steel, leading the authors to conclude that the passivation property of PPy is maintained up to a certain

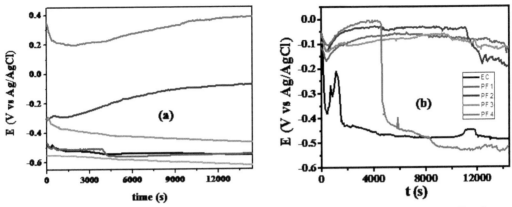

Figure 5.12: (a) The variation of Open Circuit Potential (OCP) with time for uncoated steel (i), epoxy-coated (ii) and epoxy with 1.0 wt% (iii), 2.0 wt% (iv), 3.0 wt% (v), 4.0 wt% (vi) loading of PPy/SiO$_2$ composite coated steel immersed in 3.5% NaCl solution at room temperature (30 ± 2°C). (b) Variation of OCP with time for epoxy-coated (EC), epoxy with 1.0 wt% (PF1), 2.0 wt% (PF2), 3.0 wt% (PF3), 4.0 wt% (PF4) loading of PPy/flyash composite coated steel, immersed in 3.5% NaCl solution at room temperature (30 + 2°C) (Ruhi et al., American Journal of Polymer Science, DOI: 10.5923/s.ajps.201501.03)

extent. The reported work also demonstrated the positive effect of Fe$_3$O$_4$ incorporated in PPy matrix as the potential is maintained in the anodic region throughout the exposure period. The corrosion resistance of the coatings were further strengthened by applying a top layer of hybrid coatings, and for this the OCP was maintained close to 0 V vs Ag/AgCl when the hybrid coating is applied on top of the PPy and PPy-Fe$_3$O$_4$ coatings. The reported work concluded that the time of protection exerted by the PPy coating in the chloride solution is enhanced when it is reinforced by Fe$_3$O$_4$ or protected with an external barrier coating (PPy/GPTMS or PPy-Fe$_3$O$_4$/GPTMS); the self-healing ability of PPy on carbon steel becomes evident and the performance of the coating system is better.

5.4.2 Potentiodynamic polarization (Tafel plots)

Drawing Tafel plots is one of the conventional methods of studying corrosion mechanism of metals/alloys/paints by measuring different corrosion parameters, like corrosion current density (i_{corr}), corrosion potential (E_{corr}), anodic (β_a) and cathodic (β_c) Tafel constants, Corrosion Protection Efficiency (% P.E.), Corrosion Rate (C.R.), etc. The magnitude of these parameters is measured when the exposed surface of the test specimens are polarized by sweeping the potential above and below the E_{corr} at a constant scan rate. A number of research works have been reported in which polarization technique is adopted to evaluate the corrosion resistance of the polypyrrole-based composite coatings. The reported work of B.D. Mert explains the corrosion protection of aluminum surface by electrochemically synthesized PPy/TiO$_2$ composite coatings (Mert 2016). The polarization studies of this composite coating show the shifting of corrosion potential (E_{corr}) to noble potential and the decrease of the corrosion current density (i_{corr}). The protection mechanism explained by the author is that the composite coating restricts the mass transfer between the metal surface and the chloride-containing electrolyte. The TiO$_2$ as a filler has played a very important role in maintaining the potential in the noble region. The corrosion resistance properties of bi-layer coatings on aluminum alloy (AA) 2024-T3 is reported by Kartsonakis et al. (2012). The work reports the first layer of PANI–PPy copolymer contains CeO$_2$ nano-containers loaded with 2-mercaptobenzothiazole (MBT) and the second layer is made of the sol-gel coating. The polarization studies were carried out by exposing the coated/uncoated test specimens in 0.05 M NaCl solution for 72 hours. The studies have shown a remarkable decrease in the anodic and cathodic current of the bi-layer coated Al panels as compared to the other coated samples and the bare Al substrate. The synergistic combination of the properties of copolymer having nano-containers and the top layer of ORMOSIL result in the enhanced corrosion resistance by decreasing the active area of the electrodes. The authors did not claim any self-healing properties in the coatings but the formation of a perfect barrier layer over Al substrate. An interesting explanation of polarization curves of polypyrrole/titanate nano-tubes (TiNT) coatings developed on 904 L stainless steel (SS) is given by Herrasti et al. (2011). The reported work discusses that during polarization studies, additional current is produced by the polypyrrole to the current due to the oxidation and reduction of the substrate (stainless steel). This is due to the presence of π electrons in the aromatic ring and the redox property of the polymer. In order to measure the current produced by the PPy or PPy/TiNT composite coatings, the coatings of the same thickness and composition were developed on platinum electrode. This attempt was done to measure the current during the polarization studies on the platinum electrode coated with PPy and PPy/TiNT composite. The measured current is assumed to be identical to that of the coatings on the stainless steel substrate. Thereafter, the current observed on the platinum electrode because of the presence of was subtracted from the current observed for the PPy and PPy/TiNT coatings on stainless steel. This has given a rough idea on the extent of corrosion on the stainless steel substrate. The polarization curves exhibited more positive corrosion potential and less corrosion current for the coated specimens. Specifically, the PPy/TiNT coated specimen has evidenced more positive E_{corr} as compared to the PPy coated specimen. The explanation given by the authors is that the increase in corrosion potential (E_{corr}) is due to the presence of TiNT in the polymer matrix. The TiNT in water is negatively charged and electrostatically repels the chloride ions. The authors inferred from the polarization test results that the corrosion current density (i_{corr}) of coated specimens are almost three orders of magnitude less as compared to the bare stainless steel. The polarization studies of polypyrrole/ZnO nano-composite coatings on mild steel substrate the show high value of polarization resistance (Rp) as compared to the neat polypyrrole coating and the bare metal (Hosseini et al. 2011). The introduction of ZnO nano-rods in PPy matrix has improved the corrosion resistance of the coatings. The authors concluded that the large aspect ratio, shape and size of the ZnO nano-rods attribute the excellent corrosion resistant property of the nano-composite coatings. The reported work demonstrated that the initial resistance of the PPy coatings and the PPy-ZnO composite coatings were almost same; however, the corrosion resistance of the PPy coatings decreased sharply with exposure time as compared to the PPy-ZnO composite coatings. The authors proposed that the prolonged corrosion protection mechanism of the PPy-ZnO nano-composite coating was due to the presence of nano-rods of ZnO which resulted in the formation of a compact coating having robust nature toward the diffusive chloride ions. The corrosion studies of PPy/SiO$_2$ and PPy/flyash composite coatings carried out by Ruhi et al. on the mild steel substrates also evidenced superior corrosion resistance of the composite coatings with less corrosion current density (i_{corr}) and more positive corrosion potential (E_{corr}) than the bare steel (Ruhi et al. 2014; Ruhi et al. 2015).

Tafel polarization behavior of epoxy coating (EC) and epoxy coatings with different wt% loading of PPy-flyash composite (PF1, PF2, PF3 and PF4) after 1 hour of immersion in 3.5% NaCl solution at room temperature (25 ± 2°C) were studied. The plots were drawn to derive the values of different electrochemical parameters like corrosion current density (icorr), anodic (βa) and cathodic (βc) Tafel constants by extrapolating the anodic and cathodic curve using Tafel extrapolation method (Table 5.2).

From the measured data, it can be observed that the i_{corr} of epoxy coatings with 1.0% (PF1) and 2.0% (PF2) loading of PPy-flyash are two orders of magnitude and more than two orders of magnitude less, respectively, as compared to the epoxy coating. Furthermore, the occurrence of the notably higher value of anodic and cathodic Tafel constants for specimens PF1 and PF2 imply the effective role of PPy-flyash composite in controlling the anodic and cathodic corrosion reactions. The effective corrosion protection by PPy-flyash composite, present as an additive in the epoxy resin, is the reason for the significant reduction in the values of i_{corr} for specimens PF1 and PF2. Accordingly, the Corrosion Protection Efficiency (% P.E.) as calculated from Equation 1 is observed to be 98.9 and 99.0% for specimens PF1 and PF2, respectively. The composite present in the epoxy resin acts as an effective physical barrier against the penetration of chloride ions and protects the underlying metal surface. In addition to this, flyash particles provide mechanical integrity to the coating in 3.5% NaCl

Table 5.2: Electrochemical parameters obtained from Tafel extrapolation for PPy/flyash composites in 3.5% NaCl solution (Ruhi et al., American Journal of Polymer Science, DOI: 10.5923/s.ajps.201501.03)

Substrate description	i_{corr} (A/cm²)	β_a (mV/Decade)	β_c (mV/Decade)	Protection efficiency (%)
EC	6.2×10^{-7}	359.5	96.1	.....
PF1	6.4×10^{-9}	529.8	1247	98.9
PF2	5.8×10^{-9}	748.2	1038	99.0
PF3	1.3×10^{-7}	2078	92.4	79.1
PF4	1.4×10^{-7}	1924	111.4	77.4

solution. In this way, it behaves similar to other conventional coatings which restrict the penetration of ions through them. But the basic difference between the conventional coatings and the coatings with PPy-flyash composite is the polypyrrole that provides anodic protection to the steel surface by shifting its potential to passive region. This is due to the strong oxidative property of conjugated polymers that work as an oxidant to the steel surface. Here, PPy-flyash is added as an additive in the epoxy resin. However, the role of polypyrrole to intercept electrons at the metal surface and to transport them cannot be ruled out, and it improves the corrosion resistance behavior of the coatings. Therefore, PPy-flyash composite delays the start of corrosion on the mild steel surface. In this way, it reinforces the corrosion resistance properties of the epoxy coating. However, the i_{corr} values increased with further loading of PPy-flyash composite in epoxy. The observed i_{corr} for specimens PF3 and PF4 is found to be almost five times higher as compared to epoxy coatings (EC). This could be due to the detrimental effect of the addition of PPy-flyash composite beyond 2.0 wt% in epoxy system. It is speculated that 2.0 wt% is the optimum limit of PPy-flyash loading in epoxy.

Ruhi et al. have presented elaborate explanation on the polarization curves of epoxy coatings with different wt% loading of PPy/SiO$_2$ composites as exposed in 3.5% NaCl solution. The Tafel parameters like E_{corr}, i_{corr}, β_a and β_c were extracted by extrapolating the corresponding anodic and cathodic curves of the plots. The E_{corr} of the coated specimens is found to be more positive as compared to the bare steel. Among the coated specimens, the E_{corr} of the epoxy coatings with different wt% loading of PPy/SiO$_2$ composites is more positive as compared to the neat epoxy coating. In accordance with the E_{corr} values, the values of i_{corr} are found to be less for the epoxy coatings with PPy/SiO$_2$ composites. In this reported work, the epoxy coating formulation with 3.0% loading of PPy/SiO$_2$ has shown the maximum corrosion protection efficiency (99.9%). Various electrochemical parameters obtained from Tafel Extrapolation method are tabulated in Table 5.3. The authors have mentioned that in the chloride bearing neutral solutions, corrosion rate remains very high for the bare steel. Epoxy coatings offer good corrosion protection to the active metal surfaces. However, the adhesion of epoxy coatings to its substrate is compromised with the ingress of electrolyte/ions with the lapse of time, and the coating loses its protection efficiency. The reported work emphasized that the incorporation of PPy/SiO$_2$ composite in the epoxy system reinforces its corrosion protection efficiency and delays the onset of corrosion. The authors have given an explanation about the dual protection offered by the PPy/SiO$_2$ composite; firstly, as a perfect barrier against penetration of chloride ions due to the presence of SiO$_2$ in the polymer matrix and secondly, the anodic protection due to the PPy that maintains the potential of the steel in passive region.

The corrosion test results were supported by the parameter, cross-linking density of the neat epoxy and the epoxy with different wt% loadings of PPy/SiO$_2$ composites. Authors proposed that the cross-linking density is proportional glass transition temperature (T_g). The higher the T_g value, the more will be the barrier property of the coating system. The reported work presented Differential Scanning Calorimetry (DSC) test results of the epoxy and epoxy with composites between the temperature ranges of 25°C to 150°C. The measured T_g values of neat epoxy were found to be increased with the increase of the wt% loading of PPy/SiO$_2$ composites exhibiting an increase in cross-linking density

Table 5.3: Corrosion inhibition efficiency obtained for PPy/SiO$_2$ composites in 3.5% NaCl (Ruhi et al., Progress in Org. Coatings, doi.org/10.1016/j.porgcoat.2014.04.013)

Sample name	i_{corr} (A/cm^2)	β_a (mV/decade)	β_c (mV/decade)	Protection efficiency (%)
Bare M.S.	6.2×10^{-5}	85.7	106.9	--
Epoxy-coated M.S.	8.4×10^{-7}	92.1	189.8	97.99
PPy/SiO$_2$/Epoxy (1%)	1.9×10^{-7}	333.6	97.9	99.95
PPy/SiO$_2$/Epoxy (2%)	1.8×10^{-8}	659.4	126.7	99.98
PPy/SiO$_2$/Epoxy (3%)	1.3×10^{-9}	408.7	603.5	99.99
PPy/SiO$_2$/Epoxy(4)	1.5×10^{-9}	285.1	441.7	99.99

of epoxy in presence of PPy/SiO$_2$ composites. The reported work concluded that due to the increased cross-linking density of the composite, superior corrosion resistance is evidenced from the designed composite coatings. Another work on multilayered polypyrrole-SiO$_2$ composite coatings on stainless steel substrate is reported by Grari et al. (Grari et al. 2015). In this work, a new approach of coating development is adopted to enhance the homogenous distribution of SiO$_2$ particles in the system. For this layer-by-layer technique is adopted and mainly comprises polypyrrole layer deposited by electrochemical process and SiO$_2$ particle layer deposited by electrophoretic process. The reported work designed two different sets of multi-layer coatings, namely PPy/SiO$_2$/PPy layers and SiO$_2$/PPy layers on the stainless steel substrate. The corrosion inhibition of the composite coatings is evaluated by drawing the cyclic polarization curves after 1 hour of exposure to 3.0% NaCl solution. The values of corrosion potentials (E$_{corr}$), corrosion current density (i$_{corr}$), pitting potential (E$_{pit}$) and repassivation potential (E$_{rep}$) are derived from these curves. The PPy/SiO$_2$ composite coated stainless steel specimens demonstrated low i$_{corr}$ and more positive values of E$_{corr}$, E$_{pit}$ and E$_{rep}$ as compared to neat PPy-coated specimen and bare substrate. The improvement in corrosion protection is attributed to SiO$_2$ particles which reinforce the barrier properties of the coatings. Furthermore, the corrosion resistance of both sets of composite coatings was carried out. The polarization parameters showed that the PPy/SiO$_2$ system exhibited the best corrosion resistance due to the larger amount of SiO$_2$ forming a dense layer on stainless steel. Mrad et al. studied the effect of molybdate anions (MoO$_4^{2-}$) and 8-hydroxyquinoline (8HQ) species as dopants for polypyrrole in mitigating the corrosion of AA6061-T6 aluminum alloy in the presence of chloride ions (Mrad et al. 2011). The coatings were designed by electrosynthesis of PPy on the surface of Al alloy exposing to the sulfuric acid solution doped with molybdate anions (MoO$_4^{2-}$) and 8-hydroxyquinoline (8HQ) species. The cyclic potential scan (between -0.60 and 0.80 V at 50 mV^{s-1}) was performed to evaluate the corrosion resistance of neat PPy coatings and the PPy coatings doped with MoO$_4^{2-}$ anions and 8HQ species. The results indicated that PPy coatings were more resistant toward the ingress of ions when doped with 8HQ inhibitor. However, discouraging results were obtained for PPy coatings doped with molybdates. The 8HQ doped PPy films were found to be highly protective against the localized attack of chloride ions. Interestingly, the authors correlated the electrochemical test results with the morphological evidences. The PPy/MoO$_4^{2-}$ was noticed to be porous which allowed the diffusion of chloride ions through the coating to the metal surface to initiate corrosion at coating/metal interface. However, reduced ion permeability across PPy/8HQ coatings declared the coating system to be an efficient barrier against corrosion. The reported work supported the above claims that by adding the cross-sectional views of the coatings show that the PPy coatings experience an increase in thickness when doped with MoO$_4^{2-}$ anions. This leads to an increase in porosities in the coating as compared to the PPy/8HQ coatings. The authors further explained that the bulky polyanions, like Mo$_7$O$_{24}^{6-}$, Mo$_8$O$_{26}^{4-}$ or Mo$_{36}$O$_{112}$(H$_2$O)$_{16}^{8-}$, when incorporated in the polymer matrix causes swelling and porosities in the final designed coatings. In the case of 8HQ doping, the synergistically combined effect of both organic species (Py and 8HQ) give organic entity with a size more amenable to form a more compact and thicker coating.

5.4.3 *Electrochemical Impedance Spectroscopy (EIS)*

As mentioned in the earlier section, EIS is an important technique that makes a reliable prediction about the coating service life by explaining the complex corrosion mechanism and different stages of coating degradation (Hinderliter et al. 2006; Thu et al. 2001; Kendig et al. 1990). The impedance data for a particular coating system is obtained by plotting Nyquist or Cole-Cole plots and Bode plots. In Nyquist plots, the real part of the impedance is plotted against the imaginary part. In the Bode plots, the absolute value of impedance and the phase shift are plotted with log frequency. Unlike Nyquist plots, Bode plots show frequency information. Various research works have included impedance data to discuss the corrosion kinetics mechanism of paints and coatings. An elaborate discussion on the impedance data of the mild steel substrate, the epoxy-coated steel and the epoxy with different wt% loading of PPy/SiO$_2$ composites coated steel is given by Ruhi et al. (2014). The corrosion kinetics were studied for the freshly immersed samples and for the samples immersed for a period of 30 days in 3.5% NaCl solution. The Nyquist plots of the coated steel samples showed capacitive resistive behavior with one time constant during the initial period of immersion. Pore resistance (R$_{pore}$) of the coated samples were found to be significantly high as compared to the bare steel. Accordingly, the coating capacitance (C$_c$) for all the coated samples was observed to be low, indicating low electrolyte uptake tendency of the surface coatings. While categorizing the coated samples further, the epoxy with PPy/SiO$_2$ composite coatings demonstrated a remarkably high corrosion resistance than the neat epoxy coatings. Further discussions were done on the impedance data recorded for the composite coatings with the lapse of immersion time. The effective role of the composites was observed as the composite coatings maintained high R$_{pore}$ and less C$_c$ throughout the immersion period of 30 days. The authors reported diffusion-controlled reactions for these coatings, but no under coating corrosion was claimed during this time span. Test specimens were kept at OCP conditions for 1 hour in 3.5% NaCl solution for impedance analysis. All the measurements were carried out at room temperature (25 ± 2°C). The impedance graphs obtained for epoxy coating (EC) and epoxy coatings with different wt% loading of PPy-flyash composite (PF1, PF2, PF3 and PF4) are displayed in Nyquist plots (Figure 5.13) and Bode plots (Figure 5.15).

The Nyquist plot of specimen EC shows a small semi-circle or an arc with a low value of the impedance (Figure 5.13b). Interestingly, the Nyquist plot for specimen PF1 shows a high frequency capacitive behavior followed by low-frequency diffusion-controlled behavior of the coating. The capacitive behavior shown by the specimen PF1 in the high-frequency region demonstrates the corrosion resistance property of the coating, while the diffusion-controlled behavior in the low-frequency region indicated the occurrence of diffusion process at the coating/metal interface. The Nyquist plot for specimen PF2 shows capacitive and resistive behavior with significantly high impedance. The corresponding Nyquist plots for specimens PF3 and PF4 exhibit a very small semi-circle with a very low impedance value. The Nyquist plots of specimens EC, PF2, PF3 and PF4 have shown one time constant. Therefore, a simplistic circuit (Figure 5.14a), having a resistor that is connected in series to a parallel-connected capacitor and resistor, is applied to measure parameters like pore resistance (R$_{pore}$) and coating capacitance (C$_c$). Whereas, for specimen PF1—which shows diffusion behaviour—an additional circuit element, i.e., Warburg impedance (W) is introduced in the equivalent circuit (Figure 5.14b).

The electrical resistance of a coating system is measured in terms of pore resistance (R$_{pore}$), and it signifies the performance of the surface coating. On the other hand, coating capacitance (C$_c$) is an important parameter to measure the integrity of the coating and is related to water uptake tendency of the coating. Among the test specimens, the specimen EC evidenced the occurrence of a low pore resistance (R$_{pore}$) and a high coating capacitance (C$_c$). This is due to the high water uptake by the epoxy coating due to its weak barrier property. The R$_{pore}$ increased almost by one and two orders of magnitude for specimen PF1 and PF2, respectively, as compared to specimen EC. The specimen PF2 evidenced the highest R$_{pore}$ (1.2 × 10^7 Ω) among the test specimens exhibiting its superior corrosion resistance property. The coating capacitance (C$_c$) occurred in the decreasing order as PF4, PF3, EC,

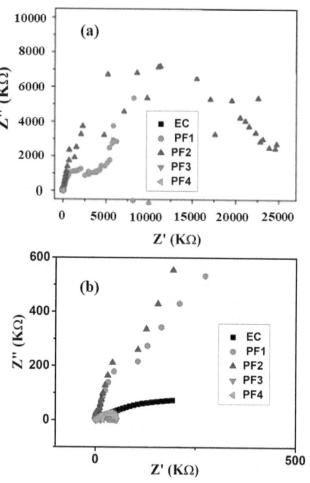

Figure 5.13: Nyquist plots of epoxy coating (EC) and epoxy coatings with 1.0 wt% (PF1), 2.0 wt% (PF2), 3.0 wt% (PF3) and 4.0 wt% (PF4) loading of PPy-flyash composite in 3.5% NaCl solution at 25 ± 2°C. A magnified view of high-frequency region is shown in figure (b) (Ruhi et al., American Journal of Polymer Science, DOI: 10.5923/s.ajps.201501.03)

PF1 and PF2. The low coating capacitance observed for epoxy coatings with 1.0 and 2.0 wt% loading of PPy-flyash composite is due to the low electrolyte uptake by the surface coating. However, the C_c increased with the further increase of the wt% loading of composite in the epoxy system. So, we can say that the increased wt% loading of PPy-flyash composite caused high electrolyte uptake and had a detrimental effect on the barrier property of the coating.

The Bode plots present the simultaneous measurement of modulus of impedance |Z| with respect to frequency as shown in Figure 5.15.

The magnitude of impedance in the low-frequency region gives an idea about the barrier property of a surface film. The high magnitude of impedance in this region signifies high pore resistance toward the diffusion of electrolyte. The Bode plots of specimens PF1 and PF2 show a straight line (slope −1) in the high-frequency region. The slope of the plot becomes −1/2 in the lower frequency region with a high value of |Z| (Figure 5.15). The values of |Z| in this region, for specimens EC, PF3 and PF4 are observed to be low, revealing a low pore resistance and easy electrolyte uptake tendency of the coatings. It can be concluded that the epoxy coatings formed with 1.0 and 2.0 wt% loading of PPy-flyash composite are compact and has superior corrosion resistance in 3.5% NaCl solution. The impedance data have shown that epoxy coatings with 1.0 and 2.0 wt% loading of PPy-flyash have excellent corrosion resistance in the initial period of immersion (1 hour) in 3.5% NaCl solution.

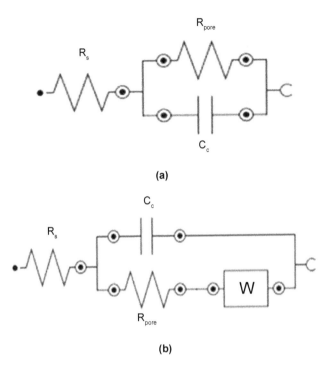

Figure 5.14: Electrical equivalent circuits of (a) intact coating in contact with 3.5% NaCl solution, (b) coatings showing diffusion of chloride ions. Here, R_s is the electrolyte resistance, R_{pore} the pore resistance and C_c the coating capacitance (Ruhi et al., American Journal of Polymer Science, DOI: 10.5923/s.ajps.201501.03)

However, evaluation of corrosion resistance of coatings for a relatively longer period of immersion is equally important. Therefore, impedance parameters of the coatings are derived up to 30 days of immersion in 3.5% NaCl solution. The Nyquist plots of coatings started exhibiting two-time constants after 24 hours of immersion. The first time constant at higher frequency region corresponds to the capacitive behavior of the coating, whereas the second time constant at low-frequency region exhibits the diffusion-controlled corrosion reaction occurring on the coating surface. The decrease of R_{pore} with time exhibits the weakening of the barrier property of the coating as the diffusion of electrolyte takes place through the coating. In accordance with the above results, the coating capacitance (C_c) is found to be low for the test specimens (Figure 5.16). However, it increased with immersion time. The C_c is an important parameter that shows the water uptake tendency of a surface coating. Water has a higher dielectric constant ($\epsilon = 80$), and the absorption of water by the coating increases the coating capacitance. So, it can be concluded that diffusion of electrolyte takes place through the coating resulting in the decrease of R_{pore} and subsequent increase of C_c. However, among the test specimens, the epoxy coating with 2.0 wt% loading of PPy-flyash composite have shown highest R_{pore} and lowest C_c, exhibiting its superior corrosion resistance even during prolonged immersion time.

Montoya et al. (2014) have given EIS data in their research work on various PPy composite coatings developed on mild steel and stainless steel substrates. The measurements were carried out by immersing the test samples in 0.1 mol l⁻¹ NaCl solution for an immersion period of 1.5 hours and 20 hours. The coatings exhibited good barrier properties during the initial hour of immersion. However, the modulus of impedance for PPy, GPTMS and PPy/GPTMS coating systems decreased significantly with time. In addition to this, the two capacitive loops are visible in the Nyquist plots of these coatings which eventually show diffusion-controlled corrosion process. The authors proposed that the barrier properties of the coatings reduced with time due to the ingress of electrolyte through the pores and defects of the coating. Other possible explanations of the coating failure are also

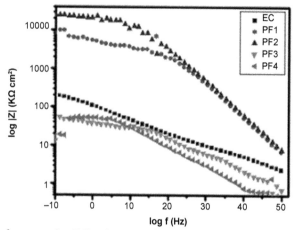

Figure 5.15: Bode plots of epoxy coating (EC) and epoxy coatings with 1.0 wt% (PF1), 2.0 wt% (PF2), 3.0 wt% (PF3) and 4.0 wt% (PF4) loading of PPy-flyash composite in 3.5% NaCl solution at 25±2°C (Ruhi et al., American Journal of Polymer Science, DOI: 10.5923/s.ajps.201501.03)

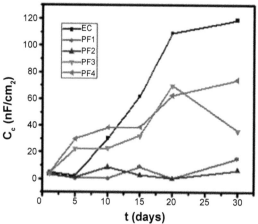

Figure 5.16: Variation of C_c of epoxy coating (EC) and epoxy coatings with 1.0 wt% (PF1), 2.0 wt% (PF2), 3.0 wt% (PF3) and 4.0 wt% (PF4) loading of PPy-flyash composite in 3.5% NaCl solution (Ruhi et al., American Journal of Polymer Science, DOI: 10.5923/s.ajps.201501.03)

mentioned in this work and are mainly due to the exchange of oxalate ions of the PPy with that of the chloride ions, the increased charge carrier mobility of the chloride ions within the polymer, the diffusion-controlled process and the formation of corrosion products. However, encouraging results were obtained for the PPy-Fe_3O_4 and PPy-Fe_3O_4/GPTMS composite coatings. They maintained high impedance modulus even after 20 hours of immersion. The conclusion drawn from the impedance data is that the incorporation of Fe_3O_4 in the PPy matrix improves the barrier properties of the coatings. The EIS discussion on composite coatings consisting of polypyrrole and alumina nanoparticles is reported by Tallman et al. (2008). The impedance data were collected after 1 hour and 168 hours of immersion in dilute Harrison solution (DHS), containing 0.5 g NaCl and 3.5 g/l $NH_4(SO_4)$. The authors have reported detailed corrosion kinetics on how the behavior of the coating changes with the absorption of moisture when exposed to the Harrison solution for prolonged period. Interesting observations were noted from the Nyquist plots for the initial period of immersion. The respective Nyquist plots of composite coatings with different wt% loading of alumina nano-particles evidenced two semi-circles, indicating the presence of two different phases in the coating. The equivalent circuit employed to extract the EIS parameters was designed with two Randles circuits connected in parallel to explain this behavior.

The authors assumed that the low-frequency semi-circle shows the behavior of alumina nano-particle. The reason behind this assumption is that the values of charge transfer resistance calculated from the radius of the semi-circle decreased with the decrease of the wt% loadings of filler (alumina nano-particles) in the coating. Furthermore, the high frequency semi-circle is mentioned to be related to the behavior of the polymer matrix as the pattern of this semi-circle. The Nyquist plots evidenced a significant change in their shape after 168 hours of immersion. Interestingly, the coatings with higher wt% loadings of filler evidenced comparatively less change in shape than the coatings with low wt% of the alumina nano-particles. The composite coatings with low loadings of fillers would have increased volume fraction of voids due to the continuous ingress of moisture across them. The Bode plots supported the findings of the Nyquist plots. The modulus of impedance ($|Z|$) dropped for the 0.5 wt% loading of the filler, for 1.0 wt% it remained same and for the coatings having 2.0 wt% of the filler the increase of $|Z|$ is evidenced with the immersion time. The authors discussed that the presence of nano-particles of alumina has some surface effects that improved the barrier properties of the coatings and slowed down the diffusion of electrolyte across it.

Valenca et al. (2015) demonstrated the corrosion resistance properties of ZnO-Nps/PPy composite coatings by plotting Nyquist plots in 3.0 wt% NaCl solution. The impedance data revealed that epoxy coatings with low loading of ZnO-Nps/PPy composite (0.2% w/w) produced large capacitive loops with high impedance value as compared to the coatings with high loadings of ZnO-Nps/PPy composite. The authors concluded that the coating with low loading of ZnO-Nps/PPy composite has significantly high efficiency against corrosion. The poor corrosion resistance of coatings with higher loadings of ZnO-Nps/PPy composite is due to the particle agglomeration which allows the electrolyte to diffuse into the coating and initiates the corrosion process at coating/metal interface. The authors also used capacitance data to assess electrolyte uptake tendency of the coatings over time. The capacitance was found to be in the range of 10^{-8} to 10^{-10} F for the freshly immersed coatings. However, it increased with the increase of the immersion time and reached around 10^{-7} to 10^{-4} F. Interestingly, the reported capacitance value of the coating with 0.2% w/w loading of ZnO-Nps/PPy composite maintained the capacitance as low as 2.5×10^{-10} F even after 24 hours of immersion. The corrosion resistance of PPy/CeO$_2$ nano-composite coatings is discussed by extracting the EIS data in 0.6 M NaCl solution (Kumar et al. 2017). The authors report that incorporation of nano-sized ceria (CeO$_2$) particles in PPy matrix has improved the corrosion resistance of the coatings, and maximum corrosion protection efficiency is measured for 3.0% loading of CeO$_2$ nano-particles in PPy matrix. The impedance test results showed high values of charge transfer resistance (R_{ct}) and lower values of double-layer capacitance (CPE_{dl}) for the PPy/CeO$_2$ nano-composite coatings as compared to the pure PPy coatings. The research work published by Niratiwongkorn et al. (2016) discussed the self-healing properties of polyvinyl butyral/polypyrrole-carbon black composite (PVB/PPyCB) coatings on carbon steel. The impedance data were extracted by immersing the composite (PVB/PPyCB) coated, PPyCB coated and neat PVB coated carbon steel specimens in 4.0% NaCl solution. The authors discussed that the low-frequency magnitude of impedance ($|Z|$) in Bode plot of 5.0% and 10% PPyCB composite coatings are almost of the same order to that of the neat PVB coating. However, the magnitude of the impedance of the 20% PPyCB coating in the lower frequency region is observed to be higher than the neat PVB coating. The above observations were explained as the effective barrier properties of the organic coatings which inhibits the ingress of oxygen and moisture across them. These types of coatings show one-time constant in the higher frequency region and have a plateau of an 90 degrees in this region. The PVB coatings showed an angle close to 90, and the nature of the plateau remained unaffected by the addition of PPyCB in the PVB matrix. The research work concluded that the PVB/PPyCB composite coatings have a superior barrier property which is sufficient to protect the underlying metal. The authors have mentioned the valid reasons of this superior corrosion protection of the composite coatings, and they are mainly due to the good dispersion of the composites in the PVB matrix and synergistic interaction between the PPyCb and

PVB matrix. The work also discussed the corrosion behavior of coatings after 2 days of immersion in 4.0% Nacl solution. The decrease in the magnitude of |Z| in the low-frequency region is noticed for the neat PVB coatings which revealed the weakening of its barrier properties. However, interesting results are obtained for the PVB/PPyCB composite coatings. The low frequency |Z| value increased after 2 days of immersion, and the values increased more significantly for the higher concentrations of the composite (PVB/PPyCB20). The authors mentioned that the corrosion protection imparted by the composites is due to the synergistic effect of the redox reaction of polypyrrole and the electrical conductivity of the PPyCB pigments.

5.4.4 Self-healing mechanism of the conducting polymer-based composite coatings

The corrosion process is a complex electrochemical phenomenon that may consist of one or more oxidation and reduction reactions occurring at regions (metal/electrolyte interface, metal/coatings interface, etc.) where the galvanic cell is established. It is a thermodynamically driven process and can be reduced by controlling the dynamics of various factors that directly influence its rate. So far the chapter discussed the application of single/multi-layered coatings, functionalized coatings, to reduce the rate of corrosion. These coatings preliminarily act as an efficient barrier between the metal surface and the corrosive electrolyte/environment; hence, they retard the rate of the ingress of water, oxygen, various corrosive ions, etc., reaching the metal surface. However, with the progress of time, the barrier nature of almost every coating diminishes drastically and paves the way for electrolyte/ions to approach metal surface and to commence the corrosion process. There are some active coatings which have the ability to sense the damage and work accordingly. In other words, these coatings are intelligent and smart with self-healing properties. Conducting polymer-based coatings are smart coatings with self-healing properties. These coatings contain materials that have the ability to interact with the environment. Thus, they offer superior barrier protection, as most conventional coatings do, but along with that they also provide anodic protection by shifting the potential of the metal in passive region. In addition to this, the redox behavior of the conducting polymers also makes them electrochemical active which can bind or expel certain dopants/corrosion inhibitors to form complexes at the anodic regions of the corroding metal surface. Therefore, the conducting polymers such as polypyrrole, polyaniline and polythiophene-based coatings may be complimented as more than just barrier coatings. These are active coatings that interact chemically or electrochemically with the metal surface to alter its corrosive behavior. These coatings are genuine alternatives to the ones containing toxic hexavalent chromium ions. The coatings containing active species like Cr(VI) or Zn interacts with the metal surface by complex chemical and electrochemical reactions. Similarly, the conducting polymers design active coatings that not only protect the metals surface from corrosion but also sense the onset of corrosion or coating failure. In the presence of an aerated electrolyte, the metal and the conducting polymer form a galvanic couple. With the commencement of corrosion, the anodic region is formed at coating/metal interface where metal is oxidized. The electrons released during the oxidation process are accepted by the conducting polymer layer which ultimately gets reduced. Here, the shifting of the cathode from the coating/metal interface to conducting polymer coating/electrolyte interface (Figure 5.17) is noticed. Another parallel process that might occur in the presence of dopants/inhibitors in the conducting polymer is during the reduction process where the dopants release from the conducting polymer and interact with the oxidized metal surface to form a metal-dopant complex at coating metal interface. These complexes are expected to block the active sites of the corroding metal surface, thus inhibiting the progress of corrosion. In these cases, the conducting polymer coatings act as inhibitor release coatings.

Ruhi et al. 2014 have discussed the self-healing behavior of polypyrrole/SiO$_2$ composite coatings by creating an artificial defect of diameter 0.75 mm. Various coating compositions like neat epoxy coating and epoxy with 1.0, 2.0, 3.0 and 4.0 wt% loadings of PPy/SiO$_2$ composite coatings were

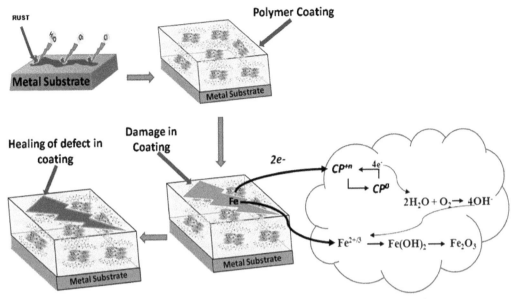

Figure 5.17: Scheme showing the self-healing property of conducting polymer coatings

immersed in 3.5% NaCl solution for 24 hours. Tafel plots were drawn for these coating compositions during an immersion period of 4 hours and 24 hours. The Tafel plots drawn after 4 hours of immersion demonstrated more positive values of corrosion potential (E_{corr}) of the epoxy coatings with PPy/SiO$_2$ composites as compared to the neat epoxy coatings (Figure 5.18A). The corrosion current density was measured to be less for the PPy/SiO$_2$ composite coatings in comparison to the neat epoxy coating. The results indicate that the coatings having PPy/SiO$_2$ composite can inhibit corrosion in the defect in a more efficient way than the neat epoxy coating. The Tafel data is complemented by the FESEM micrographs of the defect region after 4 hours of immersion (Figure 5.19). The defect region of epoxy coating showed the presence of rust with an elemental composition of Fe (55.35%), C (4.29%), O (18.84%) and Cl (0.88%). However, interesting morphological features of the defect are observed under FESEM for the composite coatings with 3.0 and 4.0% loadings of PPy/SiO$_2$. The FESEM micrographs revealed a partially healed image of the defect with an elemental composition of C (10.68%), Fe (42.53%), O (27.93%), Cl (0.59%) and Na (1.52%). In addition to this, the zoomed image of the corroded and passivated region revealed different morphologies. The difference in the elemental composition of the corroded region of the defect and the partially healed region of the defect helped authors to conclude that PPy/SiO$_2$ composite helped the coating to act as an active coating with self-healing property. Authors mentioned that the composite is not able to heal the defect completely as the dimension of the defect is appreciably big although the partial healing of the defect was noticed. An assumption was given that the conducting polymers are efficient to heal micro-defects of the coatings but could not heal significantly larger defects. Further investigations allowed the drawing of the Tafel plots after 24 hours of immersion in 3.5% NaCl solution (Figure 5.18B). The authors concluded that the barrier property of the oxide layer present in the defect further weakened with the lapse of time.

5.4.5 *Corrosion study of the PPy-based composite coatings under accelerated test conditions*

Salt spray test is an accelerated corrosion testing method using high salt content (5.0% NaCl) and high relative humidity. The coated metal panels are exposed to salt spray fog/mist under standard conditions as mentioned in ASTM B117. The progress of corrosion is noticed for a given period to

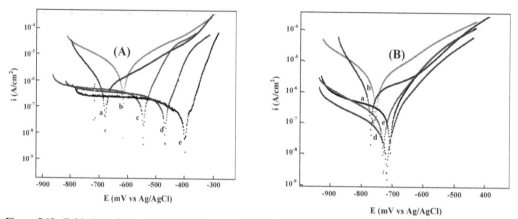

Figure 5.18: Tafel plots of coatings having an artificial defect (diameter 0.75 mm), immersed in 3.5% NaCl solution for 4 hours (A) and 24 hours (B) at temperature 25 ± 2°C. The plots shown here are for (a) neat epoxy coating and epoxy with (b) 1.0%, (c) 2.0%, (d) 3.0% and (e) 4.0% loadings of PPy/SiO$_2$ composite. (Ruhi et al., Progress in Organic Coatings DOI: 10.1016/j.progcoat. 2014.04.013)

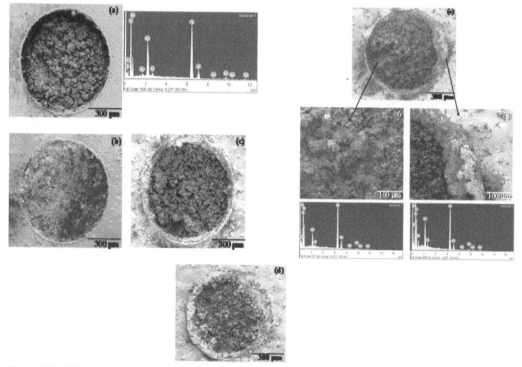

Figure 5.19: FESEM micro-structures of artificial defects created (diameter 0.75 mm) in (a) epoxy-coated mild steel surface and epoxy with (b) 1.0%, (c) 2.0%, (d) 3.0% and (e) 4.% loadings of PPy/SiO$_2$ composite. The EDS spectrum is also shown. Figure (f) and (g) show the high magnification images of the corroded region and passivated region of the defect of 4.0% loading of the composite coating (Ruhi et al., Progress in Organic Coatings DOI: 10.1016/j.progcoat.2014.04.013)

evaluate the corrosion endurance of the paints/coatings. An intentional scribe mark is introduced on the coated metal panels, and the appearance of blisters, the occurrence of undercoating corrosion, etc., are studied. PPy-based composite coatings have shown superior corrosion resistance under salt spray fog of NaCl. Ruhi et al. (2014, 2015) have discussed salt spray test results of PPy/SiO$_2$, PPy/flyash and PPy/chitosan composite coatings developed on the mild steel surface. Figure 5.20 shows

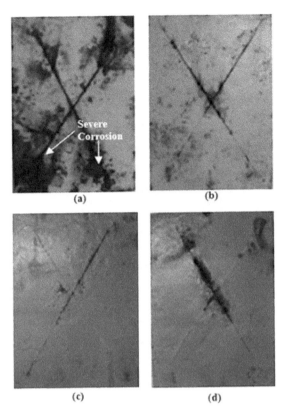

(a)   (b)

(c)   (d)

Figure 5.20: Photographs of (a) epoxy-coated and epoxy with (b) 1.0 wt%, (c) 2.0 wt%, (d) 3.0 wt% loading of PPy-flyash composite coated steel specimens exposed to salt spray fog after 150 day (Ruhi et al., American Journal of Polymer Science, DOI: 10.5923/s.ajps. 201501.03)

the photographs of epoxy-coated mild steel panels (EC) and epoxy with different wt% loading of PPy-flyash composite (PF1, PF2 and PF3) coated steel panels after exposure to salt spray fog for 150 days.

Epoxy-coated mild steel panel reveals the appearance of severe rusting and blistering along the scribe mark as shown in Figure 5.20a. The appearance of rust indicates the loss of adherence of the epoxy coating to its substrate during prolonged exposure to the salt spray fog. Several pinholes are also visible on the surface of the epoxy coating. The photographs of corroded coated steel panels revealed that the neat epoxy coating deteriorates severely under the experimental conditions. On the contrary, the epoxy coatings with various wt% loadings of composites withstood the extremely corrosive conditions even after 65 days of exposure. The appearance of blisters and coating delamination are common for the epoxy-coated steel surface. The PPy-flyash improves the corrosion resistance properties of the steel as evident from the photographs (see Figure 5.20b, c and d). The authors report almost insignificant progress of undercoating corrosion or blistering on the PPy composite coated steel panels. The explanation given for the difference in the behavior of these two coating systems is the densely packed coating formation in the presence of composites. The remarkably high corrosion protection offered by specimens PF1 and PF2, as compared to epoxy coating alone, is due to the dual protection mechanism shown by the synthesized composite. Polypyrrole passivates the underlying metal and inhibits the diffusion of corrosive ions, whereas flyash particles as reinforcing material reduce the degradation of the coating under corrosive conditions. The high cross-linking density of the cured composite coatings is superior in corrosion protection. In addition to this, the various fillers (SiO_2, Flyash, Chitosan, etc.) act as interlocking points in the cured coatings which enhance the mechanical integrity of the coating. Lenz et al. have discussed salt spray test results of polypyrrole/zinc phosphate composite coatings developed on mild steel surface (Lenz et al. 2003). The results disclosed that the steel panels without zinc phosphate pigment showed severe corrosion along the X-cut mark, whereas the panels coated with PPY/zinc phosphate evidenced almost no points of corrosion or blisters. The authors conclude that the zinc phosphate pigment makes an appreciable improvement on the corrosion resistance of the PPy coatings.

Ashassi-Sorkhabi et al. discussed the effect of the addition of nano-materials and organic additives in the given polymer matrix to design coatings for St-12 steel (Ashassi et al. 2014). Among the different combination of composites coatings mentioned in the reported work, polypyrrole/multi-walled carbon nano-tubes/chitosan (PPy-MWCNTs-chitosan) composite coatings were found to be superior in all aspects.

5.5 Conclusion

The present chapter discusses the chemistry of polypyrrole, starting from chemical oxidative polymerization of pyrrole monomer to electroactive properties of the polymer. It also mentions different synthesis routes of the polypyrrole composites. Various polypyrrole-based composites containing fillers, like SiO_2, TiO_2, clay, ZnO, Al_2O_3, MWCNTs, grapheme, etc., and their designs as corrosion-resistant coatings over ferrous and non-ferrous substrates are presented in the chapter. The designing of composite coatings by adopting a number of approaches are too discussed. The chapter has presented an elaborate discussion on different characterization techniques, like X-ray diffraction, Fourier Transform Infrared Spectroscopy (FTIR), Field Emission Scanning Electron Microscopy (FESEM) and Transmission Electron Microscopy (TEM) for compositional and micro-structural characterizations of the composites. The corrosion characterization of the polypyrrole-based composite coatings is the core content of this chapter. Electrochemical studies, like Open Circuit Potential (OCP) versus time, potentiodynamic polarization and Electrochemical Impedance Spectroscopy (EIS) are employed to explain the corrosion resistance offered by the composite coatings. The mechanism of corrosion protection shown by the coatings is also discussed in detail. An interesting section of this chapter is about the self-healing property shown by the polypyrrole/SiO_2 composite coatings. Authors have demonstrated this property of the coatings by presenting the FESEM images of the artificial defect on the coated surfaces and their self-healing tendencies with the passage of time. The chapter also discusses the physicomechanical characterization of the coated surfaces. It concludes on how polypyrrole-based composite coatings offer advanced corrosion protection to the metal surfaces. The reason is the synergistic action of conducting polymer (polypyrrole) and the fillers present in the composite coatings.

References

Ahmed, K., Kanwal, F., Ramay, S. M., Mahmood, A., Atiq, S. and Al-Zaghayer, Y. S. (2016). High dielectric constant study of TiO_2-polypyrrole composites with low contents of filler prepared by *in situ* polymerization. Advances in Condensed Matter Physics.

Al-Dulaimi, A. A., Hashim, S. and Khan, M. I. (2011). Corrosion protection of carbon steel using polyaniline composite with inorganic pigments. Sains Malaysiana, 40(7): 757–763.

Ashassi-Sorkhabi, H., Bagheri, R. and Rezaei-Moghadam, B. (2014). Sonoelectrochemical synthesis, optimized by Taguchi method, and corrosion behavior of polypyrrole-silicon nitride nanocomposite on St-12 steel. Synthetic Metals, 195: 1–8.

Ashassi-Sorkhabi, H., Bagheri, R. and Rezaei-Moghadam, B. (2014). Protective properties of PPy-Au nanocomposite coatings prepared by sonoelectrochemisty and optimized by the Taguchi method. Journal of Applied Polymer Science, 131(22).

Ashassi-Sorkhabi, H., Bagheri, R. and Rezaei-moghadam, B. (2015). Sonoelectrochemical synthesis of PPy-MWCNTs-chitosan nanocomposite coatings: characterization and corrosion behavior. Journal of Materials Engineering and Performance, 24(1): 385–392.

Ashassi-Sorkhabi, H., Bagheri, R. and Rezaei-Moghadam, B. (2016). Corrosion protection properties of PPy-ND composite coating: sonoelectrochemical synthesis and design of experiment. Journal of Materials Engineering and Performance, 25(2): 611–622.

Bumgardner, J. D., Wiser, R., Gerard, P. D., Bergin, P., Chestnutt, B., Marini, M., Ramsey, V., Elder, S. H. and Gilbert, J. A. (2003). Chitosan: potential use as a bioactive coating for orthopaedic and craniofacial/dental implants. Journal of Biomaterials Science Polymer Edition, 14 5: 423–438.

Batool, A., Kanwal, F., Imran, M., Jamil, T. and Siddiqi, S. A. (2012). Synthesis of polypyrrole/zinc oxide composites and study of their structural, thermal and electrical properties. Synthetic Metals, 161(23-24): 2753–2758.

Carneiro, J., Tedim, J., Fernandes, S. C., Freire, C. S. R., Silvestre, A. J. D., Gandini, A., Ferreira, M. G. S. and Zheludkevich, M. L. (2012). Chitosan-based self-healing protective coatings doped with cerium nitrate for corrosion protection of aluminum alloy 2024. Progress in Organic Coatings, 75: 8–13.

Cheah, K., Forsyth, M. and Truong, V. T. (1998). Ordering and stability in conducting polypyrrole. Synthetic Metals, 94(2): 215–219.

Chen, G. Z., Shaffer, M. S., Coleby, D., Dixon, G., Zhou, W., Fray, D. J. and Windle, A. H. (2000). Carbon nanotube and polypyrrole composites: coating and doping. Advanced Materials, 12(7): 522–526.

Chen, J. H., Huang, Z. P., Wang, D. Z., Yang, S. X., Wen, J. G. and Ren, Z. F. (2001). Electrochemical synthesis of polypyrrole/carbon nanotube nanoscale composites using well-aligned carbon nanotube arrays. Applied Physics A, 73(2): 129–131.

Ciric-Marjanovic, G., Dragicevic, L., Milojevic, M., Mojović, M., Mentus, S., Dojcinovic, B., Marjanovic, B. and Stejskal, J. (2009). Synthesis and characterization of self-assembled polyaniline nanotubes/silica nanocomposites. The Journal of Physical Chemistry B, 113(20): 7116–7127.

Dai, T., Yang, X. and Lu, Y. (2007). Conductive composites of polypyrrole and sulfonic-functionalized silica spheres. Materials Letters, 61(14-15): 3142–3145.

DeBerry, D. W. (1985). Modification of the electrochemical and corrosion behavior of stainless steels with an electroactive coating. Journal of The Electrochemical Society, 132(5): 1022–1026.

Dhawan, S. K. and Trivedi, D. C. (1992). Thin conducting polypyrrole film on insulating surface and its applications. Bull. Mater. SCL, Vol. 16, No. 5, October 1993, pp. 371–380.

Diaz, A. F., Kanazawa, K. K. and Gardini, G. P. (1979). Electrochemical polymerization of pyrrole. Journal of the Chemical Society, Chemical Communications, (14): 635–636.

Djebaili, K., Mekhalif, Z., Boumaza, A. and Djelloul, A. X. P. S. (2015). XPS, FTIR, EDX, and XRD analysis of Al_2O_3 scales grown on PM2000 alloy. Journal of Spectroscopy.

Eisazadeh, H. (2007). Studying the characteristics of polypyrrole and its composites. World Journal of Chemistry, 2(2): 67–74.

Fan, J., Wan, M., Zhu, D., Chang, B., Pan, Z. and Xie, S. (1999). Synthesis, characterizations, and physical properties of carbon nanotubes coated by conducting polypyrrole. Journal of Applied Polymer Science, 74(11): 2605–2610.

Fayyad, E. M., Sadasivuni, K. K., Ponnamma, D. and Al-Maadeed, M. A. A. (2016). Oleic acid-grafted chitosan/graphene oxide composite coating for corrosion protection of carbon steel. Carbohydrate Polymer, 20(151): 871–878.

Fenelon, A. M. and Breslin, C. B. (2002). The electrochemical synthesis of polypyrrole at a copper electrode: corrosion protection properties. Electrochimica Acta, 47(28): 4467–4476.

Fenelon, A. M. and Breslin, C. B. (2003). The electropolymerization of pyrrole at a CuNi electrode: corrosion protection properties. Corrosion Science, 45(12): 2837–2850.

Ferreira, C. A., Aeiyach, S., Coulaud, A. and Lacaze, P. C. (1999). Appraisal of the polypyrrole/cataphoretic paint bilayer system as a protective coating for metals. Journal of Applied Electrochemistry, 29(2): 259–263.

Fu, Y., Su, Y. S. and Manthiram, A. (2012). Sulfur-polypyrrole composite cathodes for lithium-sulfur batteries. Journal of the Electrochemical Society, 159(9): A1420–A1424.

Gao, X., Luo, W., Zhong, C., Wexler, D., Chou, S. L., Liu, H. K., Shi, Z., Chen, G., Ozawa, K. and Wang, J. Z. (2014). Novel germanium/polypyrrole composite for high power lithium-ion batteries. Scientific Reports 4: 6095.

Gergely, A., Pfeifer, É., Bertóti, I., Török, T. and Kálmán, E. (2011). Corrosion protection of cold-rolled steel by zinc-rich epoxy paint coatings loaded with nano-size alumina supported polypyrrole. Corrosion Science, 53(11): 3486–3499.

Giuliani, C., Pascucci, M., Riccucci, C., Messina, E., de Luna, M. S., Lavorgna, M., Ingo, G. M. and Di Carlo, G. (2018). Chitosan-based coatings for corrosion protection of copper-based alloys: A promising more sustainable approach for cultural heritage applications. Progress in Organic Coatings, 122: 138–146.

Goy, R. C., Morais, S. T. and Assis, O. B. (2016). Evaluation of the antimicrobial activity of chitosan and its quaternized derivative on *E. coli* and *S. aureus* growth. Revista Brasileira de Farmacognosia, 26(1): 122–127.

Grari, O., Taouil, A. E., Dhouibi, L., Buron, C. C. and Lallemand, F. (2015). Multilayered polypyrrole–SiO$_2$ composite coatings for functionalization of stainless steel: Characterization and corrosion protection behavior. Progress in Organic Coatings, 88: 48–53.

Grgur, B. N., Krstajić, N. V., Vojnović, M. V., Lačnjevac, Č. and Gajić-Krstajić, L. (1998). The influence of polypyrrole films on the corrosion behavior of iron in acid sulfate solutions. Progress in Organic Coatings, 33(1): 1–6.

Guibal, E. (2004). Interactions of metal ions with chitosan-based sorbents: A review. Separation and Purification Technology, 38: 43–74.

Hahn, B. D., Park, D. S., Choi, J. J., Ryu, J., Yoon, W. H., Choi, J. H., Kim, H. E. and Kim, S. G. (2011). Aerosol deposition of hydroxyapatite-chitosan composite coatings on biodegradable magnesium alloy. Surface and Coatings Technology, 205(8-9): 3112–3118.

Han, Y. and Lu, Y. (2007). Preparation and characterization of graphite oxide/polypyrrole composites. Carbon, 45(12): 2394–2399.

Herrasti, P., Recio, F. J., Ocon, P. and Fatás, E. (2005). Effect of the polymer layers and bilayers on the corrosion behavior of mild steel: Comparison with polymers containing Zn microparticles. Progress in Organic Coatings, 54(4): 285–291.

Herrasti, P., Kulak, A. N., Bavykin, D. V., de Léon, C. P., Zekonyte, J. and Walsh, F. C. (2011). Electrodeposition of polypyrrole–titanate nanotube composites coatings and their corrosion resistance. Electrochimica Acta, 56: 1323–1328.

Hien, N.T.L., Garcia, B., Pailleret, A. and Deslouis, C. (2005). Role of doping ions in the corrosion protection of iron by polypyrrole films. Electrochimica Acta, 50(7-8): 1747–1755.

Hinderliter, B. R., Croll, S. G., Tallman, D. E., Su, Q. and Bierwagen, G. P. (2006). Interpretation of EIS data from accelerated exposure of coated metals based on modeling of coating physical properties. Electrochimica Acta, 51: 4505–4515.

Hosseini, M. G., Sabouri, M. and Shahrabi, T. (2007). Corrosion protection of mild steel by polypyrrole phosphate composite coating. Progress in Organic Coatings, 60(3): 178–185.

Hülser, P. and Beck, F. (1990). Electrodeposition of polypyrrole layers on aluminium from aqueous electrolytes. Journal of Applied Electrochemistry, 20(4): 596–605.

Ioniță, M. and Prună, A. (2011). Polypyrrole/carbon nanotube composites: molecular modeling and experimental investigation as anti-corrosive coating. Progress in Organic Coatings, 72(4): 647–652.

Iroh, J. O. and Su, W. (1999). Characterization of the passive inorganic interphase and polypyrrole coatings formed on steel by the aqueous electrochemical process. Journal of Applied Polymer Science, 71(12): 2075–2086.

Jadhav, N., Matsuda, T. and Gelling, V. (2017). Mica/polypyrrole (doped) composite containing coatings for the corrosion protection of cold rolled steel. Journal of Coatings Technology and Research, 1–12.

Janssen, W. and Beck, F. (1989). Electrochemical codeposition of polypyrrole with polyacrylates. Polymer, 30(2): 353–359.

Jiang, S., Liu, Z., Jiang, D., Cheng, H., Han, J. and Han, S. (2016). Graphene as a nanotemplating auxiliary on the polypyrrole pigment for anticorrosion coatings. High Performance Polymers, 28(6): 747–757.

Kartsonakis, I. A., Koumoulos, E. P., Balaskas, A. C., Pappas, G. S., Charitidis, C. A. and Kordas, G. C. (2012). Hybrid organic–inorganic multilayer coatings including nanocontainers for corrosion protection of metal alloys. Corrosion Science, 57: 56–66.

Kendig M. and Scully, J. (1990). Basic aspects of electrochemical impedance application for the life prediction of organic coatings on metals. Corrosion, 46(1) 22–29.

Khor, E. and Whey, J.L.H. (1995). Interaction of chitosan with polypyrrole in the formation of hybrid biomaterials. Carbohydrate Polymers, 26: 183–187.

Koene, L., Hamer, W. J. and De Wit, J. H. W. (2006). Electrochemical behaviour of poly(pyrrole) coatings on steel. Journal of Applied Electrochemistry, 36(5): 545–556.

Konwer, S., Boruah, R. and Dolui, S. K. (2011). Studies on conducting polypyrrole/graphene oxide composites as supercapacitor electrode. Journal of Electronic Materials, 40(11): 2248.

Kowalski, D., Ueda, M. and Ohtsuka, T. (2007). Corrosion protection of steel by bi-layered polypyrrole doped with molybdophosphate and naphthalenedisulfonate anions. Corrosion Science, 49(3): 1635–1644.

Kowalski, D., Ueda, M. and Ohtsuka, T. (2007). The effect of counter anions on corrosion resistance of steel covered by bi-layered polypyrrole film. Corrosion Science, 49(8): 3442–3452.

Kowalski, D., Ueda, M. and Ohtsuka, T. (2008). The effect of ultrasonic irradiation during electropolymerization of polypyrrole on corrosion prevention of the coated steel. Corrosion Science, 50(1): 286–291.

Kowalski, D., Ueda, M. and Ohtsuka, T. (2010). Self-healing ionpermselective conducting polymer coating. Journal of Materials Chemistry, 20(36): 7630–7633.

Kubota, N., Tatsumoto, N., Sano, T. and Toya, K. (2000). A simple preparation of half Nacetylated chitosan highly soluble in water and aqueous organic solvents. Carbohydrate Research, 324: 268–274.

Kudoh, Y. (1996). Properties of polypyrrole prepared by chemical polymerization using aqueous solution containing $Fe_2(SO_4)_3$ and anionic surfactant. Synthetic Metals, 79(1): 17–22.

Kudoh, Y., Akami, K. and Matsuya, Y. (1998). Properties of chemically prepared polypyrrole with an aqueous solution containing $Fe_2(SO_4)_3$, a sulfonic surfactant and a phenol derivative. Synthetic Metals, 95(3): 191–196.

Kumar, A. M. and Rajendran, N. (2012). Influence of zirconia nanoparticles on the surface and electrochemical behaviour of polypyrrole nanocomposite coated 38 316L SS in simulated body fluid. Surface & Coatings Technology, 213: 155–166.

Kumar, A. M., Babu, R. S., Ramakrishna, S. and de Barros, A. L. (2017). Electrochemical synthesis and surface protection of polypyrrole-CeO_2 nanocomposite coatings on AA2024 alloy. Synthetic Metals, 234: 18–28.

Kumar, G. and Buchheit, R. G. (2006). Development and characterization of corrosion resistant coatings using the natural biopolymer chitosan. ECS Transactions, 1(9): 101–117.

Kumar, S. A., Bhandari, H., Sharma, C., Khatoon, F. and Dhawan, S. K. (2013). A new smart coating of polyaniline–SiO_2 composite for protection of mild steel against corrosion in strong acidic medium. Polymer International, 62(8): 1192–1201.

Le Thu, Q., Bierwagen, G. P. and Touzain, S. (2001). EIS and ENM measurements for three different organic coatings on aluminum. Progress in Organic Coatings, 42: 179–187.

Lehr, I. and Saidman, S. (2014). Bilayers polypyrrole coatings for corrosion protection of SAE 4140 steel. Portugaliae Electrochimica Acta, 32(4): 281–293.

Lei, J., Liang, W. and Martin, C. R. (1992). Infrared investigations of pristine, doped and partially doped polypyrrole. Synthetic Metals, 48(3): 301–312.

Lenz, D. M., Delamar, M. and Ferreira, C. A. (2003). Application of polypyrrole/TiO_2 composite films as corrosion protection of mild steel. Journal of Electroanalytical Chemistry, 540: 35–44.

Li, J., Huang, H., Fielden, M., Pan, J., Ecco, L., Schellbach, C., Delmas, G. and Claesson, P. M. (2016). Towards the mechanism of electrochemical activity and self-healing of 1 wt% PTSA doped polyaniline in alkyd composite polymer coating combined AFM-based studies. RSC Advances 6: 19111–19127.

Li, Y., Li, G., Peng, H. and Chen, K. (2011). Facile synthesis of electroactive polypyrrole–chitosan composite nanospheres with controllable diameters. Polymer International, 60: 647–651.

Lingappan, N., Gal, Y. S. and Lim, K. T. (2013). Synthesis of reduced graphene oxide/polypyrrole conductive composites. Molecular Crystals and Liquid Crystals, 585(1): 60–66.

Liu, B., Zhang, Z., Wan, J. and Liu, S. (2017). Enhanced corrosion protection of polypyrrole coatings on carbon steel via electrodeposition. International Journal of Electrochemical Science, 12(2): 994–1003.

Lota, K., Lota, G., Sierczynska, A. and Acznik, I. (2015). Carbon/polypyrrole composites for electrochemical capacitors. Synthetic Metals, 203: 44–48.

Mert, B. D. (2016). Corrosion protection of aluminum by electrochemically synthesized composite organic coating. Corrosion Science, 103: 88–94.

Mollahosseini, A. and Noroozian, E. (2009). Electrodeposition of a highly adherent and thermally stable polypyrrole coating on steel from aqueous polyphosphate solution. Synthetic Metals, 159(13): 1247–1254.

Mondal, J., Marandi, M., Kozlova, J., Merisalu, M., Niilisk, A. and Sammelselg, V. (2014). Protection and functionalizing of stainless steel surface by graphene oxide-polypyrrole composite coating. Journal of Chemistry and Chemical Engineering, 8: 786–793.

Montoya, P., Martins, C. R., de Melo, H. G., Aoki, I. V., Jaramillo, F. and Calderón, J. A. (2014). Synthesis of polypyrrole-magnetite/silane coatings on steel and assessment of anticorrosive properties. Electrochimica Acta, 124: 100–108.

Mrad, M., Dhouibi, L., Montemor, M. F. and Triki, E. (2011). Effect of doping by corrosion inhibitors on the morphological properties and the performance against corrosion of polypyrrole electrodeposited on AA6061-T6. Progress in Organic Coatings, 72: 511–516.

Niratiwongkorn, T., Luckachan, G. E. and Mittal, V. (2016). Self-healing protective coatings of polyvinyl butyral/polypyrrole-carbon black composite on carbon steel. RSC Advances, 6: 43237–43249.

Nogami, Y., Pouget, J. P. and Ishiguro, T. (1994). Structure of highly conducting PF6-doped polypyrrole. Synthetic Metals, 62(3): 257–263.

Ohtsuka, T., Iida, M. and Ueda, M. (2006). Polypyrrole coating doped by molybdo-phosphate anions for corrosion prevention of carbon steels. Journal of Solid State Electrochemistry 10(9): 714–720.

Omastova, M., Pionteck, J. and Trchová, M. (2003). Properties and morphology of polypyrrole containing a surfactant. Synthetic Metals, 135-136: 437–438.

Omastova, M., Trchová, M., Kovářová, J. and Stejskal, J. (2003). Synthesis and structural study of polypyrroles prepared in the presence of surfactants. Synthetic Metals, 138(3): 447–455.

Paliwoda-Porebska, G., Stratmann, M., Rohwerder, M., Potje-Kamloth, K., Lu, Y., Pich, A. Z. and Adler, H. J. (2005). On the development of polypyrrole coatings with self-healing properties for iron corrosion protection. Corrosion Science, 47(12): 3216–3233.

Pei, Q. and Qian, R. (1992). Electrochemical polymerization of pyrrole in aqueous buffer solutions. Journal of Electroanalytical Chemistry, 322(1-2): 153–166.

Petitjean, J., Aeiyach, S., Lacroix, J. C. and Lacaze, P. C. (1999). Ultra-fast electropolymerization of pyrrole in aqueous media on oxidizable metals in a one-step process. Journal of Electroanalytical Chemistry, 478(1-2): 92–100.

Radhakrishnan, S., Sonawane, N. and Siju, C. R. (2009). Epoxy powder coatings containing polyaniline for enhanced corrosion protection. Progress in Organic Coatings, 64(4): 383–386.

Ramesan, M. (2013). Synthesis, characterization, and conductivity studies of polypyrrole/copper sulfide nanocomposites. Journal of Applied Polymer Science, 128(3): 1540–1546.

Redondo, M. and Breslin, C. B. (2007). Polypyrrole electrodeposited on copper from an aqueous phosphate solution: Corrosion protection properties. Corrosion Science, 49(4): 1765–1776.

Riaz, U., Ashraf, S. M. and Ahmad, S. (2007). High performance corrosion protective DGEBA/polypyrrole composite coatings. Progress in Organic Coatings, 59(2): 138–145.

Rikhari, B., Mani, S. P. and Rajendran, N. (2018). Electrochemical behavior of polypyrrole/chitosan composite coating on Ti metal for biomedical applications. Carbohydrate Polymer, 189: 126–137.

Ruhi, G., Bhandari, H. and Dhawan, S. K. (2014). Designing of corrosion resistant epoxy coatings embedded with polypyrrole/SiO$_2$ composite. Progress in Organic Coatings, 77(9): 1484–1498.

Ruhi, G., Bhandari, H. and Dhawan, S. K. (2015). Corrosion resistant polypyrrole/flyash composite coatings designed for mild steel substrate. American Journal of Polymer Science, 5(1A): 18–27.

Sanches, E.A., Alves, S.F., Soares, J.C., Silva, A.M.D., Silva, C.G.D., Souza, S.M.D. and Frota, H.O.D. (2015). Nanostructured polypyrrole powder: a structural and morphological characterization. Journal of Nanomaterials, 16(1): 301.

Saravanan, K., Sathiyanarayanan, S., Muralidharan, S., Azim, S. S. and Venkatachari, G. (2007). Performance evaluation of polyaniline pigmented epoxy coating for corrosion protection of steel in concrete environment. Progress in Organic Coatings, 59(2): 160–167.

Selvaraj, M., Palraj, S., Maruthan, K., Rajagopal, G. and Venkatachari, G. (2010). Synthesis and characterization of polypyrrole composites for corrosion protection of steel. Journal of Applied Polymer Science, 116(3): 1524–1537.

Sharifirad, M., Omrani, A., Rostami, A. A. and Khoshroo, M. (2010). Electrodeposition and characterization of polypyrrole films on copper. Journal of Electroanalytical Chemistry, 645(2): 149–158.

Stejskal, J., Omastova, M., Fedorova, S., Prokeš, J. and Trchová, M. (2003). Polyaniline and polypyrrole prepared in the presence of surfactants: a comparative conductivity study. Polymer, 44(5): 1353–1358.

Su, N., Li, H. B., Yuan, S. J., Yi, S. P. and Yin, E. Q. (2012). Synthesis and characterization of polypyrrole doped with anionic spherical polyelectrolyte brushes. Express Polymer Letters, 6(9).

Su, W. and Iroh, J. O. (1997). Formation of polypyrrole coatings onto low carbon steel by electrochemical process. Journal of Applied Polymer Science, 65(3): 417–424.

Su, W. and Iroh, J. O. (1999). Electropolymerization of pyrrole on steel substrate in the presence of oxalic acid and amines. Electrochimica Acta, 44(13): 2173–2184.

Sugama, T. and Cook, M. (2000). Poly(itaconic acid)-modified chitosan coatings for mitigating corrosion of aluminum substrates. Progress in Organic Coatings, 38: 79–87.

Sugama, T. and Milian-Jimenez, S. (1999). Dextrine-modified chitosan marine polymer coatings. Journal of Material Science, 34: 2003–2014.

Sunderland, K., Brunetti, P., Spinu, L., Fang, J., Wang, Z. and Lu, W. (2004). Synthesis of γ-Fe$_2$O$_3$/polypyrrole nanocomposite materials. Materials Letters, 58(25): 3136–3140.

Sutha, S., Kavitha, K., Karunakaran, G. and Rajendran, V. (2013). *In vitro* bioactivity, biocorrosion and antibacterial activity of silicon integrated hydroxyapatite/chitosan composite coating on 316 L stainless steel implants. Materials Science & Engineering C, 33(7): 4046–4054.

Tallman, D. E., Levine, K. L., Siripirom, C., Gelling, V. G., Bierwagen, G. P. and Croll, S. G. (2008). Nanocomposite of polypyrrole and alumina nanoparticles as a coating filler for the corrosion protection of aluminium alloy 2024-T3. Applied Surface Science, 254(17): 5452–5459.

Tang, B. Z., Geng, Y., Sun, Q., Zhang, X. X. and Jing, X. (2000). Processible nanomaterials with high conductivity and magnetizability. Preparation and properties of maghemite/polyaniline nanocomposite films. Pure and Applied Chemistry, 72(1-2): 157–162.

Troch-Nagels, G., Winand, R., Weymeersch, A. and Renard, L. (1992). Electron conducting organic coating of mild steel by electropolymerization. Journal of Applied Electrochemistry, 22(8): 756–764.

Valença, D.P., Alves, K.G.B., Melo, C.P.D. and Bouchonneau, N. (2015). Study of the efficiency of polypyrrole/ZnO nanocomposites as additives in anticorrosion coatings. Materials Research, 18: 273–278.

Van Schaftinghen, T., Deslouis, C., Hubin, A. and Terryn, H. (2006). Influence of the surface pre-treatment prior to the film synthesis, on the corrosion protection of iron with polypyrrole films. Electrochimica Acta, 51(8-9): 1695–1703.

Varshney, S., Ohlan, A., Jain, V. K., Dutta, V. P. and Dhawan, S. K. (2014). Synthesis of ferrofluid based nanoarchitectured polypyrrole composites and its application for electromagnetic shielding. Materials Chemistry and Physics, 143(2): 806–813.

Wang, J., Li, X., Du, X., Wang, J., Ma, H. and Jing, X. (2017). Polypyrrole composites with carbon materials for supercapacitors. Chemical Papers, 71(2): 293–316.

Wencheng, S. and Iroh, J. O. (1998). Effects of electrochemical process parameters on the synthesis and properties of polypyrrole coatings on steel. Synthetic Metals, 95(3): 159–167.

Wessling, B. (1994). Passivation of metals by coating with polyaniline: corrosion potential shift and morphological changes. Advanced Materials, 6(3): 226–228.

Wu, T. M. and Lin, S. H. (2006). Synthesis, characterization, and electrical properties of polypyrrole/multiwalled carbon nanotube composites. Journal of Polymer Science Part A: Polymer Chemistry, 44(21): 6449–6457.

Xu, L., Chen, W., Mulchandani, A. and Yan, Y. (2005). Reversible conversion of conducting polymer films from superhydrophobic to superhydrophilic. Angewandte Chemie International Edition, 44(37): 6009–6012.

Yalçınkaya, S., Demetgül, C., Timur, M. and Çolak, N. (2010). Electrochemical synthesis and characterization of polypyrrole/chitosan composite on platinum electrode: Its electrochemical and thermal behaviours. Carbohydrate Polymer, 79(4): 908–914.

Yan, M., Vetter, C. A. and Gelling, V. J. (2013). Corrosion inhibition performance of polypyrrole Al flake composite coatings for Al alloys. Corrosion Science, 70: 37–45.

Yan, Q., Li, C., Huang, T. and Yang, F. (2017). Electrochemical synthesis of polypyrrole-Al_2O_3 composite coating on 316 stainless steel for corrosion protection. AIP Conference Proceedings, AIP Publishing.

Yang, S., Tirmizi, S. A., Burns, A., Barney, A. A. and Risen Jr, W. M. (1989). Chitaline materials: soluble chitosan-polypyrrole copolymers and their conductive doped forms. Synthetic Metals, 32: 191–200.

Yang, X. and Lu, Y. (2005). Hollow nanometer-sized polypyrrole capsules with controllable shell thickness synthesized in the presence of chitosan. Polymer, 46(14): 5324–5328.

Yi, Y., Liu, G., Jin, Z. and Feng, D. (2013). The use of conducting polyaniline as corrosion inhibitor for mild steel in hydrochloric acid. International Journal of Electrochemical Science, 8: 3540–3550.

Yuvaraj, H., Jeong, Y. T., Kim, H. G., Gal, Y. S., Hong, S. S. and Lim, K. T. (2010). Synthesis and property of polypyrrole/multi-walled carbon nanotube nanocomposites in supercritical carbon dioxide. Molecular Crystals and Liquid Crystals, 532(1): 72/[488]–482/[498].

Zeybek, B., Aksun, E. and Üğe, A. (2015). Investigation of corrosion protection performance of poly (N-methylpyrrole)-dodecylsulfate/multi-walled carbon nanotubes composite coatings on the stainless steel. Materials Chemistry and Physics, 163: 11–23.

Zhang, T. and Zeng, C. (2005). Corrosion protection of 1Cr18Ni9Ti stainless steel by polypyrrole coatings in HCl aqueous solution. Electrochimica Acta, 50(24): 4721–4727.

Zhang, W. H., Fan, X. D., Tian, W. and Fan, W. W. (2012). Polystyrene/nano-SiO_2 composite microspheres fabricated by Pickering emulsion polymerization: Preparation, mechanisms and thermal properties. Express Polymer Letters, 6(7).

Zhou, M. and Heinze, J. (1999). Electropolymerization of pyrrole and electrochemical study of polypyrrole: 1. Evidence for structural diversity of polypyrrole. Electrochimica Acta, 44(11): 1733–1748.

6

Polypyrrole/Biopolymer Hybrid Coatings

6.1 Introduction of biopolymers

Biopolymers are environmentally friendly, biodegradable and low-cost organic polymers that are extracted from plants or animals. The biopolymers have been present on the earth for millions of years in the form of protein, nucleic acids, carbohydrates, lipids, polysaccharides-like cellulose, starch, etc. The greater part of the human body and the biotic component of the ecosphere are composed of these biopolymers. Cellulose is the most abundant biopolymer comprising almost 33% of the plant component on the earth. The biopolymers are mainly obtained from sugar, starch, cellulose and synthetic materials (synthetic compounds obtained from petroleum). These polymers have an infinite number of applications, ranging from packaging industry to making cutlery, from agricultural applications to biomedical applications, from wastewater treatment to designing of biosensors, etc. The major advantage of these biopolymers over the synthetic polymers is their environmentally friendly nature, i.e., they are biodegradable and has no residual impact on the environment. According to Research Nester, the global biopolymer market is expected to register considerable growth over the forecast period.

Ajay Daniel from Research Nester has elaborated that biopolymers are biodegradable polymers that do not contaminate the water resources like the traditional polymers as shown in schematic Figure 6.1. They are sustainable and renewable polymers which emit less amount of carbon dioxide in the environment. The growing demand for biopolymer as eco-friendly alternatives is anticipated to drive the growth of the biopolymer market over the forecast period. Apart from the above-mentioned applications, some of the biopolymers like gum acacia, chitosan, alginate, tragacanth gum, starch, etc., exhibit good corrosion inhibition properties (Prabhu and Rao 2013). Chitosan is found in the shells of crustaceans, such as lobsters, crabs shrimps and even in many other organisms like insects and fungi. It is one of the most abundant biodegradable materials in the world. Gum acacia is a natural gum consisting of the hardened sap of various species of the acacia tree. Gum arabic (Figure 6.2) consists of a mixture of lower relative molecular mass polysaccharide and higher molecular weight hydroxyproline-rich glycoprotein.

Alginate is a polysaccharide distributed widely in the cell walls of brown algae. Gum tragacanth is a viscous, odorless, tasteless and water-soluble mixture of polysaccharides (Figure 6.3) obtained from sap that is drained from the root of the plant and later dried.

Starch is a polymeric carbohydrate consisting of a large number of glucose units joined by glycosidic bonds (Figure 6.4). This polysaccharide is produced by most green plants as energy storage.

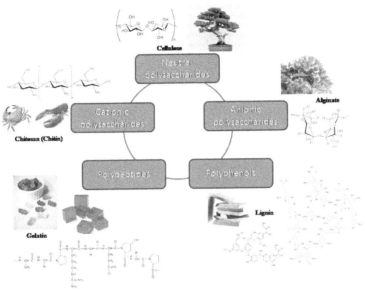

Figure 6.1: Schematic representation of different biopolymers (Daniel Ajay, Research Nester)

Color version at the end of the book

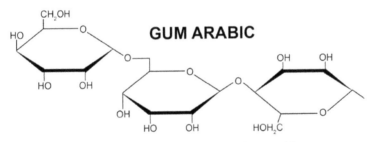

Figure 6.2: Structural representation of Gum Arabic

Gum tragacanth

Figure 6.3: Structural representation of Gum tragacanth

It is the most common carbohydrate in human diets and is contained in large amounts in staple foods like potatoes, wheat, maize (corn), rice and cassava.

These polymers have polar groups like –OH, NH$_2$, etc., which help them form strong bonds with the metallic surfaces (Waanders et al. 2002). Corrosion, as we all know, is a detrimental phenomenon which deteriorates the useful properties of metals and their alloys. Corrosion prevention is one of the serious and most difficult problems in industries where metals are prevalent. Various efforts, like

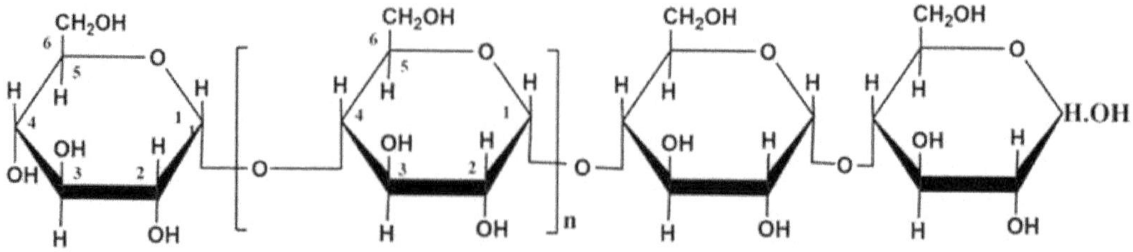

Figure 6.4: Amylose molecule in starch

anodic protection, cathodic protection, application of corrosion-resistant coatings and use of corrosion inhibitors, are done to tackle the problems of corrosion. When it comes to the use of corrosion inhibitors, chemical inhibitors like chromate, phosphate, molybdates, arsenates, etc., are prominent and certainly show superior corrosion properties but they are also known to be highly toxic for the marine life. Thus they do not pass the environmental regulations that exist around the world. Apart from this, their non-biodegradable nature and bioaccumulation property have discouraged their use as a corrosion inhibitor. On the other hand, biopolymers with good corrosion inhibition have an edge over these chemical-based corrosion inhibitors and are considered as safer and greener options. Their biocompatibility, biodegradable nature and non-toxicity have paved the way for them to be exploited as the potential alternative for corrosion inhibition purpose. These polymers possess oxygen, nitrogen or sulfur atoms which are held in a conjugated ring structure. These atoms render adsorption of the polymer on the metal surface to form an inert barrier layer. Interestingly, these polymers govern anodically and cathodically controlled corrosion reactions on the steel surface (Farag et al. 2015; Ramesh Kumar et al. 2015; Murulana et al. 2016). Charitha et al. (2017) have discussed the use of starch as an eco-friendly green inhibitor for corrosion protection of 6061 Al alloy in 0.1 M hydrochloric acid medium. The potentiodynamic polarization data for different concentrations of the inhibitor show that the corrosion current density (i_{corr}) and corrosion rate (CR) decreased with the increase of the concentration of the inhibitor. The authors proposed that the adsorption of inhibitor on the metal surface forms a barrier layer separating it from the aggressive electrolyte. The corrosion protection mechanism offered by the starch is explained based on the structure of starch which has two polymeric units, amylose (a non-branched, helical polymer having α-1,4 linked D-glucose unit) and amylopectin (a branched polymer having α-1,4 and α-1,6 linked D-glucose units). Starch is a big molecule having a number of –OH groups present along the chain. Here, the presence of lone pair of electrons and π-electrons of oxygen atoms are responsible for the protection of Al surface. The authors explained that coordinate bonds are formed between the inhibitor and the Al surface by the donation of electrons to the vacant p-orbitals of Al. This helps in the formation of a protective layer of chemically adsorbed starch on the substrate to control the corrosion process. In another interesting work, Obot et al. have discussed the effective corrosion protection offered by Sodium alginate (a polysaccharide biopolymer) to the API X60 steel exposed in neutral 3.5% NaCl medium (Obot et al. 2017). The authors have highlighted the structure of alginate and correlated it with their corrosion inhibition properties. The biopolymer, alginate is an anionic polysaccharide obtained from seaweed. It possesses carboxylate (–COONa) and hydroxyl (–OH) groups along with oxygen (a heteroatom). The presence of heteroatoms and the availability of lone pairs of electrons favor the adsorption of sodium alginate in the steel substrate, whereas its bulky size and large surface area cover the metal surface efficiently to form a perfect barrier layer. It is mentioned in the reported work that sodium alginate is a mixed inhibitor and is reported to protect mild steel substrates against corrosion in acidic solutions (Al-Bonayan 2014; Tawfik 2015). However, Obot et al. opted to study the corrosion kinetics of sodium alginate in chloride-containing neutral solution. The corrosion studies were carried out by employing weight loss techniques, like OCP versus time measurements, Electrochemical Frequency Modulation (EFM) and impedance measurements. The steel specimens covered with inhibitor evidenced less weight loss as compared to the bare steel. The OCP values were found to be more positive for the steel specimens covered with sodium alginate, exhibiting the well-passivated metal surface. The authors have taken support from the reported work of (Umoren et al. 2008) which says there is a synergistic interaction between halide ions and the polysaccharides that result in superior corrosion inhibition by the sodium alginate layer. On the other hand, the same halide ions cause a detrimental effect on the oxide layer present on the bare steel surface that results in the progress of severe corrosion. The impedance test results demonstrated the appearance of one time constant (Nyquist plots) and high impedance modulus (Bode plots) for the steel specimens with inhibitor. High charge transfer resistance (R_{ct}) and low capacitance (C_c and C_{dl}) for these specimens claimed superior corrosion resistance. The authors beautifully defended the obtained results that organic inhibitors displace adsorbed water molecules at metal/electrolyte interface. Since these inhibitors have low dielectric constant as compared to water, thus they reduce capacitance and rate of transfer of oxidants across the interface. The reported work claimed that the corrosion inhibition increased with the increase of the concentration of sodium alginate in the solution. Apart from this, the contact angle measurement results indicated that the water contact angle increased with the increase of the concentration of sodium alginate and approached 120° with 1,000 ppm concentration, concluding that the surface of the steel is less wetted with sodium alginate. Shih-Chen Shi et al. investigated the corrosion characteristics of the films of hydroxyl propyl methylcellulose derivatives, namely, hydroxy propyl methylcellulose phthalate (HPMCP) and hydroxyl propyl methyl cellulose acetate succinate (HPMCAS) films on high-speed steel substrate (Shi and Su 2016). The parent biopolymer, HPMC is biodegradable (Ehrich et al. 1990; Jiménez et al. 2010; Falguera et al. 2011), biocompatible and has a good film-forming ability (Byun et al. 2012). Various research works of Arukalam et al. have discussed its application as a corrosion inhibitor (Arukalam 2014; Okechi Arukalam et al. 2014; Arukalam et al. 2015). However, Shih-Chen Shi et al. revealed that HPMC has good solubility in water, therefore it is not suitable for designing solid films for aqueous medium or media containing a humid environment. The main reason for designing the derivatives of HPMC is to overcome this problem as both the derivatives of HPMC, i.e., HPMCP and HPMCAS are water-insoluble and perform well in acidic medium. The corrosion studies of the steel substrate (HSS), HPMCP coated and HPMCAS coated high-speed steel (HSS) substrate were carried out by Tafel plots and impedance analysis. The Tafel plots demonstrated more positive corrosion potential (E_{corr}) and less corrosion current density (i_{corr}) for the HPMCP and HPMCAS-coated steel specimens as compared to the bare steel (HSS), indicating the reduced corrosion rates of the coated test specimens. Interestingly, the corrosion rates decreased with the increase in the film thickness, showing a positive correlation between the thickness of the film and corrosion properties of the coatings. The authors further compared the corrosion properties of the HPMCP and HPMCAS coatings and noticed that i_{corr} values are significantly less in the presence of phthalate functional group which renders the hydrophobic property to the fabricated coatings. Therefore, HPMCP coatings are found to be more efficient in inhibiting corrosion on HSS as compared to HPMCAS coatings. The reported work also correlated the corrosion characteristics based on their viscosities. The reported viscosity values for HPMCP and HPMCAS are 100 and 200 mPas, respectively. The authors inferred that the high viscosity of HPMCAS has resulted in the formation of inferior quality of films with pores and defects, hence caused their poor corrosion resistance. On the other hand, the low material viscosity of HPMCP and its hydrophobic property resulted in the formation of a compact uniform film of superior corrosion resistance properties. Samyn (Samyn 2014) has designed a polymer coating having nano-particulate organic compound (styrene maleimide nano-particles) with chemically bonded vegetable oil for corrosion protection of aluminum metal. The author reported that imidization of poly(styrene-co-maleic anhydride) is carried out in the presence of different vegetable oils to form styrene maleimide nano-particles. The reported work has evaluated the overall performance of the coating based on the homogeneity of the coatings, a high degree of imidization, a high water contact angle of the coated surface, low free oil content and good stability of the coatings. Various types of refined vegetable oils like soy oil, castor oil, high-oleic sunflower oil, rapeseed oil, corn oil and hydrogenated castor oil were used to synthesize organic nano-particles encapsulating vegetable oils (SMI/oil). This is then mixed with carnauba wax and water-based emulsion of styrene butadiene (SB) latex to form the final coating compositions. The substrate is subjected to metallographic and chemical pre-treatment steps, followed by dip-coating into the aqueous dispersion of SMI/oil nano-particles, carnauba wax and SB latex to fabricate the coated surfaces. Authors have given a detailed explanation about the significance of surface pre-treatment on the corrosion resistance of the developed coatings. They quoted that the degreasing the aluminum surface resulted in the formation of a more homogenous coating as compared to the non-degreased surfaces. It is well-known that the aluminum has a naturally occurring layer of aluminum oxides and hydroxides and has cracks and inclusions. Therefore, alkaline etching of the substrate was done to remove the reactive top layer, to fabricate rough surface for proper adhesion of the coatings and to provide good wettability for coating with aqueous polymer dispersion. The corrosion test of the coatings was performed by salt spray tests for 480 hours. The visual inspection of the coated Al samples indicated that the alkaline etched polymer-coated samples has significantly high corrosion resistance among the test samples. Authors concluded that appropriate surface treatment followed by deposition of compact/defect-free polymeric coating on the substrate is the key for the excellent corrosion resistance properties. In another work, Roux et al. studied the corrosion inhibition of EPS 180 exopolysaccharides which were the biopolymers used as a concrete admixture for the steel rebars (Roux et al. 2010). For this study, C15 rebars were inserted in CEMI and CEMV cement paste containing EPS 180 and were exposed to seawater. The corrosion behavior of the rebars was studied using OCP versus time measurements and Electrochemical Impedance Spectroscopy (EIS). Talking about EPS biopolymer, some bacterial species like *Lactobacillus reuteri* secretes extracellular polymeric substances (EPS-biopolymer). Among the various EPs biopolymers, EPS 180 exopolysaccharides show corrosion inhibition to mild steel. There are reported studies which evidenced that coatings containing EPS 180 protect steel structures exposed to seawater (Vincke et al. 2002; Stadler et al. 2008). The authors highlighted the role of this biopolymer in reducing the porosities in the concrete as a result of which the penetration of aggressive ions through the concrete is hindered. The electrochemical test results indicated that an addition of EPS 180 solution in the cement alters the oxygen reduction cathodic reaction of the rebars but has no effect on the passive layer. The biopolymer clogged the pores in the concrete for an initial few hours as the seawater is noticed to reach the rebars after certain exposure time. An eco-friendly biopolymer such as starch inhibitor is studied to control corrosion of 6,061 aluminum alloy in 0.25 M hydrochloric acid solution (Charitha and Rao 2015). The reported work extracted the electrochemical data by running potentiodynamic polarization scans and using EIS studies. The different Tafel parameters like E_{corr}, i_{corr}, corrosion rate (CR), percentage inhibition efficiency (% IE), anodic (β_a) and cathodic (β_c) Tafel constants were carefully compiled to correlate the effect of concentration of inhibitor and the values of Tafel parameters. The authors inferred that the corrosion current and the corrosion rate decreased with an increase in the concentration of the inhibitor. The corrosion protection mechanism was given as the formation of a protective barrier layer of inhibitor on the metal surface. From the catholic and anodic slopes, it was suggested that hydrogen evolution is activated and controlled, and it does not change the mechanism of the reduction reaction. Authors mentioned that starch is a mixed type inhibitor. The Nyquist plots produced capacitive loops indicating that the corrosion is controlled by the charge transfer process. It is basically due to the chemisorption of starch on the metal surface to form a barrier layer which protects it from corrosion. Recent research work has highlighted tragacanth gum as an eco-friendly green inhibitor for corrosion protection of carbon steel in 1.0 M HCl medium (Mobin et al. 2017). In this work, the authors have mentioned that carbon steel is extensively used in various industrial applications because of its good strength, easy availability and low cost. But this also true that this particular steel is very prone to acid corrosion which limits its service life in various engineering applications. Biopolymers are promising eco-friendly alternatives for corrosion protection of active metals like carbon steel. The research work mentions that tragacanth gum is the dried exudate obtained from stems and branches of plant Astragalus. Tragacanth gum contains a biopolymer named arabinogalactan (AG) which has corrosion inhibition properties. AG is a highly branched anionic polysaccharide composed of L-arabinose, D-galactose, L-rhamnose, D-xylose,

L-fucose, D-glucose and D-galacturonic acid. When the gum is solubilized in water, it is divided into soluble tragacanthin part and insoluble bassorin part. The authors mentioned that the AG extracted from tragacanth gum is water-soluble and generally produces low viscosity solutions. Some unique structural features of AG, like the presence of hydroxyl groups (–OH), the presence of heteroatom O, the unshared pair of electrons on the oxygen atom, its large size, etc., help it adsorb effectively on the surface of steel substrate and inhibit corrosion at even low concentrations. The corrosion studies are performed by drawing potentiodynamic polarization curves. The reported work explained that the E_{corr} values shifted anodically (as compared to the blank sample) with the addition of AG in 1.0 M HCl. In addition to this, the i_{corr} decreased progressively indicating the presence of an efficient barrier AG layer that protects the metal surface from the corrosive electrolyte. Interestingly, the authors noticed that both the reactions, i.e., cathodic hydrogen evolution and anodic metal dissolution are inhibited by this layer. The maximum corrosion inhibition is reported at a concentration of 500 ppm of AG. The EIS study evaluated the corrosion kinetics of carbon steel in 1.0 M HCl in the presence and absence of AG. The Nyquist and Bode plots are drawn for different concentrations of AG in 1.0 M HCl solution indicating high-frequency capacitive behavior. The different EIS parameters like charge transfer resistance (R_{ct}) increases and double-layer capacitance (C_{dl}) decreases with the increase of AG concentration. The conclusion drawn from the above observations is that the protective film accumulated on the steel surface is effective in protecting the metal against corrosion. The authors also confirmed from the UV-Visible spectroscopic measurements that there exists an interaction between AG with Fe^{2+} to form a complex in the HCl solution.

6.2 Chitosan: properties and applications

Among the various biopolymers utilized for corrosion inhibition purposes, the name of chitosan comes on the top of the list because of its amazing blend of properties. Chitosan is a natural polysaccharide derived from partial deacetylation of chitin which is a highly abundant low-cost biopolymer found in the exoskeletons of the marine crustaceans. Chitosan is a straight-chain copolymer consisting of beta 1-4 linked D-glucosamine units and N-acetyl-D-glucosamine units (Figure 6.5). It is one of the most plentiful, biocompatible and biodegradable natural polymers with remarkably good combination of properties.

Chitosan dissolves in an aqueous medium having pH < 6. At this pH, chitosan becomes a polycationic electrolyte having a high positive charge density on amine groups that are present along the chain. This property of chitosan helps in its interaction with various anionic species. Apart from this, they have an excellent physicochemical film-forming ability. Its polycationic nature, biocompatibility, excellent film-forming ability, non-toxic nature and anti-microbial property have made it a valuable polymer for various applications including biomedical, pharmaceutical, corrosion inhibition, etc. (Sutha et al. 2013; Goy et al. 2016). However, chitosan has one major limitation, i.e., its non-solubility in water, organic solvents and aqueous bases. Dilute acids like hydrochloric acid, acetic acid, phosphoric acid and nitric acid are best suitable to dissolve chitosan (Kubota et al. 2000; Guibal 2004). Coming to its corrosion resistance properties, it is popular due to its attractive physicochemical property. This makes chitosan a sustainable and potential candidate for designing corrosion-resistant coatings for active metals (Sugama and Cook 2000; Kumar and Buchheit 2006). They form complexes with metal ions and adhere strongly on negatively charged surfaces. However, they are prone to absorb atmospheric moisture because of the presence of free –OH and –NH$_2$ groups along the polymeric chain. After absorbing moisture it turns into the hydrogel, causing biodegradation of the coating followed by easy ingress of the electrolyte which results in complete failure of the coating (Sugama and Milian-Jimenez 1999). Various research works have been carried out to reduce the hydrophilic nature of chitosan and to improve its bonding characteristics with the metal surface (Bumgardner et al. 2003; Hahn et al. 2011).

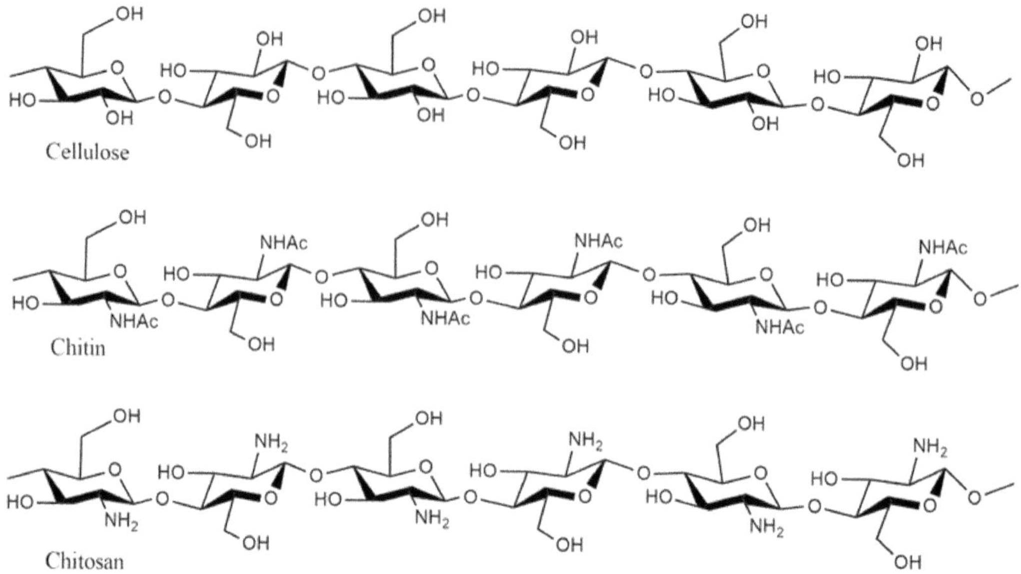

Figure 6.5: Chemical structures of cellulose, chitin and chitosan

6.3 Chitosan-based composite coatings

The reported works mentioned in the earlier section show that the neat chitosan coatings were not sufficient to provide a long term corrosion protection to the metal surface. To overcome its limitations and to exploit its film-forming properties, attempts were made to synthesize various chitosan-based hybrid/composite systems. This section of the chapter will focus on some of the reported works done in this field. Giuliani et al. have synthesized chitosan-based active coatings having corrosion inhibitors like benzotriazole (BTA) and mercaptobenzothiazole (MBT) for copper-based alloy (Giuliani et al. 2018) that are used for indoor artifacts. The present work allows renewable and eco-friendly polymeric material like chitosan to form corrosion-resistant barrier coating which also acts as a reservoir of corrosion inhibitors. The authors explained the synergistic combination of the chemical properties of the inhibitors and the physical properties of the chitosan matrix to obtain a coating system with superior barrier properties. Carneiro et al. declared chitosan as a promising candidate with an exceptionally good combination of properties, like film-forming ability, able to easily functionalize and its good solubility to design green protective coatings (Carneiro et al. 2012). The present work discusses the synthesis of cerium nitrate doped chitosan coatings for aluminum alloys. The amino groups present along with the chitosan chain form complex with the Ce^{3+} ions which act as a corrosion inhibitor. The composite coatings demonstrated self-healing behavior with the release of Ce^{3+} ions from the polymeric matrix. In another interesting work, oleic acid-grafted chitosan/graphene oxide composite coating is developed for the corrosion protection of mild steel substrate (Fayyad et al. 2016). The synthesized composite coating system comprises of a green polymeric matrix having graphene oxide as nano-fillers and oleic acid to impart a best possible combination of anti-corrosive properties. In present work, the authors conclude that the composite coatings improved the corrosion resistance of the substrate hundred folds as compared to the neat chitosan coating.

6.4 Conducting polymer/chitosan composite coatings

Apart from these works, several researchers have reported the synthesis of conducting polymer/ chitosan composites with advanced properties (Yang et al. 1989; Khor and Whey 1995; Yang and Lu 2005). Studies report that the chemical oxidative polymerization of pyrrole in the presence of chitosan reduces the size and improves the solubility of the polypyrrole (Yingmei et al. 2011). Yalchikaya et al. have studied the electrochemical and thermal behavior of polypyrrole/chitosan composite coatings on the platinum electrode (Yalçınkaya et al. 2010). Apart from this, the incorporation of chitosan in the polypyrrole matrix is observed to improve the bonding characteristics of the fabricated coatings. Rikhari et al. have discussed the corrosion properties of electrochemically synthesized polypyrrole/ chitosan (PPy/CHI) composite coatings on titanium substrate for bio-implant applications (Rikhari et al. 2018). The reported work has done extensive corrosion studies on coated metal specimens using Electrochemical Impedance Spectroscopy (EIS), Dynamic Electrochemical Impedance Spectroscopy (DEIS) and polarization studies in the simulated body fluid. The polarization data showed significantly less corrosion current density (i_{corr}) and high polarization resistance (R_p) for the composite coated Ti specimens as compared to the neat PPy coated and bare metal surface, indicating the passive state of the metal surface underneath the composite coating. The authors mentioned that the high corrosion resistance of the chitosan composite coating due to the structural compatibility of both the polymers (chitosan and polypyrrole) that results in the synergistic interaction between them. The possible explanation given about the interaction between chitosan and PPy is due to the presence of free –OH groups in the chitosan chain that forms hydrogen bonds with the nitrogen atom of PPy. This results in the non-availability of the free –OH groups to form hydrogen bonds with water/moisture. Thus, the composite coating showed superior corrosion protection in SBF. Furthermore, the composite layer also facilitates the growth of HAp in the stipulated time of 168 hours in the simulated body fluid. Thus, the composite coating system improves the corrosion resistance as well as the biocompatibility of the Ti implants. The micro-structural features of the composite coated Ti substrate evidenced in the presence of rough surface having the morphology of cauliflower with fiber-like patterns. The rough surface is expected to be beneficial in enhancing bioactivity and bone-bonding ability (Madhan et al. 2012). Another important investigation on corrosion-resistant properties of PPy/chitosan-based composite coatings was carried out by Ruhi et al. (2015). Figure 6.6 shows the scheme of the synthesis of chitosan-polypyrrole-SiO$_2$ composites.

The authors presented elaborate explanation about the superior corrosion resistance offered by the epoxy coatings having chitosan-polypyrrole-SiO$_2$ composites by providing morphological evidence, electrochemical data and compositional details (Ruhi et al. 2015). As the scheme shows, the chemical oxidative polymerization of the pyrrole monomer occurs in the chitosan solution in the presence of sodium salt of p-toluene sulphonate (*p*-TS) and FeCl$_3$. The authors proposed chemical interaction between the chitosan, PPy and *p*-TS. The synthesized composites were reported to be introduced into the epoxy resin in different wt% loadings to develop composite coatings of desired properties. The synergistic combination of the film-forming ability of chitosan, redox behavior of polypyrrole and mechanical integrity provided by the SiO$_2$ particles are the main reasons for the improvement of the overall performance of the designed composite coatings. The compositional analysis using FTIR and XRD techniques have revealed the interaction between the two polymeric chain and the SiO$_2$ particles. The micro-structural examination of the composites indicated morphological changes as evidenced by Figure 6.7. The FESEM micrographs (Figure 6.7a and c) and the TEM micrographs (Figure 6.7b and d) of the chitosan-polypyrrole and chitosan-polypyrrole-SiO$_2$ composites, respectively, revealed a clustered structure composed of spherical nano-particles of chitosan and polypyrrole (diameters ~ 40–50 nm) arranged in a regular pattern. The presence of SiO$_2$ particles in the polymer matrix is also noticed in the FESEM (Figure 6.7c) and TEM (Figure 6.7d) micrographs of chitosan-polypyrrole-SiO$_2$ composites.

The reported work discussed that the polymerization of pyrrole in the absence of chitosan form polypyrrole particles having diameters 100–150 nm with aggregated cauliflower morphology (Yingmei et al. 2011). However, the *in situ* chemical oxidative polymerization of pyrrole monomer in the presence of chitosan forms much smaller nano-spheres of polypyrrole (40–60 nm). The possible explanation given by the authors is the chitosan acts as a steric stabilizer and discourages the formation of large size polypyrrole particles (Yalçınkaya et al. 2010). The corrosion-resistant behavior of the composite coatings is discussed by drawing Open Circuit Potential (OCP) versus time curves, Tafel plots, Nyquist

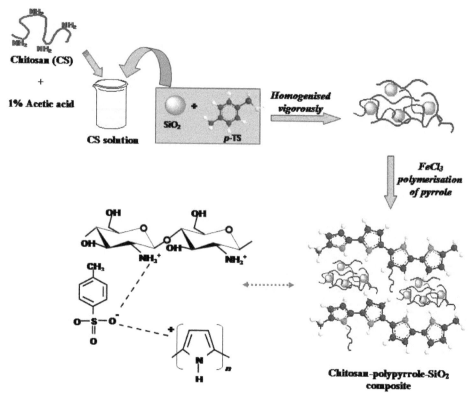

Figure 6.6: The schematic description of various steps of the synthesis of chitosan-polypyrrole-SiO$_2$ composite (Ruhi et al., Synthetic Metals doi.org/10.1016/ j.synthmet. 2014.12.019)

Color version at the end of the book

and Bode plots for the epoxy coatings with and without chitosan-polypyrrole-SiO$_2$ composites. The interpretation of the electrochemical data of the coated specimens in 3.5% NaCl solution indicated the superior corrosion resistance of the epoxy coatings with lower loadings of the composite. The authors gathered a preliminary idea that the designed composite system has an effective barrier property and a self-passivation tendency by interpreting the OCP versus time curves of coated and uncoated mild steel specimens while immersed in the corrosion potential under freely corroding conditions of the composite coatings, demonstrating passivation behavior with gradual shift of OCP toward positive potential.

Whereas, the neat epoxy coatings showed a gradual shift of OCP toward negative potential demonstrating the commencement of diffusion of electrolyte across the epoxy layer. The reported work has presented detailed data of Tafel plots and impedance study to support the advanced corrosion resistance behavior of the chitosan-polypyrrole-SiO$_2$ composite coatings. The low corrosion current density (i$_{corr}$) and pore positive corrosion potential (E$_{corr}$) of composite coatings, as compared to the neat epoxy coatings, have proved its superior corrosion protection ability. The authors have given a possible explanation of the corrosion protection mechanism of the composites (Figure 6.8).

The synergistic combination of the film-forming ability of chitosan—strong oxidative property of polypyrrole and SiO$_2$ particles as a perfect reinforcing material—in the composite results in the formation of a composite system with excellent corrosion resistance properties. Polypyrrole provides anodic protection to the metal by consuming the electrons released during the metal oxidation. Hence, it shifts the cathodic corrosion reaction from the metal/coating interface to coating/electrolyte interface. The Tafel parameters are mentioned in Table 6.1 for the coated and uncoated mild steel specimens.

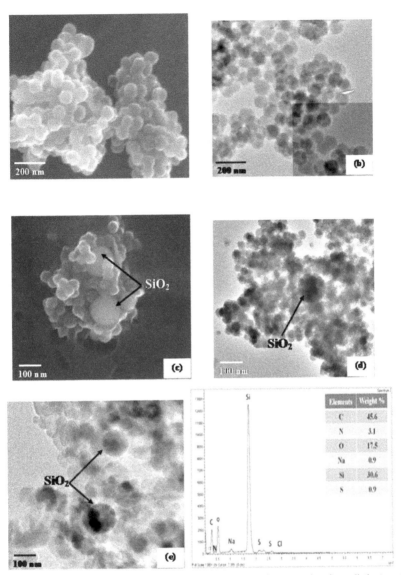

Figure 6.7: FESEM (a) and TEM (b) micrographs of chitosan-polypyrrole composite show distinct particles having spherical morphology. FESEM (c) and TEM (d) micrographs of chitosan-polypyrrole-SiO$_2$ composite show embedded SiO$_2$ particles in the polymer matrix. TEM micrograph of (e) polypyrrole-SiO$_2$ composite (without chitosan) (Ruhi et al., Synthetic Metals doi.org/10.1016/ j.synthmet.2014.12.019)

6.5 Electrochemical impedance studies of PPy/chitosan composites

Impedance measurements were carried out by immersing the uncoated and coated steel specimens in 3.5% NaCl solution at OCP conditions for 1 hour at room temperature. Nyquist and Bode plots were drawn for uncoated steel (BS), epoxy coating (EC) and epoxy coatings with different wt% loadings of chitosan polypyrrole-SiO$_2$ composite (CsPC1, CsPC2 and CsPC3), polypyrrole-SiO$_2$ composite (PS1 and PS2) and chitosan-polypyrrole composite (CP). The Nyquist plots are displayed in Figure 6.9.

The Bode plots (|Z| and phase angle versus frequency plots) of the composite coatings have shown significantly high impedance modulus in the low-frequency region which is equivalent to an undamaged coating (Figure 6.9). Bode plots obtained for epoxy coating (EC), epoxy with different wt% loadings of chitosan-polypyrrole-SiO$_2$ composite (CsPC1, CsPC2 and CsPC3) and

$$CH_3COOH + H_2O \Longleftrightarrow CH_3COO^- + H_3O^+$$

$$Chit\text{-}NH_2 + H_3O^+ \Longleftrightarrow Chit\text{-}NH_3^+ + H_2O$$

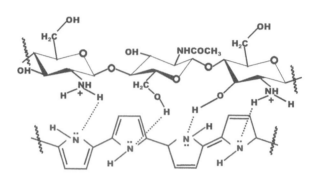

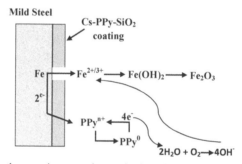

Figure 6.8: A scheme showing the corrosion protection mechanism of PPy-chitosan-SiO$_2$ composite (Ruhi et al., Synthetic Metals doi.org/10.1016/j.synthmet.2014.12.019)

Table 6.1: Tafel parameters extracted by drawing potential versus current data for the coated and uncoated mild steel specimens immersed in 3.5% NaCl solution.

Sample name	E_{corr} (V)	i_{corr} (A/cm^2)	β_a (mV/dec)	β_c (mV/dec)	Protection efficiency (% P.E.)
Bare steel	−0.62	6.2×10^{-5}	85.7	106.9	---
Neat epoxy-coated steel	−0.54	8.4×10^{-7}	101.8	220.8	98.64
PPy/SiO$_2$/Epoxy-coated steel (1%)	−0.51	1.9×10^{-7}	343.3	98.2	99.69
PPy/SiO$_2$/Epoxy-coated steel (2%)	−0.42	1.8×10^{-8}	710.1	129.0	99.97
CP	−0.41	4.4×10^{-9}	324.5	585.6	99.99
PPy/SiO$_2$/Chitosan/Epoxy-coated steel (1%)	0.03	0.9×10^{-9}	317.8	390.3	99.99
PPy/SiO$_2$/Chitosan/Epoxy-coated steel (2%)	0.02	0.7×10^{-9}	406.4	1950	99.99
PPy/SiO$_2$/Chitosan/Epoxy-coated steel (3%)	−0.46	5.5×10^{-9}	366.8	1330	99.99

Note: BS = bare steel, EC = neat epoxy-coated steel, PS1 and PS2 are epoxy with 1.0 and 2.0 wt% loadings of polypyrrole-SiO$_2$ coated steel, respectively, CP = epoxy with chitosan-polypyrrole coated steel and CsPC1, CsPC2 and CsPC3 are epoxy with 1.0, 2.0 and 3.0 wt% loadings of chitosan-polypyrrole-SiO$_2$ coated steel specimens

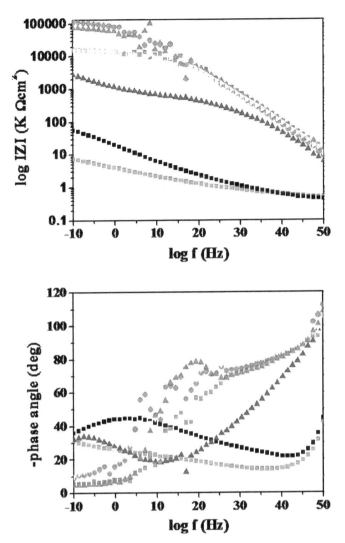

Figure 6.9: Bode plots of epoxy-coated steel specimen (■) and epoxy with 1.0% (■), 2.0% (▲) loadings of polypyrrole-SiO₂ composite coated steel specimens. Plots for epoxy with 1.0% (▲), 2.0% (●) and 3.0% (■) loadings of chitosan-polypyrrole-SiO₂ composites coated and 2.0% (▼) loading of chitosan-polypyrrole composite (CP) coated steel specimens immersed in aqueous solution of 3.5% NaCl for 1 hour at room temperature (25± 2°C). (Ruhi et al., Synthetic Metals doi.org/10.1016/j.synthmet.2014.12.019)

polypyrrole-SiO₂ composite (PS) coated steel specimens are displayed. Bode plot is informative as it gives a simultaneous measurement of modulus of impedance |Z| for the frequency. Bode plot in low-frequency region is particularly interesting as the barrier property of the system is represented by high impedance modulus (high pore resistance, R_{pore}) in this region. As shown in Figure 6.9, a significantly high magnitude of impedance (|Z|) with a slope of –1 at frequency ~ 10 Hz for specimens CsPC1 and CsPC2 signify an excellent barrier property of the surface film. The respective region for the epoxy-coated steel specimen shows a significantly less |Z|, revealing its weak barrier property against corrosive ions. The phase angle plots of the epoxy coatings with 2.0 wt% loadings of chitosan-polypyrrole-SiO₂ composite (CsPC2) show a wide capacitive behavior with a high value of phase angle 81 in the middle frequency region. It is reported that the capacitive behavior in this region shows the insulating properties of the surface film. Furthermore, the measurement of phase angle in low-frequency region is quite important to measure the compactness of the coating. The occurrence

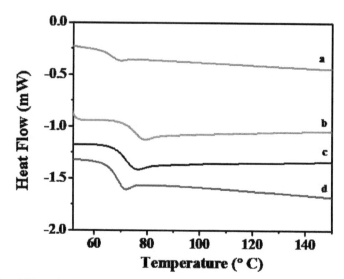

Figure 6.10: Differential Scanning Calorimetry (DSC) thermograms of (a) epoxy resin and epoxy resin with (b) 2.0% loadings of chitosan-polypyrrole-SiO$_2$ (c) 2.0% loadings of chitosan-polypyrrole and (d) 2.0% loadings of polypyrrole-SiO$_2$ composites. The plots are recorded after second heating in a nitrogen atmosphere with a heating rate of 10°C/min (Ruhi et al., Synthetic Metals doi.org/10.1016/ j.synthmet.2014.12.019)

of very low phase angle value for specimen CsPC2 demonstrates the existence of a compact coating on the steel surface. This further compliments the superior barrier nature of the chitosan-polypyrrole-SiO$_2$ composite coatings.

The reported work correlated the superior barrier property of the composite coatings with their high cross-linking density which is measured by Differential Scanning Calorimetry (DSC) as shown in Figure 6.10. The glass transition temperature (T$_g$) of the epoxy resin increased appreciably with the wt% loading of the composites, indicating the increase of the cross-linking density of the cured composite coatings.

The authors emphasized that the synergistically combined properties of chitosan, polypyrrole and SiO$_2$ particles are responsible for the improved corrosion performance of the composite coatings. The designed composite coatings are claimed to show superior adhesion to its metal surfaces as shown in Figure 6.11. The bend test (deformation angle 175°) results show cracked epoxy-coated surface (Figure 6.11a). However, adhesion is shown to be improved with the incorporation of chitosan-polypyrrole-SiO$_2$ composites in epoxy resin as no crack appeared on the coated surfaces (Figure 6.11b, c and d).

Ashassi-Sorkhabi et al. discussed the effect of the addition of nano-materials and organic additives in the given polymer matrix to design coatings for St-12 steel (Ashassi-Sorkhabi et al. 2015). The synthesized coatings were PPy, PPy-chitosan, PPy-MWCNTs and PPy-MWCNTs-chitosan composite coatings. Among the different coating systems, polypyrrole/multi-walled carbon nano-tubes/chitosan (PPy-MWCNTs-chitosan) composite coatings were found to be superior in all aspects. The morphological evidence in form of SEM micrographs revealed a compact layer of PPy-MWCNTs-chitosan on the steel surface. The authors supported their findings by explaining that MWCNTs are the main component in the nano-composite on which the oxidation of pyrrole monomer takes place. Additionally, the -OH and -NH$_2$ groups of chitosan form bonds with the N-atom of polypyrrole, and the said composite grows around the MWCNTs. The polycationic nature of chitosan also helps in forming a compact composite structure on MWCNTs surface. Coming to corrosion studies, the OCP versus time plots obtained by exposing the test specimens in 3.5% NaCl solution exhibit a sudden shift potential toward negative potential with time. This suggests the commencement of the diffusion of chloride ions across the coatings. However, the PPy-MWCNTs-chitosan composite coatings maintained the significantly high positive potential throughout the immersion period indicating its

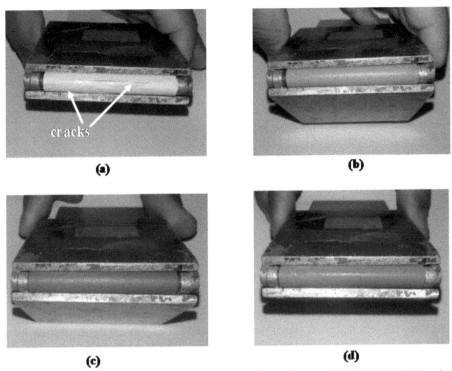

Figure 6.11: Photographs of the surface of (a) epoxy-coated steel sheet and epoxy with (b) 1.0%, (c) 2.0% and (d) 3.0% loading of chitosan-polypyrrole-SiO$_2$ coated steel sheets subjected to bend test (at 175°) (Ruhi et al., Synthetic Metals doi. org/10.1016/j.synthmet.2014.12.019)

good corrosion protection ability. The polarization curves also evidenced more positive E$_{corr}$ values and high polarization resistance for the PPy-MWCNTs-chitosan composite coatings. Authors have taken support from literature to explain the corrosion inhibition and barrier property of chitosan (Gebhardt et al. 2012; El-Haddad 2013). Chitosan forms a compact and dense film in the presence of polypyrrole and MWCNTs. The p-electrons of the MWCNTs act as templates for the growth of the polymer chains which ultimately allows the formation of an effective barrier film. A brief quantitative discussion of the impedance data was done to study the corrosion kinetics of all the coating compositions. For these coatings, the high-frequency semi-circle is attributed to the coating resistance, and the low-frequency region signifies the charge transfer reaction taking place at coating/metal interface. The Nyquist plots indicate that the PPy coatings show high-frequency capacitive behavior followed by the low-frequency diffusion-controlled corrosion behavior. The diffusion behavior is explained by adding the Warburg element in the equivalent circuit. Authors mentioned different impedance parameters like the coating resistance (R$_{coat}$), the coating constant phase element (CPE-1), the constant phase element of double-layer (CPE-2) for the corrosion process, the charge transfer resistance (R$_{corr}$) and the Warburg element in Table 6.2.

In Table 6.2, the W-R, W-T and W-P are the three characters of the finite-length Warburg element. A Warburg element occurs when charge carrier diffuses through a material (mass transport). The W-T and W-R value represent the Warburg coefficient and Warburg resistance, respectively, and W-P is exponent which set in 0.5. Table 6.2 demonstrates that the PPy-MWCNTs-chitosan nano-composite coating has a higher coating resistance (Rcoat) as compared to the PPy, PPy-MWCNTs and PPy-chitosan coatings. In addition to this, the absence of the Warburg element at lower frequency region for the nano-composite coating (Figure 6.7) indicates that the coating can limit the diffusion process. The authors assumed that PPy homogenously coats the surface of MWCNTs and chitosan blocks the ingress of corrosive species resulting in the effective corrosion protection of the substrate.

Table 6.2: Impedance data obtained by fitting the equivalent circuits in the Nyquist using Z view (II) software (Ashassi-Sorkhabi, H., Bagheri, R. and Rezaei-moghadam, B. J. of Materi. Eng. and Perform. (2015) 24: 385. doi.org/10.1007/s11665-014-1297-9)

Sample	CPE1				CPE2			W-R	W-T	W-P
	R_s, Ωcm^2	Yo, $\Omega^{-1} cm^{-2}$	n_1	R_{coat}, Ωcm^2	Yo x 10^{-5} $\Omega^{-1} cm^{-2}$	n_2	R_{corr}, Ωcm^2			
St-12 steel	7	...	...	...	63	0.75	1962	...	...	...
PPy	10	9.02	0.82	176	69	0.50	243	14710	213	0.5
PPy-chitosan	6.34	0.00016	0.80	236	210	0.60	274	...	...	...
PPy-MWCNTs	6.33	0.011	0.62	70	4.00	0.90	518	...	...	...
PPy-MWCNTs-chitosan	6.75	0.0013	0.64	861	2.45	0.80	23446	...	...	...

6.6 Brief discussion on polyaniline/chitosan composite coatings

Another interesting work on corrosion inhibition of conducting copolymer/chitosan/SiO$_2$ composite coatings is discussed by Sambyal et al. (Sambyal et al. 2018). The reported work has discussed the synthesis of corrosion-resistant poly(aniline-anisidine)/chitosan/SiO$_2$ composite coatings for mild steel substrate. The reported work claims that the composite coatings are designed in such a manner that useful property of every component is synergistically combined to achieve the desired results. Authors have highlighted the surface passivation behavior of polyaniline (PANI), a conducting polymer with unique redox properties. In the synthesis steps, the monomers (aniline and anisidine) were adsorbed on the SiO$_2$ surface followed by chemical oxidative copolymerization of the monomers in the chitosan medium at –2°C which is shown in Figure 6.12.

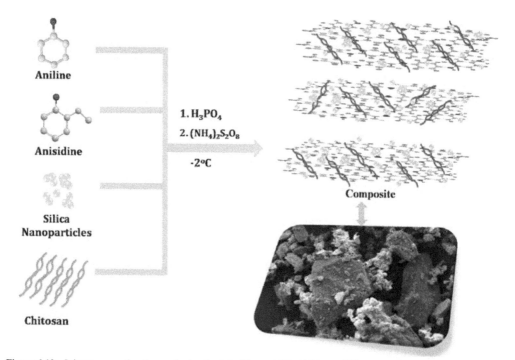

Figure 6.12: Scheme presenting the synthesis of poly(aniline-anisidine)/chitosan/SiO$_2$ composite (Sambyal et al., Progress in Organic Coatings doi.org/10.1016/j.porgcoat.2018.02.014)

The synthesized bluish-green fine powdered composite was blended with epoxy to form different coating compositions. The final coating compositions were spray coated onto the mild steel panels. The synthesized composite was subjected to compositional characterization using XRD and ATR-FTIR, micro-structural characterization using SEM and thermal characterization using TGA. Furthermore, the designed coatings were also subjected to different physico-mechanical characterization like Taber abrasion test, scratch hardness test, bend test and corrosion tolerance tests like Tafel polarization study, EIS study, salt spray test. A detailed electrochemical data of Tafel polarization and EIS study are discussed for the coated and uncoated test specimens for an exposure period of 20 days in 3.5% NaCl solution. The ATR-FTIR spectrum of the composite revealed the presence of characteristic peaks of SiO_2, polyaniline and chitosan. In addition to this, peak showing the presence of the methoxy group of anisidine was also noticed. The reported work presented the broad XRD diffraction pattern of the composite indicating its semi-crystalline nature. The composite is also mentioned to be thermally more stable as compared to the neat polymers. The presence of fillers is concluded to be responsible for the increased thermal stability of the composite. The Tafel polarization data of the coated and uncoated test specimens were presented up to 20 days of immersion in 3.5% NaCl solution at room temperature. Table 6.3 mentions the Tafel parameters of the exposed specimens. The first impression gained from the compiled data is that the i_{corr} of the coated test specimens remained almost three to four orders of magnitude less than the bare steel throughout the immersion period of 20 days. This indicates the presence of an effective barrier coating of poly(aniline-anisidine)/chitosan/SiO_2 composite on the metal surface.

The redox behavior of poly(aniline-anisidine) helps to passivate the micro-sized defects and pores that appear during long exposure duration in a corrosive medium. Furthermore, chitosan helps the formation of homogenous and gives flexible coating with good adhesion properties, whereas SiO_2 particles provide mechanical integrity to the composite coatings. The authors depend on EIS and salt spray test results for the detailed corrosion characteristics of the composite coating.

Nyquist plots of epoxy with 1.0% (APCS1), 2.0% (APCS2), 3.0% (APCS3), 4.0% (APCS4) and 5.0% (APCS5) loadings of poly(aniline-anisidine)/chitosan/SiO_2 composite coated steel substrate immersion in 3.5% NaCl is shown in Figure 6.13. The reported work evaluated the corrosion-resistant performance of the surface coatings by compiling the EIS data and then comparing the pore resistance (R_{pore}) and coating capacitance (C_c) values of the uncoated with the poly(aniline-anisidine)/chitosan/SiO_2 composite coated steel specimens. The R_{pore} appeared to be remarkably high for the polymer composite coated specimens as compared to the uncoated steel. The polymer composite coated surfaces have less affinity toward the electrolyte as the coating capacitance (Cc) is found to be less as compared to the uncoated steel surface. Table 6.4 indicates that the corrosion resistance behavior of the coatings changed with the passage of immersion time as the barrier property of the surface coating weakened.

However, corrosion resistance remains superior to the uncoated steel surface. The authors emphasized that the polymer composite can replenish the mechanical integrity of the coating during the exposure period of 20 days and hence able to passivate the metal surface beneath the coating. Apart from the corrosion studies, the physico-mechanical testing was also conducted to study the robustness of the surface coatings. The Taber Abrasion Test of the neat epoxy and epoxy with polymer composite coated steel panels were carried out to evaluate the wear resistance property of the coating. The polymer composite coatings evidenced less weight loss (as mentioned in Table 6.5).

The result shows that the presence of polymer composite enhances the wear resistance ability in the coating formulation (Figure 6.14a). Furthermore, in the scratch resistance test, a scratch length of 150 mm is scribed with a tip of 1 mm diameter stainless steel ball by increasing the load to 1 Kg at a speed of 100 mm/s. No scratch was visible on the polymer composite coated steel surfaces (Figure 6.14b) indicating its excellent scratch resistance.

The reported work claimed that the polymer composite coated steel panels were superior when subjected to Mandrels Bend Test by bending it with an angle of 175°. No detachment of the coating

Table 6.3: The Tafel parameters obtained for poly(aniline-anisidine)/chitosan/SiO$_2$ composite after exposing the test specimens in 3.5% NaCl solution up to 20 days

Tafel parameters	Exposure in days			
	1	5	10	20
BS				
βa (mV/decade)	300.5	249.2	239.2	185.8
βc (mV/decade)	80.3	87.8	92.9	85.2
E$_{corr}$ (mV)	−557.7	−653.5	−631.5	−628.7
i$_{corr}$ (A/cm^2)	2.1×10^{-5}	2.5×10^{-5}	3.4×10^{-5}	3.7×10^{-5}
C.R. (mm/year)	0.23	0.30	0.39	0.40
APCS1				
βa (mV/decade)	1,950.0	122.7	153.5	348.8
βc (mV/decade)	1,64.1	107.9	238.8	494.3
E$_{corr}$ (mV)	−489.7	−687.0	−629.8	−435.5
i$_{corr}$ (A/cm^2)	2.5×10^{-8}	1.1×10^{-7}	3.7×10^{-7}	4.2×10^{-8}
C.R. (mm/year)	3.0×10^{-4}	1.3×10^{-3}	3.3×10^{-3}	4.9×10^{-4}
APCS2				
βa (mV/decade)	883.8	1,200.0	191.7	391.8
βc (mV/decade)	400.0	200.0	215.3	827.6
E$_{corr}$ (mV)	−457.2	−517.7	−605.2	−709.9
i$_{corr}$ (A/cm^2)	3.5×10^{-9}	1.4×10^{-7}	1.5×10^{-7}	1.5×10^{-8}
C.R. (mm/year)	4.0×10^{-5}	1.7×10^{-3}	1.8×10^{-3}	1.8×10^{-4}
APCS3				
βa (mV/decade)	604.2	112.5	155.2	152.1
βc (mV/decade)	73.5	153.5	151.7	120.3
E$_{corr}$ (mV)	−645.0	−751.0	−717.3	−715.2
i$_{corr}$ (A/cm^2)	12.1×10^{-8}	1.7×10^{-7}	6.8×10^{-7}	2.1×10^{-6}
C.R. (mm/year)	1.3×10^{-3}	2.0×10^{-3}	7.9×10^{-3}	2.4×10^{-2}
APCS4				
βa (mV/decade)	224.2	225.5	2,500.0	203.3
βc (mV/decade)	102.5	477.5	923.0	675.5
E$_{corr}$ (mV)	−585.8	−613.5	−571.2	−661.0
i$_{corr}$ (A/cm^2)	4.7×10^{-9}	5.1×10^{-8}	2.0×10^{-6}	2.2×10^{-6}
C.R. (mm/year)	5.4×10^{-5}	6.0×10^{-4}	2.3×10^{-2}	2.5×10^{-2}
APCS5				
βa (mV/decade)	902.8	555.7	328.3	2,080.0
βc (mV/decade)	1,400.0	195.7	297.2	859.5
E$_{corr}$ (mV)	−433.3	−612.0	−396.3	−655.5
i$_{corr}$ (A/cm^2)	5.3×10^{-8}	3.6×10^{-8}	1.0×10^{-8}	6.3×10^{-8}
C.R. (mm/year)	6.2×10^{-4}	4.2×10^{-4}	1.2×10^{-4}	7.3×10^{-4}

was observed, whereas the neat epoxy-coated steel surface evidenced the appearance of cracks (Figure 6.14c).

Another research work by Sambyal et al. 2018 reported on the corrosion resistance of poly(aniline-co-o-toluidine)–chitosan–SiO$_2$/epoxy composite (POCS) coatings for mild steel substrate. It presented a detailed comparison of the electroactive properties of homopolymers (polyaniline and polytoluidine) and the copolymer poly(aniline-co-o-toluidine) by chemical oxidative polymerization as well as using cyclic voltametry (CV).

The poly(aniline-co-o-toluidine)/chitosan/SiO$_2$ composites were synthesized by copolymerization of aniline and o-toluidine (monomers are taken in equimolar ratio) using ammonium persulfate (APS) as oxidant as given in Figure 6.15. First of all, SiO$_2$ nano-particles were synthesized taking tetra-ethylorthosilicate (TEOS) as the precursor. The hydrolysis of TEOS was carried out by ethanol and

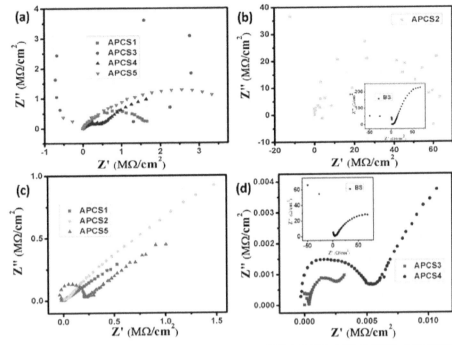

Figure 6.13: (a and b) Nyquist plots of epoxy with 1.0% (APCS1), 2.0% (APCS2), 3.0% (APCS3), 4.0% (APCS4) and 5.0% (APCS5) loadings of poly(aniline-anisidine)/chitosan/SiO$_2$ composite coated steel substrate immersion in 3.5% NaCl solution for 24 hours at room temperature 25 ± 2°C. Figure 6 (c & d): The respective Nyquist plots after 20 days of immersion in 3.5% NaCl solution (Sambyal et al., Progress in Organic Coatings doi.org/10.1016/j.porgcoat.2018.02.014)

Table 6.4: The EIS parameters obtained after exposing the test specimens in 3.5% NaCl solution up to 20 days

CS	Exposure days	EIS parameters			
		1	5	10	20
BS	R_{pore} (Ω)	71 ± 8.1%	61.7 ± 3.1%	55.20 ± 9.4%	37.5 ± 11.4%
	C_c (F/cm^2)	$2.5 \times 10^{-8} \pm 5.6\%$	$1.7 \times 10^{-7} \pm 9.1\%$	$1.5 \times 10^{-7} \pm 11.2\%$	$1.15 \times 10^{-3} \pm 12.2\%$
	W (Mho)	-------	---------	$8.3 \times 10^{-3} \pm 7.0\%$	--------
APCS1	R_{pore} (Ω)	$4.9 \times 10^{7} \pm 12.2\%$	4596.1 ± 4.9%	1895 ± 4.4%	752 ± 12.3%
	C_c (F/cm^2)	$2.05 \times 10^{-9} \pm 14.0\%$	$2.8 \times 10^{-9} \pm 8.5\%$	$8.2 \times 10^{-9} \pm 7.2\%$	$5.7 \times 10^{-8} \pm 2.1\%$
	W (Mho)	--------	$3.3 \times 10^{-5} \pm 7.1\%$	$7.4 \times 10^{-5} \pm 5.9\%$	$2.3 \times 10^{-4} \pm 8.6\%$
APCS2	R_{pore} (Ω)	$1.6 \times 10^{6} \pm 7.2\%$	65123 ± 13.0%	16709 ± 6.4%	37243 ± 8.3%
	C_c (F/cm^2)	------	$4.2 \times 10^{-9} \pm 7.5\%$	$1.06 \times 10^{-8} \pm 10.7\%$	$8.5 \times 10^{-11} \pm 9.3\%$
	W (Mho)	------	$1.17 \times 10^{-6} \pm 8.4\%$	$1.74 \times 10^{-5} \pm 14.5\%$	$1.9 \times 10^{-6} \pm 5.3\%$
	CPE (µMho s cm^{-2})	$7.49 \times 10^{-8} \pm 10.2\%$	--------	-------	-------
APCS3	R_{pore} (KΩ)	17951 + 12.8%	697.4 ± 7.3%	1686 ± 12.1%	1052.5 ± 8.2%
	C_c (F/cm^2)	$2.34 \times 10^{-6} + 8.9\%$	$1.9 \times 10^{-7} \pm 6.5\%$	$5.8 \times 10^{-8} \pm 13.4\%$	$1.9 \times 10^{-5} \pm 11.3\%$
	W (Mho)	------	$4.9 \times 10^{-5} \pm 11.6\%$	$8.61 \times 10^{-5} \pm 11.5\%$	$7.2 \times 10^{-4} \pm 12.9\%$
APCS4	R_{pore} (Ω)	$2.1 \times 10^{5} \pm 11.3\%$	12401 ± 8.8%	15360 ± 4.5%	3397.4 ± 3.18%
	C_c (F/cm^2)	$1.67 \times 10^{-9} \pm 9.7\%$	$1.17 \times 10^{-9} \pm 10.8\%$	$1.63 \times 10^{-9} \pm 7.7\%$	$1.73 \times 10^{-9} \pm 7.7\%$
	W (Mho)	$6.5 \times 10^{-7} \pm 11.5\%$	$3.5 \times 10^{-6} \pm 7.3\%$	$3.4 \times 10^{-5} \pm 11.6\%$	$2.7 \times 10^{-4} \pm 8.7\%$
APCS5	R_{pore} (Ω)	$3.7 \times 10^{6} \pm 7.6\%$	$5.4 \times 10^{5} \pm 6.6\%$	$206 \times 10^{5} \pm 10.8\%$	27847.49 ± 7.1%
	C_c (F/cm^2)	-------	------	$6.7 \times 10^{-9} \pm 5.4\%$	$9.4 \times 10^{-10} \pm 14.7\%$
	W (Mho)	-------	-------	$8.6 \times 10^{-7} \pm 8.9\%$	$9.8 \times 10^{-6} \pm 3.7\%$
	CPE (µMho s cm^{-2})	$1.04 \times 10^{-7} \pm 12.7\%$	$2.91 \times 10^{-7} \pm 9.1\%$	-------	------

Table 6.5: Weight loss in coating specimens (CS) during the abrasion resistance test

Sr. No.	Sample name	Coating loss in mg/1,000 cycle
1.	Epoxy	68
2.	APCS1	53
3.	APCS2	52
4.	APCS3	54
5.	APCS4	56
6.	APCS5	57

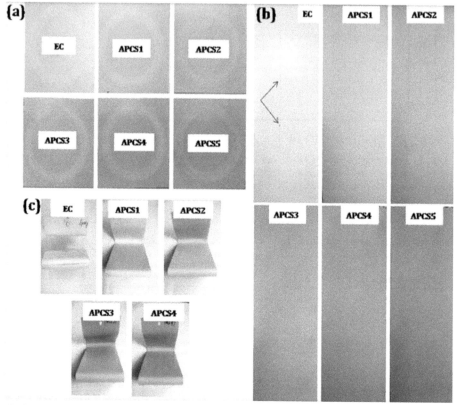

Figure 6.14: Photographs of (a) Taber abrasion test (b) Scratch hardness test, (c) Bend test of 1.0% (APCS1), 2.0% (APCS2), 3.0% (APCS3), 4.0% (APCS4) and 5.0% (APCS5) loading of polyaniline/anisidine/chitosan/SiO$_2$ composite coated specimens (Sambyal et al., Progress in Organic Coatings doi.org/10.1016/j.porgcoat.2018.02.014)

water in aqueous ammonia medium. The appearance of white turbidity shows the formation of hydrated SiO$_2$ nano-particles. These suspended SiO$_2$ particles were retrieved by centrifugation. This step was followed by calcination of the particles in a muffle furnace at 600°C. In the next step, the monomers were adsorbed on the surface of synthesized SiO$_2$ nano-particles with continuous stirring at 50°C.

Here, SiO$_2$ nano-particles were taken as 20 wt% to the monomers. The obtained slurry was added to the pre-cooled (0–5°C) solution (1.0% solution) of chitosan in dilute ortho-phosphoric acid. The copolymerization of aniline and o-toluidine was carried out in this medium with the dropwise addition of APS (0.1 M) with constant stirring for 4–5 hours as shown in Figure 6.15. The composites were

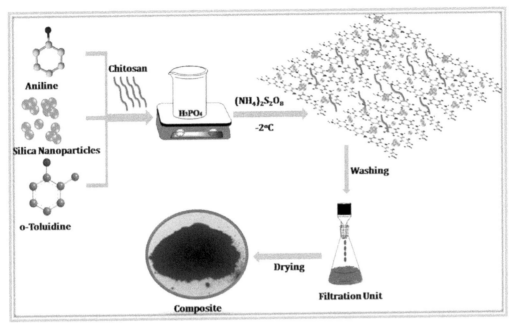

Figure 6.15: Schematic of the synthesis of poly(aniline-co-o-toluidine)/chitosan/SiO₂ composite by chemical oxidative copolymerization process (Sambyal et al., Materials & Corrosion (2018) DOI: 10.1002/maco.201709554)

retrieved by vacuum filtration followed by washing under vacuum till the filtrate shows neutral pH. In the final step, the composite powder was dried under vacuum at 60°C.

Electrochemical polymerization of aniline and its derivatives takes place in pH < 4 (Stilwell 1988; Ganash 2014) by sweeping the potential between –0.20 to 1.2 V versus Ag/AgCl at a scan rate of 20 mVs−1. The reported work mentioned the use of H_3PO_4 as a medium for polymerization for the polymers. The occurrence of the anodic peak at 0.8 V in the first cycle is due to the oxidation of aniline to anilinium radical cation (Figure 6.16a inset). The anodic and cathodic peaks of the subsequent scans are positioned at 0.41 and 0.31 V, respectively, showing the progress of the polymerization process (Figure 6.16a). The intensity of the anodic and cathodic peaks reduced with the progress of polymerization indicating a slight decrease in the conductivity of the surface films. However, the appearance of anodic and cathodic peaks even after 20 scans evidenced the electroactive property of the polyaniline film. For o-toluidine, the anodic/cathodic peaks appeared at the same potential (0.41 V), indicating its perfect reversible redox behavior (Figure 6.16b). The o-toluidine has a methyl group at ortho-position, and the reversibility of the redox process could be due to the steric effect produced by the group. The electrochemical polymerization of the co-monomer (aniline and o-toluidine) is carried out and the cyclic voltametry curve is compared with the homopolymers. Figure 6.16d shows the comparative CV curves of the 20th scan of polyaniline, poly o-toluidine and aniline-o-toluidine copolymer. The anodic and cathodic currents during electrochemical deposition of the copolymer are observed to be higher than o-toluidine but are lower than that of the electrochemical aniline. This trend shows that the better electroactive property of the copolymer as compared to the homopolymer. The XPS analysis evidenced the chemical composition of the coated surfaces before and after exposure to the electrolyte (3.5% NaCl solution). The XPS spectra of carbon, oxygen and nitrogen are obtained. Figure 6.17a presents the high-resolution carbon C (1s) spectra of epoxy coating showing different chemical states of carbon. The deconvoluted four peaks for carbon is at 284.5 eV, assigned to aliphatic carbon (C—C or C—H), 285.3 eV assigned to carbon bonded to oxygen (—C—O—), 286.2 eV assigned to carbon bonded to nitrogen (amine —C—NR₂, or isocyanate —CN=O) and 289.1 eV due to carbon carboxyl (C—COO−) species (Rinaudo 2006). The cured epoxy coating contains a carboxylic acid

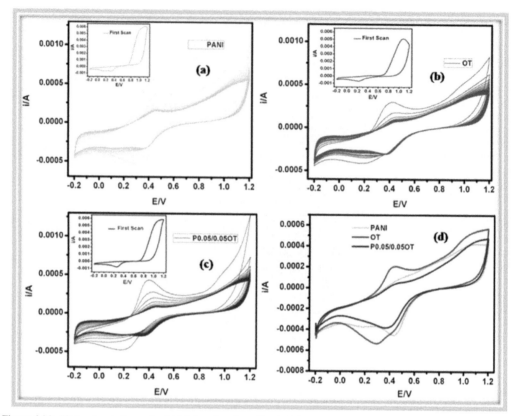

Figure 6.16: Electrochemical growth of (a) 0.1 mol per lt. aniline in 0.3 mol per lt. H$_3$PO$_4$ (b) 0.1 mol per lt. o-toluidine in 0.3 mol per lt. H$_3$PO$_4$ (c) aniline 0.05 mol per lt. +/ o-toluidine 0.05 mol per lt. in 0.3 mol per lt. H$_3$PO$_4$ in the potential range –0.2 V to 1.2 V versus Ag/AgCl at a scan rate of 20 mV per second (d) Cyclic voltammogram of PANI, o-toluidine and copolymer (Sambyal et al., Materials & Corrosion (2018) DOI: 10.1002/maco.201709554)

group which is basically due to the oxidation of the epoxy ring during curing at elevated temperatures (Bell et al. 1974). During the five days of exposure to NaCl solution, the concentrations of C—C and C—H bonding decreased indicating the degradation of the polymeric structure of epoxy resin with time (Figure 6.17b).

Figure 6.17c shows C (1s) spectra of the poly(aniline-co-o-toluidine)–chitosan–SiO$_2$/epoxy composite (POCS) coating containing deconvoluted peaks of —C—C— or —C—H— of the aromatic ring (284.5 eV) and carbon bonded to a neutral nitrogen atom (amine or imine) in the polymeric chain (285.2 eV). The peaks at 286.2 and 289.1 eV ascribed to the carbon bonded to (radical cation) nitrogen (—C—N$^+$—) and positive charge nitrogen (—C—N$^+$—). The N (1s) spectra (Figure 6.17e) of POCS3 coating shows four major components at 398.4 eV (quinonoid imine)(=N—), 399.5 eV (benzenoid amine) (—NH—), 400.2 eV (radical cation —N$^{+\bullet}$—) and 401.1 eV (positively charged —N$^+$—) (Zhong et al. 2006; Shao et al. 2009). The spectra of POCS3 (before and after exposure to NaCl solution revealed an increase in the proportion of the cation nitrogen (N+•) and positively-charged nitrogen (N+) from 33.28 to 47.53% and 12.14 to 29.10%, respectively. This shows that the copolymer consists of the emeraldine salt form. The electrochemical studies of the copolymer composite coatings for 30 days of exposure to 3.5% NaCl solution evidenced superior corrosion protection and have also shown repassivation properties. The reported work concluded that the combination of conducting copolymer, chitosan and SiO$_2$ particles has resulted in the formation of coating formulations with advanced corrosion protection properties for mild steel substrates.

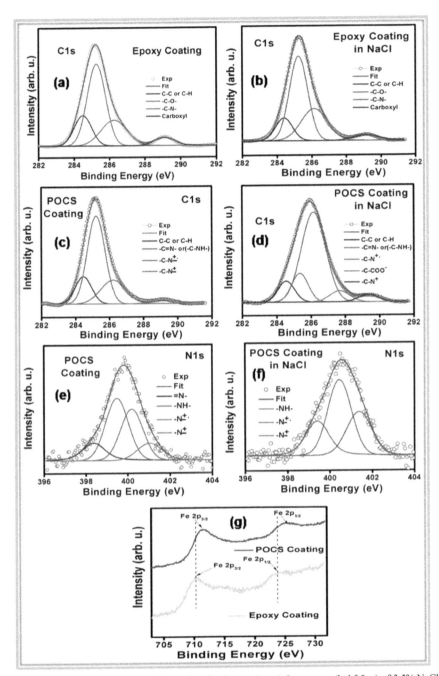

Figure 6.17: (a) XPS spectra of epoxy & POCS coatings in absence (a,c,e) & presence (b,d,f & g) of 3.5% NaCl solution (Sambyal et al., Materials & Corrosion (2018) DOI: 10.1002/ maco.201709554)

6.7 Polypyrrole/gum acacia corrosion inhibitive composite coatings

Plant-based green inhibitors are naturally occurring compounds having complex molecular structures and are gifted with various useful properties. These compounds are green as they are biodegradable, do not contain heavy metals and other toxic compounds, are easily available and are of low cost. There is a growing trend of utilizing the corrosion inhibition properties of these green inhibitors for the protection of active metals in acidic and alkaline media (El-Etre 1998; Kliškić et al. 2000; El-

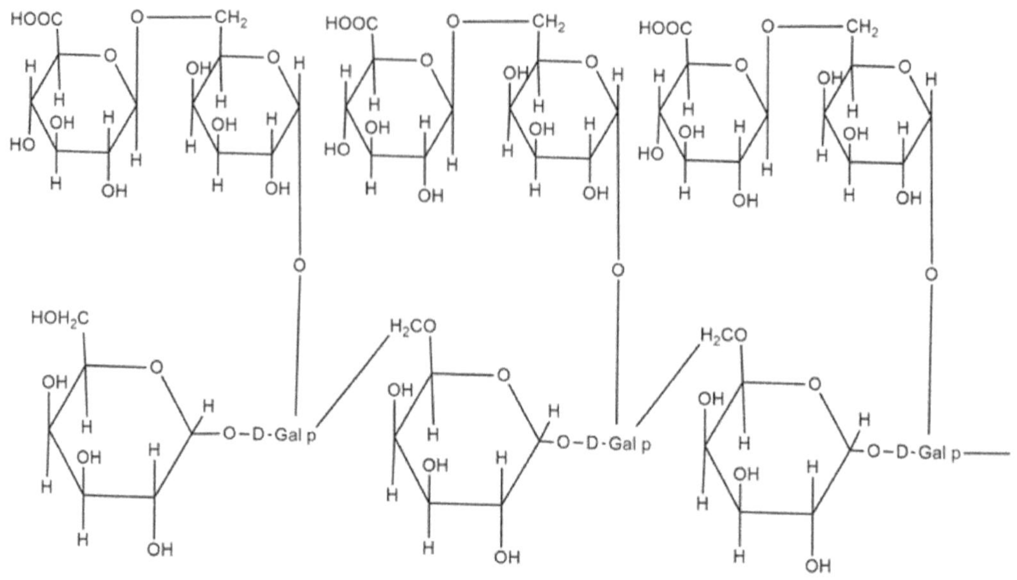

Figure 6.18: Chemical structure of acacia gum

Etre 2003; Orubite and Oforka 2004; Abiola et al. 2007). Plant-based by-products like gum exudate from Acacia seyal var. seyal as corrosion inhibitor for mild steel in freshwater have been reported (Buchweishaija and Mhinzi 2008). Acacia gum is a natural polysaccharide, more specifically a complex mixture of arabinogalactan, oligosaccharides, polysaccharides and glucoproteins (Abdallah 2004). Generally, gums are molecular structures having high molecular mass with colloidal properties, hence can produce gels or suspensions of high viscosity that tend to absorb water. Acacia gum is obtained from the stems of leguminosae tree, grown in the central part of Tanzania. Acacia gum consists of surface-active functional groups like carboxyl (–COOH), hydroxyl (–OH) and amino (–NH$_2$) groups which make it electrochemically active (Figure 6.18). These groups help acacia gum to interact with the metal surface and inhibit corrosion. Through their functional group, they form complexes with metal ions and on the metal surfaces.

Acacia gum/metal complexes possess large surface area, thus they blanket the surface and shield the metal from the corrosive environment. Various reported works highlighted the corrosion inhibition properties of acacia gum in acidic mediums like HCl and H$_2$SO$_4$ on mild steel substrates (Umoren et al. 2008; Abu-Dalo et al. 2012). The obtained results in terms of weight loss, hydrogen evolution and polarization studies claimed that acacia gum is a potential green corrosion inhibitor.

Buchweishaija et al. (2008) have discussed the corrosion inhibition performance of the acacia gum in the concentration range from 0 to 1,000 ppm (v/v) by drawing potentiodynamic polarization curves and EIS studies. The polarization curves revealed that the acacia gum blocked the electrochemical process occurring on the metal surface and hence reduced both anodic and cathodic reactions. In addition to this, the corrosion potential (E$_{corr}$) also shifted toward less negative values as compared to the bare steel surface. The observations suggested that Acacia Seyal var. seyal inhibitor acts as a mixed type inhibitor with predominant anodic effect. The reported work also concluded that the corrosion inhibition efficiency increased with the increase of the concentration of the inhibitor. The EIS studies also complimented the above observations as the R$_{ct}$ values increase while the double-layer capacitance (C$_{dl}$) decreases with the increase of concentration of acacia gum. The authors concluded that the inhibitor gets adsorbed on the metal surface forming a well adherent compact layer that inhibits the commencement of corrosion on the metal surface. Umoren et al. have studied the corrosion inhibition of gum Arabic on the aluminum metal surface exposed to alkaline NaOH medium (Umoren et al. 2006). The work reported that acacia gum exhibited remarkable corrosion inhibition effect on aluminum in NaOH medium. Furthermore, the inhibition efficiency is reported to increase with an increase in inhibitor concentration. The oligosaccharides, polysaccharides, glucoproteins and arabinoglactan constituents of acacia gum get adsorbed on the metal surface causing corrosion inhibition. These constituents contain oxygen and nitrogen atoms which are the centers of adsorption. The authors claimed that the adsorption of the inhibitor is chemical as it followed the Langmuir and Freundlich adsorption isotherms. The corrosion mechanism of the aluminum surface in NaOH solution is discussed in detail. The bare aluminum surface evidenced metal dissolution the moment it came in contact with NaOH solution.

$$Al + OH^- + H_2O \rightarrow AlO^{2-} + 3/2\ H_2$$

$$Al_2O_3 2OH^- \rightarrow 2AlO_2^- + H_2O$$

As soon as the aluminum surface comes in contact with NaOH, the OH⁻ ions get adsorbed on the surface oxide (Al_2O_3) film. The OH⁻ ions chemically attack the oxide film and cause its subsequent dissolution.

The reaction is exothermic, so the heat evolved from the above reaction favors further dissolution of the oxide film and hence aluminum metal corrosion progresses. The hydrogen evolution reaction is the indication of corrosion process here, and the volume of hydrogen evolved increases with an increase in the concentration of the NaOH. However, in the presence of inhibitor, there is a marked reduction in the volume of hydrogen evolved and it is further reduced with the increase in the concentration of the inhibitor. It indicated that acacia gum inhibited the corrosion on the aluminum surface by getting adsorbed strongly and forming a compact layer. This section of the chapter discussed the acacia gum as a potential green corrosion inhibitor. However, limited literature is available on corrosion-resistant composites containing acacia gum. Solomon et al. 2018 had studied the gum Arabic/silver nano-particle composites as a green formulation for the corrosion protection of mild steel in strongly acidic medium. The research work has reported an environmentally friendly and low-cost approach to improve the corrosion inhibition properties of gum acacia. Here, natural honey is used as a reducing agent and capping agent during the synthesis of gum Arabic/Ag nano-particles (GA/AgNPs) composite. The corrosion resistance of the composites is studied using gravimetric studies, surface micro-structural analysis, Tafel polarization studies and EIS studies. The gravimetric studies were carried out in 15% HCl and H_2SO_4 solutions with and without GA/AgNPs for the steel specimens. The results demonstrated that the rate of corrosion reduced significantly in GA/NPs containing the acidic environment. Furthermore, the concentration of GA/NPs and the temperature of the system influenced the corrosion resistance efficiency of the substrate. The metal surface was found to be well protected with high concentrations of GA/NPs at low temperature. Authors emphasized that introducing AgNPs in GA matrix enhanced the corrosion inhibition property of polymer, especially in H_2SO_4 solution. The corrosion protection efficiency of neat GA (500 ppm) is 21.84% in 0.1 M H_2SO_4 solution, whereas 500 ppm of GA/NPs offers 70.30% corrosion protection in 15% H_2SO_4 solution (a concentration much higher than 0.1 M). Authors proposed that the interactive nature of AgNPs facilitates easy adsorption of the composite on the metal surface which results in the formation of the compact barrier layer. The corrosion kinetics is studied by drawing Nyquist and Bode plots in HCl and H_2SO_4 solutions at 25°C. The Nyquist plots for the blank sample and the sample with GA/AgNPs exhibited single semi-circle, revealing a corrosion process controlled by a charge transfer mechanism. However, the Nyquist plot of GA/AgNPs system has a bigger arc, revealing its superior corrosion protection property. The size of the arc increased with the increase of the concentration of the GA/AgNPs. The conclusion drawn by the authors is that the GA/AgNPs adsorbed on the metal surface and concealed the catholic and anodic sites in such a manner that the charge transfer process was either delayed or inhibited. The authors mentioned that the Nyquist plots were not perfect semi-circle because of surface heterogeneities, grain boundaries, impurities, the formation of the porous layer, etc. The Bode plots complimented the Nyquist plots findings and evidenced one time constant with high modulus of impedance for the system containing GA/AgNPs.

This further points that the effective corrosion inhibition of higher concentration of GA/AgNPs. The authors used Tafel polarization method to evaluate whether the GA/AgNPs composite suppresses the anodic reaction, cathodic reaction or both. Interesting observations were noticed in the acid media. In H_2SO_4 solution, the composite suppressed both cathodic and anodic corrosion reactions but more effectively anodic oxidation reaction. However, in HCl medium it suppressed mainly anodic reactions. In addition to this, the i_{corr} was observed to reduce significantly in the presence of composite. The above observations indicated that the corrosion mechanism is not changed in the presence of GA/AgNPs composite but simply induced blocking effect on the anodic and cathodic sites of the metal surface. Another novel approach to design conducting polymer/Gum acacia composite coatings for mild steel was given by Sambyal et al. (2018). In this work, authors focused on the synthesis of the composite which possessed synergistically combined properties of polypyrrole (PPy) and Gum acacia (GA). Figure 6.19 shows the micro-structural features of the composite in which the GA particles are seen to be dispersed in the polymer matrix. It can be noticed from the micrograph that the oxide layer is porous and has an irregular structure with numerous clusters of iron oxide. Therefore, it permits easy diffusion of chloride ions to the metal surface. The presence of elements like Fe = 85.8 wt%, Na = 0.98 wt%, Cl = 0.22 wt%, C = 4.24 wt% and O = 8.41 wt% is proven by EDS profiling (Figure 6.19c) of the corroded epoxy-coated specimen. The presence of high wt% of Fe further confirms the presence of iron oxide layer on the surface. On the contrary, the surface morphology of epoxy coating with 2.0 wt% loading of PPy/GA composite (PA2) evidenced a homogenous and compact appearance (Figure 6.19d). This featureless surface morphology at high magnification is appeared to contain nano-sized plate-like structures (Figure 5.19e). The EDS analysis (Figure 6.19f) confirmed the presence of elements C = 43.1 wt%, O = 33.0 wt%, Si = 8.6 wt%, S = 3.9 wt%, Na = 2.8 wt%, Cl = 3.8 wt% and very less amount of Fe (1.8 wt%). The presence of very low wt% of Fe highlights that the compact layer present on the surface of PA2 is different from the regular rust layer.

The reported work discusses the formation of final coating formulation by blending the synthesized composite with epoxy resin which is spray-coated on the metal surface. The corrosion behavior of the composite coatings was studied by drawing OCP versus time curves, Tafel polarization curves and EIS curves in 3.5% NaCl solution at 25°C. OCP versus time curves of composite coated, neat epoxy-coated and bare steel sample gave a clear indication about the self-healing ability of the composite. The OCP versus time curves mentioned in the research article demonstrated a positive shift of potential with time for the composite coated steel samples (Figure 6.20). Authors proposed that the high positive potential maintained for the composite coated sample is due to the effective barrier property of the surface coating which maintains the steel surface passive. It also indicates that the micro-porosities/defects/flaws that appear in the coating during long exposure to the electrolyte are also healed due to electroactive properties of polypyrrole and corrosion inhibition tendency of Gum acacia.

The authors presented a detailed Tafel polarization data in 3.5% NaCl for an immersion period of 30 days (Table 6.6).

The corrosion potential (E_{corr}) of the composite coated steel surfaces were found to be more noble as compared to neat epoxy-coated surface. Furthermore, the i_{corr} values of the composite coated samples maintained low throughout the immersion period of 30 days. Authors claimed that the synergistic effect of the corrosion inhibition properties of polypyrrole and Gum Acacia are the basic reasons for the superior corrosion resistance of the coatings. The redox property of polypyrrole helps it in intercepting the electron released from the metal and utilizing them to the oxygen reduction at the coating/electrolyte interface. The reaction facilitates the formation of a passive oxide layer at the polymer/metal interface which results in the shifting of corrosion potential of the mild steel toward the noble direction. They further explained that Gum acacia (GA) successfully assisted the formation of the oxide layer on the metal surface. Gum acacia is a branched complex polysaccharide having good corrosion inhibition property. It easily adsorbs on the metal surface through the oxygen and nitrogen atoms and blocks the cathodic and anodic reaction sites. The above explanation received support from the surface topographical evidence that was obtained from the epoxy-coated and epoxy with PPy/GA composite coated steel surfaces after 30 days of immersion in 3.5% NaCl solution, as seen

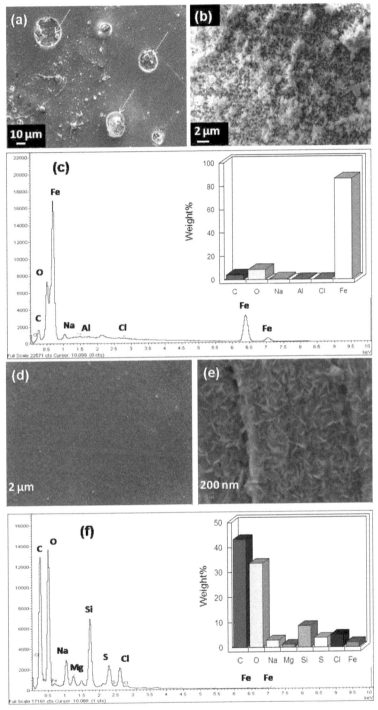

Figure 6.19: (a) The surface morphology of epoxy coating (EC) after 30 days of immersion in 3.5% NaCl solution. (b) High magnification image showing needle-shaped rust present on the coated surface. (c) EDS spectrum of the corroded coated surface. (d) The surface morphology of epoxy with 2.0 wt% loading of PPy/GA composite (PA2) coating after 30 days of immersion in 3.5% NaCl solution. (e) High magnification surface image showing plate-like nano structures. (f) EDS spectrum of the coat (Ruhi et al., Advanced Material Letters, doi. 10.5185/amlett.2018.7007)

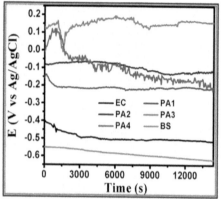

Figure 6.20: The OCP versus time curves for uncoated steel (BS), epoxy coating (EC) and epoxy coatings with 1.0 wt% (PA1), 2.0 wt% (PA2), 3.0 wt% (PA3), 4.0 wt% (PA4) loading of polypyrrole/Gum Acacia (PPy/GA) composite immersed in 3.5% NaCl solution at room temperature (30 + 2°C) (Ruhi et al., Advanced Material Letters, DOI: 10.5185/amlett.2018.7007)

Table 6.6: Different electrochemical parameters derived from Tafel extrapolation for the test specimens immersed in 3.5% NaCl solution for 30 days

Tafel parameters	Exposure in days					
	1	2	5	10	20	30
EC						
E_{corr} (mV)	−638.9	−654.2	−654.9	−668.8	−672.1	−650.3
i_{corr} (A/cm^2)	3.7×10^{-8}	4.6×10^{-8}	7.2×10^{-8}	6.5×10^{-7}	1.4×10^{-6}	2.1×10^{-6}
C.R. (mm/year)	2.0×10^{-4}	5.4×10^{-4}	8.4×10^{-4}	7.5×10^{-3}	1.6×10^{-2}	2.4×10^{-2}
PA1						
E_{corr} (mV)	-------	------	−411.4	−719.7	−676.2	−629.7
i_{corr} (A/cm^2)	-------	------	1.1×10^{-7}	9.1×10^{-8}	5.5×10^{-7}	4.6×10^{-7}
C.R. (mm/year)	-------	------	8.3×10^{-4}	1.0×10^{-3}	6.4×10^{-3}	5.4×10^{-3}
PA2						
E_{corr} (mV)	−623.8	−588.7	−381.6	−81.0	−286.4	−626.9
i_{corr} (A/cm^2)	3.0×10^{-9}	8.5×10^{-9}	4.9×10^{-9}	0.5×10^{-9}	6.0×10^{-9}	5.3×10^{-8}
C.R. (mm/year)	3.5×10^{-5}	9.9×10^{-5}	5.7×10^{-5}	6.6×10^{-6}	7.0×10^{-5}	6.2×10^{-4}
PA3						
E_{corr} (mV)	−610.6	−578.4	−542.3	−584.2	−692.3	−622.8
i_{corr} (A/cm^2)	1.8×10^{-8}	4.4×10^{-8}	1.5×10^{-8}	1.3×10^{-8}	2.5×10^{-7}	3.9×10^{-7}
C.R. (mm/year)	1.9×10^{-4}	5.2×10^{-4}	1.7×10^{-4}	1.5×10^{-4}	2.9×10^{-3}	4.5×10^{-3}
PA4						
E_{corr} (mV)	−740.2	−687.8	−595.2	−708.4	−652.8	−650.0
i_{corr} (A/cm^2)	3.9×10^{-8}	9.8×10^{-9}	6.7×10^{-8}	2.6×10^{-7}	2.2×10^{-7}	7.2×10^{-7}
C.R. (mm/year)	2.4×10^{-4}	1.1×10^{-4}	1.6×10^{-4}	3.1×10^{-4}	2.5×10^{-3}	8.4×10^{-3}

Note: EC = epoxy-coated mild steel sample, PA1, PA2, PA3 and PA4 are the epoxy with 1.0%, 2.0%, 3.0% and 4.0% PPy/GA composite coated mild steel samples, respectively.

in Figure 6.14. The epoxy-coated steel evidenced non-homogenous patchy surface having needle-shaped corrosion product (hydrated iron oxide). The EDS profile proved the presence of elements like Fe = 85.8 wt%, Na = 0.98 wt%, Cl = 0.22 wt%, C = 4.24 wt% and O = 8.41 wt%, which are mainly corrosion product. On the other hand, PPy/GA composite coated sample (PA2) evidenced a homogenous and compact nano-sized plate-like structure. The EDS data revealed very less amount of Fe (1.8%) on the composite coated surface indicating the absence of corrosion product.

Table 6.7: Different EIS parameters obtained by fitting data to suitable equivalent circuits for an immersion period of 30 days in 3.5% NaCl solution

EIS Parameters	Exposure in days					
	1/6	1	4	10	20	30
EC						
R_{pore} (KΩ)	21.9 ± 12.7%	77.5 ± 12.8%	3.0 ± 8.5%	2.8 ± 12.5%	0.9 ± 10.0%	0.7 ± 11.7%
C_c (nF/cm²)	2.6 ± 5.1%	3.7 ± 7.3%	8.1 ± 7.6%	15.6 ± 5.3%	50.0 ± 13.2%	307 ± 12.2%
W (μΩ)	0.66 ± 9.1%	0.14 ± 5.1%	0.04 ± 3.4%	0.01 ± 11.7%	0.009 ± 7.3%	0.007 ± 6.5%
PA1						
R_{pore} (KΩ)	2330 ± 4.1%	5150 ± 9.3%	2210 ± 4.1%	23.4 ± 9.8%	1.4 ± 10.5%	3.6 ± 13.2%
C_c (nF/cm²)	0.19 ± 9.3%	0.21 ± 6.7%	0.21 ± 7.2%	22.8 ± 12.9%	30.5 ± 7.5%	157.0 ± 5.2%
W (μΩ)	-------	------	------	0.02 ± 5.2%	0.01 ± 7.1%	0.03 ± 7.6%
PA2						
R_{pore} (KΩ)	1858 ± 4.3%	222 ± 6.0%	2010 ± 3.3%	650 ± 10.7%	5950 ± 13.0%	280 ± 3.5%
C_c (nF/cm²)	0.04 ± 7.9%	0.43 ± 4.1%	0.56 ± 14.4%	0.46 ± 3.3%	1.8 ± 10.3%	0.54 ± 3.8%
W (μΩ)	-------	-------	------	--------	--------	0.32 ± 10.5%
PA3						
R_{pore} (KΩ)	1680 ± 6.7%	83.8 ± 3.6%	500.5 ± 13.7%	548 ± 9.8%	7.8 ± 3.3%	4.0 ± 5.8%
C_c (nF/cm²)	3.03 ± 12.0%	0.52 ± 5.1%	0.93 ± 13.8%	0.03 ± 15.7%	1.0 ± 7.3%	1.5 ± 11.4%
W (μΩ)	-------	0.22 ± 5.1%	-------	3.33 ± 14.2%	0.02 ± 5.2%	0.01 ± 7.6%
PA4						
R_{pore} (KΩ)	50.6 ± 14.5%	42.9 ± 3.6%	10.1 ± 8.7%	305 ± 4.2%	9.5 ± 7.9%	3.2 ± 12.2%
C_c (nF/cm²)	0.81 ± 6.4%	0.48 ± 4.7%	0.48 ± 5.9%	0.5 ± 4.7%	1.1 ± 6.8%	2.8 ± 9.2%
W (μΩ)	-------	0.43 ± 3.9%	0.27 ± 10.7%	0.28 ± 15.4%	0.06 ± 9.5%	0.02 ± 10.5%

An elaborate explanation on the corrosion kinetics was given in the EIS section where Nyquist and Bode plots were drawn for the epoxy-coated and epoxy with PPy/GA composite coated steel surface. All the composite coated surfaces had shown similar corrosion mechanism (high-frequency capacitive behavior) in the initial hours of immersion in 3.5% NaCl solution, whereas the epoxy-coated surface shows high-frequency capacitive behavior followed by low-frequency diffusion tail, indicating the diffusion of chloride ions through the coating. Table 6.7 mentions the EIS parameters for neat epoxy-coated and epoxy with PPy/GA composite coated steel samples up to 30 days of immersion in 3.5% NaCl solution.

The Nyquist curves of neat epoxy coatings (EC) and epoxy coatings with different wt% loadings of polypyrrole/gum acacia composites with different loadings in epoxy (PA1, PA2, PA3 and PA4) immersed in 3.5% NaCl solution for 24 hours and 96 hours are shown in Figure 6.21.

The reported work clearly shows that the neat epoxy coating has low pore resistance (R_{pore}) as compared to the epoxy with PPy/GA composite coatings, indicating its less resistance toward the ingress of the electrolyte. In addition to this, the coating capacitance (C_c) of epoxy coating is found to be high throughout the immersion period indicating its water uptake property. The modulus of impedance ($|Z|$) for all the PPy/GA composite coated steel samples is observed to be high. From the EIS data authors concluded that the PPy/GA composite coatings are efficient in protecting the steel surface from corrosion even during long immersion periods. The synergistically combined properties of polypyrrole and Gum acacia are concluded to be the main reason behind the superior corrosion resistance shown by the composite coatings.

Figure 6.22a–f show the salt spray test results of epoxy coating and epoxy with 1.0 wt% (PA1), 2.0 wt% (PA2), 3.0 wt% (PA3) and 4.0 wt% (PA4) loadings of PPy/GA composite coatings exposed to salt spray fog for 120 days.

The epoxy-coated steel panel evidenced an extension of corrosion along the scribe mark (Figure 6.22a). Additionally, severe blistering is also noticed in several regions of the coatings (Figure 6.22b). It is well known that epoxy coatings exhibit good bonding properties with the metal

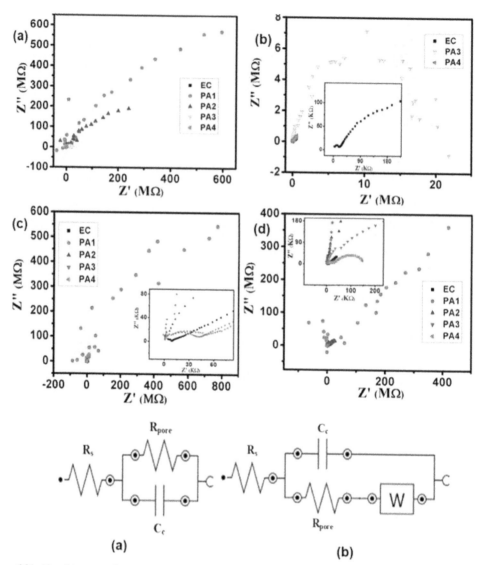

Figure 6.21: Nyquist curves of epoxy coatings (EC) and epoxy coatings with 1.0 wt% (PA1), 2.0 wt% (PA2), 3.0 wt% (PA3), 4.0 wt% (PA4) loading of polypyrrole/Gum Acacia (PPy/GA) composite after (a–b) 4 hours, (c) 24 hours and (d) 4 days of immersion in 3.5% NaCl solution at room temperature (30 ± 2°C). Electrical equivalent circuits of (a) intact coating in contact with electrolyte (b) coatings show the diffusion of electrolyte. Here, Rs (electrolyte resistance), R_{pore} (pore resistance), Cc (coating capacitance) and W (Warburg impedance) (Ruhi et al., Advanced Material Letters, DOI: 10.5185/ amlett.2018.7007)

surface because of the polar groups present along the epoxy chain. However, ingress of corrosive ions at coating/metal interface breaks the bonding and fails these epoxy coatings. On the other hand, no blistering or extended corrosion is noticed for epoxy with PPy/GA composite coated steel panels (Figure 6.22c–f). PPY/GA composites present in the epoxy system acts as a reinforcing material in the coating. Their synergy toward corrosion inhibition enhances the overall corrosion resistance of the epoxy coating system. The composite effectively inhibits the progress of undercoating corrosion and formation of oxide scale at coating/metal interface.

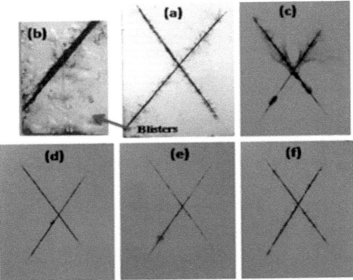

Figure 6.22: Photographs of (a) epoxy-coated steel panel (EC), epoxy with (c) 1.0 wt% (PA1), (d) 2.0 wt% (PA2), (e) 3.0 wt% (PA3) and (d) 4.0 wt% (PA4) loading of polypyrrole/Gum Acacia (PPy/GA) composite after 120 days of exposure to 5.0 % NaCl in salt spray fog at room temperature ($30 \pm 2°C$). (b) Zoomed photographs of epoxy-coated steel panel showing presence of blisters in the coating (Ruhi et al., Advanced Material Letters, DOI: 10.5185/amlett.2018.7007)

6.8 Conclusion

The present chapter discussed the development of corrosion-resistant composite coatings that are green alternatives with less residual impact on the environment. The chapter mentioned various composite coatings contain animal-based and plant-based corrosion inhibitors. The corrosion protection mechanism of these coatings is discussed in detail. An elaborate discussion on two different composite coating systems, namely chitosan/polypyrrole/SiO_2 and polypyrrole/gum acacia are presented in the present chapter. For this, chemical oxidative polymerization of pyrrole was carried out using a suitable oxidant in the presence of chitosan/gum acacia/SiO_2. The synthesized polymer composites were blended with epoxy resin and spray coated onto the mild steel substrates followed by curing at 150°C. The chapter presented an elaborate discussion on various corrosion testing techniques to evaluate the corrosion-resistant performance of the synthesized composite coatings. Open circuit potential (OCP) versus time curves, Tafel polarization curves, Nyquist and Bode curves were drawn to extract various corrosion parameters like corrosion potential (E_{corr}), corrosion current density (i_{corr}), polarization resistance (R_p), coating capacitance (C_c), double layer capacitance (C_{dl}), Warburg impedance (W), etc. Salt spray test and physico-mechanical tests were conducted to check the mechanical integrity of the coatings under accelerated corrosive conditions and the influence of physical stresses. The corrosion test results of chitosan-polypyrrole-SiO_2 composite coatings evidenced significant improvement in the corrosion-resistant properties of the steel in 3.5% NaCl solution. The composite coatings were found to able to recuperate the artificial defect created on the coated surface to some extent. The bend test results of the composite coated steel panels showed an absence of cracks under mechanical stress. Another composite coating system, polypyrrole/gum acacia also demonstrated superior corrosion protection under saline conditions. The composite coatings were observed to effective in inhibiting corrosion for even prolonged periods of immersion. The chapter concluded that the synergistic combination of the properties of plant/animal-based corrosion inhibitors, fillers and conjugated polymers resulted in the formation of robust composite coatings with superior corrosion resistant properties.

References

Abdallah, M. (2004). Guar gum as corrosion inhibitor for carbon steel in sulfuric acid solutions. Portugaliae Electrochimica Acta, 22(2): 161–175.

Abiola, O. K., Oforka, N. C., Ebenso, E. E. and Nwinuka, N. M. (2007). Eco-friendly corrosion inhibitors: The inhibitive action of Delonix Regia extract for the corrosion of aluminium in acidic media. Anti-Corrosion Methods and Materials, 54(4): 219–224.

Abu-Dalo, M. A., Othman, A. A. and Al-Rawashdeh, N. A. F. (2012). Exudate gum from acacia trees as green corrosion inhibitor for mild steel in acidic media. International Journal of Electrochemical Science, 7(10): 9303–9324.

Al-Bonayan, A. M. (2014). Sodium aliginate as corrosion inhibitor for carbon steel in 0.5 M HCl solutions. International Journal of Scientific & Engineering Research, 5(4): 611–618.

Arukalam, I. (2014). Durability and synergistic effects of KI on the acid corrosion inhibition of mild steel by hydroxypropyl methylcellulose. Carbohydrate Polymers, 112: 291–299.

Arukalam, I. O., Madufor, I. C., Ogbobe, O. and Oguzie, E. E. (2015). Inhibition of mild steel corrosion in sulfuric acid medium by hydroxyethyl cellulose. Chemical Engineering Communications, 202(1): 112–122.

Ashassi-Sorkhabi, H., Bagheri, R. and Rezaei-moghadam, B. (2015). Sonoelectrochemical synthesis of PPy-MWCNTs-chitosan nanocomposite coatings: characterization and corrosion behavior. Journal of Materials Engineering and Performance, 24(1): 385–392.

Buchweishaija, J. and Mhinzi, G. (2008). Natural products as a source of environmentally friendly corrosion inhibitors: the case of gum exudate from Acacia seyal var. seyal. Portugaliae Electrochimica Acta, 26(3): 257–265.

Bumgardner, J. D., Wiser, R., Gerard, P. D., Bergin, P., Chestnutt, B., Marini, M., Ramsey, V., Elder, S. H. and Gilbert, J. A. (2003). Chitosan: potential use as a bioactive coating for orthopaedic and craniofacial/dental implants. Journal of Biomaterials Science, Polymer Edition, 14(5): 423–438.

Byun, Y., Ward, A. and Whiteside, S. (2012). Formation and characterization of shellac-hydroxypropyl methylcellulose composite films. Food Hydrocolloids, 27(2): 364–370.

Carneiro, J., Tedim, J., Fernandes, S. C., Freire, C. S. R., Silvestre, A. J. D., Gandini, A., Ferreira, M. G. S. and Zheludkevich, M. L. (2012). Chitosan-based self-healing protective coatings doped with cerium nitrate for corrosion protection of aluminum alloy 2024. Progress in Organic Coatings, 75(1-2): 8–13.

Charitha, B. and Rao, P. (2015). Ecofriendly biopolymer as green inhibitor for corrosion control of 6061-aluminium alloy in hydrochloric acid medium. International Journal of Chem. Tech. Research, 8(11): 330–342.

Charitha, B. and Rao, P. (2017). Starch as an ecofriendly green inhibitor for corrosion control of 6061-Al alloy. Journal of Materials and Environmental Science, 8(1): 78–89.

Ehrich, W., Höh, H. and Kreiner, C. F. (1990). Biocompatibility and pharmacokinetics of hydroxypropyl methylcellulose (HPMC) in the anterior chamber of the rabbit eye. Klinische Monatsblatter fur Augenheilkunde, 196(6): 470–474.

El-Etre, A. (1998). Natural honey as corrosion inhibitor for metals and alloys. I. Copper in neutral aqueous solution. Corrosion Science, 40(11): 1845–1850.

El-Etre, A. (2003). Inhibition of aluminum corrosion using Opuntia extract. Corrosion Science, 45(11): 2485–2495.

El-Haddad, M. N. (2013). Chitosan as a green inhibitor for copper corrosion in acidic medium. International Journal of Biological Macromolecules, 55: 142–149.

Falguera, V., Quintero, J. P., Jiménez, A., Muñoz, J. A. and Ibarz, A. (2011). Edible films and coatings: Structures, active functions and trends in their use. Trends in Food Science & Technology, 22(6): 292–303.

Farag, A. A., Ismail, A. S. and Migahed, M. A. (2015). Inhibition of carbon steel corrosion in acidic solution using some newly polyester derivatives. Journal of Molecular Liquids, 211: 915–923.

Fayyad, E. M., Sadasivuni, K. K., Ponnamma, D. and Al-Maadeed, M. A. A. (2016). Oleic acid-grafted chitosan/graphene oxide composite coating for corrosion protection of carbon steel. Carbohydrate Polymers, 151: 871–878.

Ganash, A. (2014). Effect of current density on the corrosion protection of Poly(o-toluidine)-coated stainless steel. International Journal of Electrochemical Science, 9: 4000–4013.

Gebhardt, F., Seuss, S., Turhan, M. C., Hornberger, H., Virtanen, S. and Boccaccini, A. R. (2012). Characterization of electrophoretic chitosan coatings on stainless steel. Materials Letters, 66(1): 302–304.

Giuliani, C., Pascucci, M., Riccucci, C., Messina, E., de Luna, M. S., Lavorgna, M., Ingo, G. M. and Di Carlo, G. (2018). Chitosan-based coatings for corrosion protection of copper-based alloys: A promising more sustainable approach for cultural heritage applications. Progress in Organic Coatings, 122: 138–146.

Goy, R. C., Morais, S. T. and Assis, O. B. (2016). Evaluation of the antimicrobial activity of chitosan and its quaternized derivative on *E. coli* and *S. aureus* growth. Revista Brasileira de Farmacognosia, 26(1): 122–127.

Guibal, E. (2004). Interactions of metal ions with chitosan-based sorbents: a review. Separation and Purification Technology, 38(1): 43–74.

Hahn, B. D., Park, D. S., Choi, J. J., Ryu, J., Yoon, W. H., Choi, J. H., Kim, H. E. and Kim, S. G. (2011). Aerosol deposition of hydroxyapatite–chitosan composite coatings on biodegradable magnesium alloy. Surface and Coatings Technology, 205(8-9): 3112–3118.

Jiménez, A., Fabra, M. J., Talens, P. and Chiralt, A. (2010). Effect of lipid self-association on the microstructure and physical properties of hydroxypropyl-methylcellulose edible films containing fatty acids. Carbohydrate Polymers, 82(3): 585–593.

Khor, E. and Whey, J. L. H. (1995). Interaction of chitosan with polypyrrole in the formation of hybrid biomaterials. Carbohydrate Polymers, 26(3): 183–187.

Kliškić, M., Radošević, J., Gudić, S. and Katalinić, V. (2000). Aqueous extract of Rosmarinus officinalis L. as inhibitor of Al–Mg alloy corrosion in chloride solution. Journal of Applied Electrochemistry, 30(7): 823–830.

Kubota, N., Tatsumoto, N., Sano, T. and Toya, K. (2000). A simple preparation of half N-acetylated chitosan highly soluble in water and aqueous organic solvents. Carbohydrate Research, 324(4): 268–274.

Kumar, G. and Buchheit, R. G. (2006). Development and characterization of corrosion resistant coatings using the natural biopolymer chitosan. ECS Transactions, 1(9): 101–117.

Mobin, M., Rizvi, M., Olasunkanmi, L. O. and Ebenso, E. E. (2017). Biopolymer from Tragacanth gum as a green corrosion inhibitor for carbon steel in 1 M HCl solution. ACS Omega, 2(7): 3997–4008.

Murulana, L. C., Kabanda, M. M. and Ebenso, E. E. (2016). Investigation of the adsorption characteristics of some selected sulphonamide derivatives as corrosion inhibitors at mild steel/hydrochloric acid interface: Experimental, quantum chemical and QSAR studies. Journal of Molecular Liquids, 215: 763–779.

Obot, I. B., Onyeachu, I. B. and Kumar, A. M. (2017). Sodium alginate: A promising biopolymer for corrosion protection of API X60 high strength carbon steel in saline medium. Carbohydrate Polymers, 178: 200–208.

Okechi Arukalam, I., Chimezie Madufor, I., Ogbobe, O. and Oguzie, E. (2014). Hydroxypropyl methylcellulose as a polymeric corrosion inhibitor for aluminium. Pigment & Resin Technology, 43(3): 151–158.

Orubite, K. and Oforka, N. (2004). Inhibition of the corrosion of mild steel in hydrochloric acid solutions by the extracts of leaves of Nypa fruticans Wurmb. Materials Letters, 58(11): 1768–1772.

Prabhu, D. and Rao, P. (2013). Coriandrum sativum L.—A novel green inhibitor for the corrosion inhibition of aluminium in 1.0 M phosphoric acid solution. Journal of Environmental Chemical Engineering, 1(4): 676–683.

RameshKumar, S., Danaee, I., RashvandAvei, M. and Vijayan, M. (2015). Quantum chemical and experimental investigations on equipotent effects of (+) R and (−) S enantiomers of racemic amisulpride as eco-friendly corrosion inhibitors for mild steel in acidic solution. Journal of Molecular Liquids, 212: 168–186.

Rikhari, B., Mani, S. P. and Rajendran, N. (2018). Electrochemical behavior of polypyrrole/chitosan composite coating on Ti metal for biomedical applications. Carbohydrate Polymers, 189: 126–137.

Rinaudo, M. (2006). Chitin and chitosan: Properties and applications. Progress in Polymer Science, 31(7): 603–632.

Roux, S., Bur, N., Ferrari, G., Tribollet, B. and Feugeas, F. (2010). Influence of a biopolymer admixture on corrosion behavior of steel rebars in concrete. Materials and Corrosion, 61(12): 1026–1033.

Ruhi, G., Sambyal, P., Bhandari, H and Dhawan, S. K. (2018). Corrosion protection of mild steel by environment friendly polypyrrole/gum acacia composite coatings. Advanced Materials Letters, 9(3): 158–168.

Sambyal, P., Ruhi, G., Dhawan, S. K., Bisht, B. M. S. and Gairola, S. P. (2018). Enhanced anticorrosive properties of tailored poly(aniline-anisidine)/chitosan/SiO$_2$ composite for protection of mild steel in aggressive marine conditions. Progress in Organic Coatings, 119: 203–213.

Sambyal, P., Ruhi, G., Mishra, M., Gupta, G. and Dhawan, S. K. (2018). Conducting polymer/bio-material composite coatings for corrosion protection. Materials and Corrosion, 69(3): 402–417.

Samyn, P. (2014). Corrosion protection of aluminum by hydrophobization using nanoparticle polymer coatings containing plant oil. Journal of the Brazilian Chemical Society, 25(5): 947–960.

Shao, Y., Huang, H., Zhang, T., Meng, G. and Wang, F. (2009). Corrosion protection of Mg–5Li alloy with epoxy coatings containing polyaniline. Corrosion Science, 51(12): 2906–2915.

Shi, S. -C. and Su, C. -C. (2016). Corrosion inhibition of high speed steel by biopolymer HPMC derivatives. Materials, 9(8): 612.

Solomon, M. M., Gerengi, H., Umoren, S. A., Essien, N. B., Essien, U. B. and Kaya, E. (2018). Gum Arabic-silver nanoparticles composite as a green anticorrosive formulation for steel corrosion in strong acid media. Carbohydrate Polymers, 181: 43–55.

Stadler, R., Fuerbeth, W., Harneit, K., Grooters, M., Woellbrink, M. and Sand, W. (2008). First evaluation of the applicability of microbial extracellular polymeric substances for corrosion protection of metal substrates. Electrochimica Acta, 54(1): 91–99.

Stilwell, D. E. and Su Moon Park. (1988). Electrochemistry of conductive polymers II. Electrochemical studies on growth properties of polyaniline. Journal of the Electrochemical Society, 135(9): 2254–2262.

Sugama, T. and Milian-Jimenez, S. (1999). Dextrine-modified chitosan marine polymer coatings. Journal of Materials Science, 34(9): 2003–2014.

Sugama, T. and Cook, M. (2000). Poly (itaconic acid)-modified chitosan coatings for mitigating corrosion of aluminum substrates. Progress in Organic Coatings, 38(2): 79–87.

Sutha, S., Kavitha, K., Karunakaran, G. and Rajendran, V. (2013). *In-vitro* bioactivity, biocorrosion and antibacterial activity of silicon integrated hydroxyapatite/chitosan composite coating on 316 L stainless steel implants. Materials Science and Engineering: C, 33(7): 4046–4054.

Tawfik, S. M. (2015). Alginate surfactant derivatives as an ecofriendly corrosion inhibitor for carbon steel in acidic environments. RSC Advances, 5(126): 104535–104550.

Umoren, S. A., Obot, I. B., Ebenso, E. E., Okafor, P. C., Ogbobe, O. and Oguzie, E. E. (2006). Gum arabic as a potential corrosion inhibitor for aluminium in alkaline medium and its adsorption characteristics. Anti-Corrosion Methods and Materials, 53(5): 277–282.

Umoren, S. A., Ogbobe, O., Igwe, I. O. and Ebenso, E. E. (2008). Inhibition of mild steel corrosion in acidic medium using synthetic and naturally occurring polymers and synergistic halide additives. Corrosion Science, 50(7): 1998–2006.

Vincke, E., Van Wanseele, E., Monteny, J., Beeldens, A., De Belie, N., Taerwe, L., Van Gemert, D. and Verstraete, W. (2002). Influence of polymer addition on biogenic sulfuric acid attack of concrete. International Biodeterioration & Biodegradation, 49(4): 283–292.

Waanders, F. B., Vorster, S. W. and Geldenhuys, A. J. (2002). Biopolymer corrosion inhibition of mild steel: electrochemical/mössbauer results. Hyperfine interactions, 139(1-4): 133–139.

Yalçınkaya, S., Demetgül, C., Timur, M. and Çolak, N. (2010). Electrochemical synthesis and characterization of polypyrrole/chitosan composite on platinum electrode: its electrochemical and thermal behaviors. Carbohydrate Polymers, 79(4): 908–913.

Yang, S., Tirmizi, S. A., Burns, A., Barney, A. A. and Risen Jr, W. M. (1989). Chitaline materials: soluble chitosan-polyaniline copolymers and their conductive doped forms. Synthetic Metals, 32(2): 191–200.

Yang, X. and Lu, Y. (2005). Hollow nanometer-sized polypyrrole capsules with controllable shell thickness synthesized in the presence of chitosan. Polymer, 46(14): 5324–5328.

Yingmei, L., Guicun, L., Hongrui, P. and Kezheng, C. (2011). Facile synthesis of electroactive polypyrrole–chitosan composite nanospheres with controllable diameters. Polymer International, 60: 647–651.

Zhong, L., Zhu, H., Hu, J., Xiao, S. and Gan, F. (2006). A passivation mechanism of doped polyaniline on 410 stainless steel in deaerated H_2SO_4 solution. Electrochimica Acta, 51(25): 5494–5501.

7

Future Scope and Directions

In order to develop highly durable and efficient conducting copolymer-based anti-corrosion material, a significant effort has been presented in this book. The present book includes the synthesis of conducting copolymers and its composites with SiO_2 and ZrO_2 which can effectively be used for protection of mild steel in a corrosive environments such as 1.0 M HCl and 3.5% NaCl solution. The present work is carried out in two parts: the first part includes the development of highly soluble conducting copolymer based on o-toluidine (OT) and 2-amino-5-naphthol-7-sulfonic acid (ANS) and its evaluation for protection of iron in 1.0 M HCl solution which is found to be an effective corrosion inhibitor. The second part of the work includes the development of conducting copolymer nano-composites based on poly(aniline-co pentafluoro aniline)/SiO_2 and poly(aniline-co pentafluoro aniline)/ZrO_2. The synthesized copolymer nano-composites modified epoxy coating was successfully used for the protection of mild steel in 3.5% NaCl medium as an anti-corrosive coating.

Highly soluble conducting copolymers based on o-toluidine (OT) and 2-amino-5-naphthol-7-sulphonic acid (ANS) has been successfully synthesized by chemical oxidative copolymerization method. Synthesized copolymers were characterized by FTIR, TGA and SEM analysis. Corrosion inhibition behavior of the copolymer was evaluated by Tafel polarization, electrochemical impedance measurements and weight loss studies using a different concentration of inhibitors in 1.0 M HCl solution. The corrosion inhibition efficiency was found to be 95% by using 200 ppm concentration of copolymer (POT-co-ANS1) in acidic solution. The electrochemical impedance of iron in acidic medium with addition of these inhibitors revealed that these copolymers reduced the rate of corrosion of iron metal through adsorption mechanism. These studies reveal that the synthesized copolymer with a higher concentration of ANS unit has excellent corrosion inhibition properties and it can be considered as a potential material for the protection of iron in corrosive medium like .1.0 M HCl at low concentration.

In one of the chapters, the work has been carried out by developing epoxy-based poly(aniline-co-pentafluoro aniline)/SiO_2 and poly(aniline-co-pentafluoro aniline)/ZrO_2 nano-composites. Conductive copolymer composites based on poly(AN-co-PFA)/SiO_2 were prepared by *in situ* chemical oxidative polymerization. The study of FTIR, TGA, SEM and TEM confirmed the formation of copolymer composites. Synthesized copolymer composites were successfully formulated with epoxy resin in different loading and their coating was carried out on mild steel substrate by powder coating technique. The corrosion test studies in 3.5% NaCl medium have shown that the epoxy coating containing poly(AN-co-PFA)/SiO_2 had higher corrosion protection property as compared to epoxy and epoxy formulated poly(AN-co-PFA) coating. The Tafel parameters indicate low corrosion current for coatings with 3.0 and 4.0 wt.% loading of copolymer composite in 3.5 wt.% NaCl solution. These coatings have shown superior corrosion protection even for prolonged exposure to 3.5 wt.% NaCl solution. The results of mechanical testing such as abrasion resistance, scratch resistance, bend test and cross-cut tape test for epoxy formulated poly(AN-co-PFA)/SiO$_2$ coating were found to be superior to epoxy coating and epoxy formulated poly(AN-co-PFA) coating. Apart from the prevention of corrosion, it was observed that the epoxy formulated with (AN-co-PFA)/SiO$_2$ coating showed better mechanical properties as compared to epoxy and epoxy with (AN-co-PFA) coating. It was observed that the mechanical properties of the epoxy-based poly(AN-co-PFA)/SiO$_2$ coatings were found to significantly improve with the loading of poly(AN-co-PFA)/SiO$_2$ composites in the coating due to the presence of SiO$_2$ particles in a polymer matrix which reinforced the polymer by providing additional mechanical strength. To enhance the mechanical and anti-corrosion properties of the epoxy coating in saline medium, copolymer-based on poly(AN-co-PFA)/ZrO$_2$ nano-composites were synthesized. Conductive copolymer nano-composites based on poly(AN-co-PFA)/ZrO$_2$ were prepared by *in situ* chemical oxidative polymerization. FTIR, TGA, SEM and TEM studies confirmed the formation and interaction of ZrO$_2$ in copolymer nano-composites. The average dimension of poly(AN-co-PFA)/ZrO$_2$ nano-composites was found to be 20 nm. Synthesized copolymer nano-composites were successfully formulated with epoxy resin in different loading and the coating was carried out on mild steel substrate by powder coating technique. The coating showed the uniform thickness and was devoid of cracks; therefore, it was found to be an excellent corrosion-resistant material under harsh chlorine ion environment which may disclose a new opportunity in various technological applications for the marine engineering materials that require very high salt resistance ability. The corrosion test studied of epoxy coating based on poly(AN-co-PFA)/ZrO$_2$ showed superior corrosion protection performance as compared to poly(AN-co-PFA) and poly(AN-co-PFA)/SiO$_2$ based coating. Impedance spectra of poly(AN-co-PFA)/ZrO$_2$ nano-composite modified epoxy coating at different time interval also indicates its self-healing property. Tafel parameters of coated mild steel samples show the significant reduction of corrosion current density for coatings containing 3.0 and 4.0 wt.% loading of poly(AN-co-PFA)/ZrO$_2$ nano-composite at prolonged exposure to 3.5 wt.% NaCl solution. The results of mechanical testing such as abrasion resistance, scratch resistance, bend test and cross-cut tape test for poly(AN-co-PFA)/ZrO$_2$ modified epoxy coating was found to be superior to epoxy and poly(AN-co-PFA) and poly(AN-co-PFA)/SiO$_2$ modified epoxy coating. Apart from the prevention of corrosion, it was observed that the (AN-co-PFA)/ZrO$_2$ modified epoxy coating showed better mechanical properties as compared to epoxy, epoxy with (AN-co-PFA) and poly(AN-co-PFA)/ SiO$_2$ coating. Mechanical properties of poly(AN-co-PFA)/ZrO$_2$ modified epoxy coatings were found to significantly improve with the loading of poly(AN-co-PFA)/ZrO$_2$ nano-composites in the coating due to the presence of ZrO$_2$ particles in a polymer matrix which reinforced the polymer by providing additional mechanical strength. The synergistic interaction between poly(AN-co-PFA) and ZrO$_2$ NPs in epoxy coating resulted in the better thermal, mechanical, hydrophobic and anti-corrosive properties of the coating. Hence, such a system could be used as an effective anti-corrosive coating on mild steel where the probability of corrosion is greater.

Polypyrrole is a well-known conducting polymer which has been widely studied because of its better environmental stability, biocompatibility, excellent electrical properties and doping level can be tuned as per the conductivity scale. Polypyrrole has been extensively used in sensors, as a battery electrode material, ESD and as a compatible coating material for corrosion protection purpose. However, mechanical compatibility and thermal stability of these polymeric coatings in harsh climatic conditions are still not properly studied. The objective of this book is to also elaborate on the use of polypyrrole-flyash (PPy-flyash) composite coatings for corrosion protection of mild steel substrate. The corrosion inhibition property of polypyrrole and reinforcing ability of flyash are utilized to design coatings with superior corrosion resistance for saline conditions. Electrochemical Impedance Spectroscopy (EIS) and Tafel plots were employed for the electrochemical characterizations of the test specimens. The coated specimens were exposed to a salt spray fog chamber to evaluate the corrosion resistance under accelerated test conditions. How the biopolymers incorporated with conducting polymer can be utilized as a suitable matrix for coating is also elaborated in the book. Gum Acacia is a natural polymer mainly consisting of high molecular weight polysaccharides and high concentrations of calcium, magnesium and potassium salts. It is a green alternative for corrosion protection of mild steels in both acidic and corrosive mediums. It is a cost effective, environmentally friendly and is free from toxic side products. Literature has shown that there exists an interaction between Gum Acacia and steel surface. The coordination type bonding is assumed to occur between the ferrous ions and the oxygen atoms present in the backbone of the polymer. However, most of the studies on the corrosion inhibition of Gum Acacia are conducted under ambient conditions, and the achieved corrosion protection efficiency is very low. *In situ* polymerization of polypyrrole was done along with chitosan and Gum acacia in the presence of a suitable medium, and the resultant composite was found to have better corrosion prevention response than the epoxy coating. The present work is expected to produce coating formulations that will significantly improve the corrosion resistance of commercially available epoxy coatings. The synthesized composites are environmentally friendly and have almost no residual impact on environment. The coatings are developed using a powder coating technique which is an environmentally viable technique with no VOC (volatile organic compound) emission. Additionally, the proposed work in the book will pave the way to design different coating formulations by combining different plant-based corrosion inhibitors with the conjugated polymers. The principle of this work is to exploit the synergistic combination of the properties of natural (plant-based) inhibitors and conducting polymers.

Future Directions

In the present book, an attempt has been made to present data on highly durable, self-healing and hydrophobic conducting copolymers based nano-composites and evaluation of such copolymers nano-composites for the use of corrosion protection of iron in a corrosive environment. Future work would be focused on the development of super-hydrophobic and smart coating of conducting copolymer nano-composites by incorporation of different inorganic nano-materials. It is planned to design conducting polymer-based coating having better adhesive and self-healing ability in highly aggressive conditions which can be used commercially for protection of metals from corrosion. Moreover, emphasis will also be focussed on the development of eco-friendly and cost-effective green polymer-based coating.

Annexure 1
List of Abbreviations

AOT	Dioctyl sodium sulphosuccinate
APS	Ammonium persulphate
AN	Aniline
ASTM	American Society of Testing & Materials
Ag/AgCl	Silver/Silver chloride reference electrode
B	Benzenoid
C_c	Coating capacitance
C_{dl}	Double layer capacitance
CTAB	Cetyl trimethyl ammonium bromide
CPs	Conducting polymers
C.R.	Corrosion rate
DMA	N,N-dimethylacetamide
DMF	Dimethyl formamide
DMSO	Dimethylsulfoxide
DTG	Differential Thermogram
DBSA	Dodecylbenzene sulphonic acid
EA	o-Ethylaniline
EB	Emeraldine base
EDX	Energy Dispersive X-Ray Analyser
EIS	Electrochemical impedance spectroscopy
EMF	Electrochemical series
EMI	Electromagnetic interference
ESD	Electrostatic discharge
E_{corr}	Corrosion Potential
FESEM	Field Emission Scanning Electron Microscopy
E_g	Band gap
E_{SCE}	Potential vs. the saturated calomel electrode
$FeCl_3$	Ferric chloride
$Fe_2(SO_4)_3$	Ferric sulphate
FePTS	Ferric p-toluene sulphonate
FRA	Frequency Response Analyser
FTIR	Fourier transform infrared spectroscopy
FWHM	Full width at half maximum
GA	Gum Acacia
GO	Graphene oxide

HRTEM	High resolution transmission electron microscopy
HOMO	Highest occupied molecular orbital
I.E.	Inhibition efficiency
ICPs	Intrinsically conducting polymers
i_{corr}	Corrosion current density
KPS	Potassium persulphate
LB	Leucoemeraldine base
LPR	Linear polarization resistance
LS	Leucoemeraldine salt
LUMO	Lowest unoccupied molecular orbital
MWCNT	Multi walled Carbon nanotubes
NMP	N-methyl pyrrolidinone
NPs	Nanoparticles
OCP	Open circuit potential
OPF	poly(aniline-co-o-toluidine)/flyash
OT	o-Toluidine
PANI	Polyaniline
PFOA	per-fluoro octaonic acid
PFA	2,3,4,5,6 pentafluoro aniline
PF	poly(aniline-co-pentafluoroaniline)
PFS	poly(aniline-co-pentafluoroaniline)/SiO_2
PFZ	poly(AN-co-PFA)/ZrO_2
POEA	Poly(o-ethoxyaniline)
PPS	poly(aniline-co-phenetidine)/SiO_2
PPy	Polypyrrole
PPV	Polyphenylene vinylene
PTh	Polythiophene
PEDOT	Poly(3,4-ethylene dioxythiophene)
PA	Poly acetylene
PPP	Poly p-phenylene
ppm	Parts per million
Pt	Platinum
PTSA	*p*-toluene sulphonic acid
Q	Quinoid
R_{ct}	Charge transfer resistance
R_P	Polarization resistance
RGO	Reduced graphene oxide
RGO/PPy	Reduced graphene oxide/polypyrrole
R	Resistance
R_s	Electrolyte resistance
SCE	Saturated calomel electrode
SEM	Scanning electron microscopy
SiO_2	Silicon dioxide
S/cm	Siemens per centimeter
SHE	Standard hydrogen electrode
TEM	Transmission electron microscopy
TEOS	Tetra-ethylorthosilicate

TGA	Thermo gravimetric analysis
V	Volt
w/w	Weight by weight
XRD	X-ray diffraction
ZrO_2	Zirconia
α_a	Charge transfer coefficients of Anodic Reactions
α_c	Charge transfer coefficients of cathodic Reaction
β_a	Anodic Tafel Constant
β_c	Cathodic Tafel Constant
2θ	Diffraction angle
θ	Surface coverage
ϕ	Phase angle
Ω	ohm
I_{pa}	Anodic peak current
I_{pc}	Cathodic peak current
$^\circ C$	Degree Celsius
μA	micro ampere
λ	Wavelength
Z	Impedance
Z'	Real part of impedance
Z"	Imaginary part of impedance

Index

About the Authors

Dr. S.K. Dhawan is an Emeritus Scientist in CSIR-National Physical Laboratory, New Delhi, India and is working on design of conducting polymers for Smart Self Healing Coatings, EMI shielding and ESD. Dr. Dhawan has published 160 papers, filed 12 US & Indian Patents. He has received DST-Lockheed Innovation Award in 2014 for Smart Coating Research.

Dr. Hema Bhandari is Assistant Professor in Department of Chemistry, Maitreyi College, University of Delhi. She obtained her Ph.D. Degree from Indian Institute of Technology, New Delhi and CSIR-NPL, New Delhi. Dr. Hema did her Ph.D. on the topic 'Conducting Copolymers for Corrosion Inhibition' and has published 21 research papers, three US patents and published two book chapters.

Dr. Gazala Ruhi is Assistant Professor in Department of Chemistry, Maitreyi College, University of Delhi. She did her Ph.D. from Barkatullah University, Bhopal in the area of high temperature corrosion and has research experience in synthesis of conjugated polymer based nanocomposites for designing of smart coatings for corrosion protection of steel substrates in saline conditions and accelerated corrosive conditions. She has published 16 papers, two patents and published a book chapter.

Dr. Brij Mohan Singh Bisht is Scientist in National Test House, Ghaziabad under Ministry of Consumer Affairs, Government of India. He obtained his Ph.D. Degree from Uttaranchal University, Dehradun, India and CSIR-NPL, New Delhi, India. Dr. Bisht has done his Ph.D. in the area of Corrosion and published eight research papers and filed two patents in the field of Conducting polymer nanocomposites for protection of mild steel against corrosion.

Dr. Pradeep Sambyal received his Ph.D. in chemical sciences from CSIR-National Physical Laboratory, India (2017). Currently, he is working as a postdoctoral researcher at KIST, South Korea. Dr. Sambyal has published 14 papers and filed three patents. His research focuses on nanomaterials for anticorrosive coatings and EMI shielding.

Color Plate Section

Chapter 3

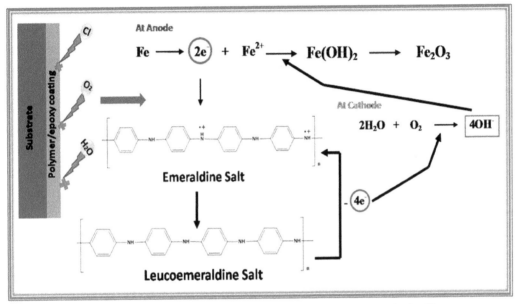

Figure 3.28: Schematic representation of corrosion inhibition mechanism of polyaniline

Chapter 4

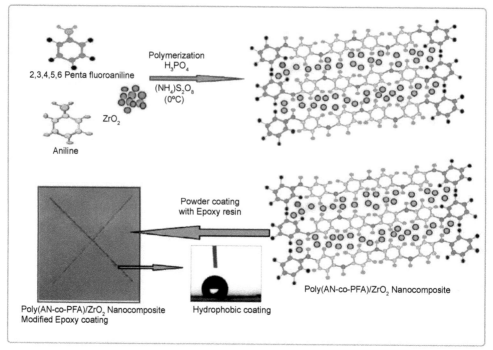

Scheme 4.2: Schematic diagram of the formation of poly(AN-co-PFA)/ZrO$_2$ nanocomposite (Bisht et al. Current Smart Materials, DOI:10.2174/2405465802666170707163408)

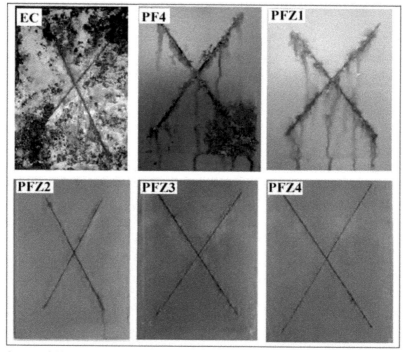

Figure 4.12: Images of (a) epoxy coated and epoxy modified with the copolymer and different loading of copolymer-zirconia nanocomposite coated mild steel panels exposed to salt spray fog for 180 days (Bisht et al. (2017) Current Smart Materials, DOI:10.2174/2405465802666170707163408)

Chapter 5

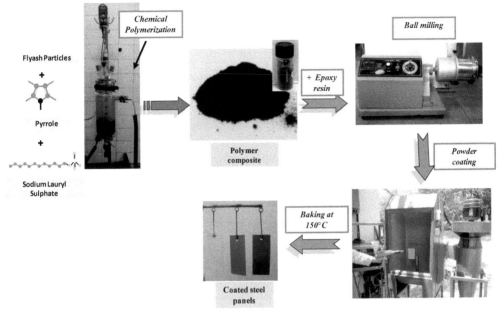

Figure 5.3: Development of PPy/Flyash composite coatings on mild steel substrate

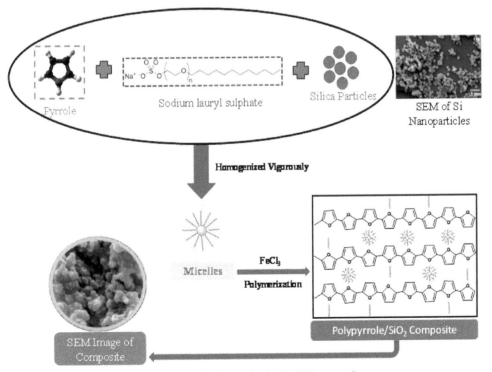

Figure 5.4: Synthesis steps for the PPy/SiO$_2$ composite

Chapter 6

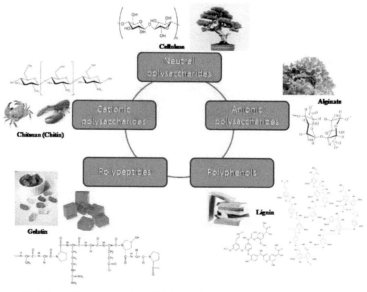

Figure 6.1: Schematic representation of different biopolymers (Daniel Ajay, Research Nester)

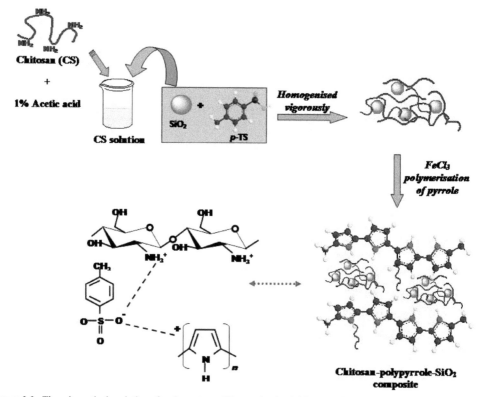

Figure 6.6: The schematic description of various steps of the synthesis of chitosan-polypyrrole-SiO$_2$ composite (Ruhi et al., Synthetic Metals doi.org/10.1016/ j.synthmet. 2014.12.019)

Milton Keynes UK
Ingram Content Group UK Ltd.
UKHW051949071024
449327UK00026B/2240

9 780367 504564